Exploring the Intersection of Forensics and Biometrics

Sarah Benziane
University of Science and Technology in Oran, Algeria

IGI Global
Scientific Publishing
Publishing Tomorrow's Research Today

Vice President of Editorial	Melissa Wagner
Director of Acquisitions	Mikaela Felty
Director of Book Development	Jocelynn Hessler
Production Manager	Mike Brehm
Cover Design	Jose Rosado

Published in the United States of America by
IGI Global Scientific Publishing
701 East Chocolate Avenue
Hershey, PA, 17033, USA
Tel: 717-533-8845 | Fax: 717-533-7115
Website: https://www.igi-global.com E-mail: cust@igi-global.com

Library of Congress Cataloging-in-Publication Data

LCCN: 2025034521 (CIP Data Pending)
ISBN13: 979-8-3373-4972-5
Isbn13Softcover: 979-8-3373-4973-2
EISBN13: 979-8-3373-4974-9

British Cataloguing in Publication Data
A Cataloguing in Publication record for this book is available from the British Library.

Table of Contents

Detailed Table of Contents

This chapter discusses the principles that underlie biometric systems, summarizes the usual strategies employed in digital identity frauds, such as deepfakes, spoofing and creating synthetic identity and reviews the strengths and weaknesses associated with the use of biometric in cyber investigation. Case studies in the real world depict the application ofbiometrics in criminal justice, border control as well as verification on mobile identity. Meanwhile the chapter discusses ethical, legal and privacy concerns of biometric data capture and application with examples of international laws like GDPR and HIPAA. Last, it discusses emerging trends such as AI-based biometric security, decentralized digital identity and privacy-respecting biometric protocols, and provides an in-depth perspective on the future of the cross-section of biometrics, cybersecurity, and biometric-based identity protection.

Forensic audit trails combined with biometric-based authentication represent a critical convergence in modern cybersecurity infrastructure. This chapter examines the technical implementation, forensic methodologies, and investigative frameworks for biometric authentication systems. The integration of immutable audit trails with biometric verification creates comprehensive forensic evidence chains essential for digital investigations. We analyze forensic challenges including spoofing attacks, presentation attacks, and biometric template security. The chapter explores multimodal biometric systems, liveness detection mechanisms, and blockchain-based audit trail implementations. Critical examination of privacy-preserving forensic techniques,

Knuckle crease patterns, formed by natural skin folds over finger joints, are emerging as a valuable biometric trait in forensic science and digital identity. Often visible in photographs and surveillance footage, they aid identification when fingerprints or DNA are unavailable. This chapter reviews their anatomical foundations, persistence, and distinctiveness, and examines automated recognition methods from handcrafted descriptors to deep learning, alongside manual classification approaches emphasizing reproducibility. Forensic applications include criminal investigations, child sexual abuse material analysis, disaster victim identification, and multimodal biometrics. Ethical, legal, and evidentiary issues are discussed, and future challenges for validation and responsible implementation are highlighted.

Recordings of the human voice are found in many contexts today, such as musical recordings, social media videos, television shows, and movies. Voice recordings used for proving identity are most often stored in databases as voiceprints, and can be the cause of some concern, however. They are often employed for accessing personal accounts but can also potentially be used in illegal or unethical ways. It is helpful to understand some of the basics of the technology and the human voice, and to explore laws from different parts of the world in regard to this biometric. Some historical and recent cases assist in seeing the progression of this technology. Cloning (deepfakes), masking, financial fraud, loss of jobs or reputation, and a concern for public figures' safety, along with the fact that the purpose of collecting some voice data is unknown is included, with surveillance and human rights issues surveyed. Recommendations include thoughtful application in the courtroom, focused legislation, continued technical research, and education for best protection.

This research aims to analyze the applicability and performance of face recognition technologies using different Python-based libraries. Face recognition and comparison tests were performed using the face_recognition, OpenCV and DeepFace libraries, respectively . First, basic face detection application was successfully implemented with the face_recognition library. Then, face recognition was performed using traditional methods using OpenCV. But limited accuracy results were obtained. Finally, deep learning-based face recognition was performed with the DeepFace library, high accuracy rates were achieved, but high processing power was required. All three methods were systematically compared in terms of hardware requirements, ease of use, processing time and accuracy rates. The findings revealed that each library has advantages and limitations for certain application scenarios. In this context, it was emphasized that the selection of the appropriate algorithm for the needs in the design of face recognition systems is critical.

This chapter summarizes, analyses and discusses the current maturity of deep learning technology regarding the development and use of multimodal biometric fusion for forensic identity verification. The chapter identifies several challenges in relation to the collection and processing of wild data (low quality/resolution images, noise due to environmental factors and/or partial occlusion). The chapter discusses advantages of using deep neural networks (DNNs) to replace traditional single-mode systems for the processing and recognition of wild data sets. Finally, the chapter describes three primary methods of fusion (feature-level, score-level and attention-based) that can be effectively used to combine disparate data sets; stresses the importance of cross-modal transformers for achieving accurate identification despite severe degradation caused by the use of dissimilar data sets; presents and emphasizes several advanced preprocessing methods, including the use of generative adversarial networks (GAN) for super-resolution.

Preface

The rapid evolution of digital technologies has profoundly reshaped both the nature of crime and the methods used to investigate it. As societies become increasingly reliant on interconnected systems, digital identities, and data-driven decision-making, forensic science has expanded beyond traditional physical evidence to encompass complex biometric, behavioral, and cyber artifacts. This volume explores the intersection of forensics and biometrics, presenting a multidisciplinary perspective on how human traits—physiological, behavioral, and multimodal—are being leveraged to strengthen identity verification, enhance investigative reliability, and ensure evidentiary integrity in modern digital environments.

Biometrics now occupy a central role in cybersecurity, criminal justice, border control, healthcare, and cloud-based infrastructures. At the same time, their widespread adoption introduces new challenges related to spoofing, deepfakes, synthetic identities, privacy, legal admissibility, and ethical governance. Advances in artificial intelligence, deep learning, blockchain, and privacy-preserving computation offer powerful tools to address these challenges, but they also raise critical questions about accountability, transparency, and human rights. This book brings together technical innovations, forensic methodologies, and legal considerations to provide a comprehensive view of current capabilities and future directions in biometric-enabled forensics.

The chapters progress from foundational concepts and threat landscapes to advanced biometric traits, multimodal fusion, and secure evidence management frameworks. Collectively, they highlight how emerging technologies—ranging from behavioral biometrics and bio-inspired optimization algorithms to blockchain audit trails and cloud forensics—are converging to redefine digital investigations. By integrating theoretical insights, empirical results, and real-world case studies, this volume aims to support researchers, practitioners, policymakers, and students seeking to understand and responsibly deploy biometric systems within forensic contexts.

CHAPTER OVERVIEW

1. An introduction to biometric intelligence as a strategic tool against cybercrime and identity fraud, outlining core biometric principles, common attack vectors such as spoofing and deepfakes, real-world applications, and the ethical, legal, and privacy frameworks shaping future biometric identity protection.

2. A behavioral biometrics perspective centered on keystroke dynamics, demonstrating how typing patterns, enhanced by machine learning, can strengthen authentication systems against traditional password-based attacks.

3. An exploration of ECG-based biometric authentication using bio-inspired optimization and feature fusion techniques, highlighting how nature-inspired algorithms improve accuracy, robustness, and security in physiological biometrics.

4. An analysis of knuckle crease patterns as an emerging forensic biometric trait, discussing anatomical foundations, automated and manual recognition methods, forensic use cases, and evidentiary and ethical considerations.

5. A deep learning–driven approach to stress recognition through facial expressions, presenting a modified VGG framework with applications in healthcare, mental wellness, and human–computer interaction.

6. A comprehensive overview of voice and speaker recognition in criminal justice, addressing technical foundations, historical and contemporary cases, risks such as cloning and masking, and legal and human rights implications.

7. A comparative performance evaluation of Python-based face recognition libraries, analyzing trade-offs among accuracy, computational cost, ease of use, and suitability for different forensic application scenarios.

8. An in-depth discussion of deep learning–based multimodal biometric fusion for forensic identification in challenging environments, emphasizing wild data, cross-modal transformers, and advanced preprocessing techniques such as GAN-based super-resolution.

9. A forensic-focused treatment of multimodal biometric fusion techniques, highlighting deep learning–based feature extraction, explainable AI, privacy governance, and demonstrated improvements in identification confidence.

10. A hybrid AI framework for detecting fake biometric fingerprints, combining generative adversarial models, image quality analysis, and machine learning classifiers to achieve high detection accuracy against spoofed data.

11. A secure biometric evidence management framework integrating homomorphic encryption, blockchain audit trails, and decentralized storage, ensuring privacy, integrity, and accountability throughout the forensic evidence lifecycle.

12. An exploration of digital forensics in cloud environments, addressing technical tools, evidentiary challenges, jurisdictional and legal complexities, and best practices for investigating data breaches and cybercrimes.

13. An IoT-focused security and attack detection study, evaluating an integrated monitoring system's ability to identify anomalies and cyberattacks, and underscoring the importance of behavior-based and temporal analysis in IoT forensics.

Together, these contributions present a cohesive and forward-looking examination of how biometrics and forensic science are co-evolving in response to contemporary digital threats, offering both technical depth and critical reflection on responsible and lawful implementation.

Sarah Benziane
University of Science and Technology, Oran, Algeria

Chapter 1

Biometric Intelligence in the War Against Cybercrime and Identity Fraud

Kavita Kanwar
https://orcid.org/0009-0000-6661-891X
Poornima University, Jaipur, India

Nikhil Kumar Goyal
https://orcid.org/0009-0007-4532-8033
Poornima University, Jaipur, India

ABSTRACT

This chapter discusses the principles that underlie biometric systems, summarizes the usual strategies employed in digital identity frauds, such as deepfakes, spoofing and creating synthetic identity and reviews the strengths and weaknesses associated with the use of biometric in cyber investigation. Case studies in the real world depict the application ofbiometrics in criminal justice, border control as well as verification on mobile identity. Meanwhile the chapter discusses ethical, legal and privacy concerns of biometric data capture and application with examples of international laws like GDPR and HIPAA. Last, it discusses emerging trends such as AI-based biometric security, decentralized digital identity and privacy-respecting biometric protocols, and provides an in-depth perspective on the future of the cross-section of biometrics, cybersecurity, and biometric-based identity protection.

DOI: 10.4018/979-8-3373-4972-5.ch001

1. INTRODUCTION

1.1 Biometrics in the Digital Age

The digital era has brought about an unparalleled number of demands of technology on identity authentication, information retrieval and safe transaction. Through this progression, biometric technologies have assumed prime integration into verifying persons through basis of their respective physiological and behavioral characteristics. Biometrics (such as fingerprints, facial characteristics, patterns in the iris, or even signs in the voice) can be used as a source of high confidence in identity confirmation to a greater extent than the usual passwords, ID cards, etc. They are especially useful when applied to the domain of digital access control and security because of their non-transportability properties, in addition to their infeasibility to forge (Adjerid & Kelley, 2018).

The increased dependence on biometrics in smartphones, banks, airport security, and even national identification systems is evidence of the increased faith in technologies. Biometrics does not only facilitate the authentication of its users, but it also plays a larger role in the field of cyber forensics and investigation. These two capabilities (preventative and investigational) mean that biometrics can be considered a staple to most digital security efforts today.

Nonetheless, the growth of the use of biometrics also brings along difficult questions dealing with the storage of data, surveillance and the consent of individuals. Due to the expected promotion of biometric systems in the common part of infrastructure, the risks of data losses, abuses, and unlawful surveillance rise. What makes the biometric identifiers so unique and undetermined is the absolute cause of their notoriety when compromised (Alsaadi & Tubaishat, 2015).

1.2 Rise of Cybercrime and Identity Theft

Cybercrime has became an advanced, unified menace which takes advantage of weak spots in digital networks, systems and even human nature. Digital identity theft is one among its numerous forms and is one of the most common and destructive forms. Hackers sell the stolen credentials to perpetrators so that they can utilize them to perpetrate fraud, gain access to sensitive information, or pretend to be a victim, which has long-term effects. As more and more personal information is kept on the Internet, the surface area of such attacks also increases.

Conventional authentication—username and passwords in this case have been found insufficient to counter the current threats in cyberspace. These certificates are simple to be phished, guessed, or leaked by means of information breaches. Because of this, even low-tech attackers are able to get unauthorized access to systems. The rise of the social engineering techniques and credential stuffing also increases the risk of the digital identities being compromised greatly (Bansal & Jain, 2019).

Biometrics have moved in to break the stage as a dominant alternative. Their connection of authentication to something that can never be changed (biological characteristics) lowers these possibilities of impersonation and stealing of credentials. Nonetheless, even the so-called biometric systems have not been spared of being hacked into although methods like spoofing, deepfake altering and manufacture of synthetic biometrics have become new threats.

1.3 Investigative Structure of Biometrics in Digital Identity Forensics

This chapter is intended to consider the application of the biometric technologies to the prevention and investigation of cybercrime, and specifically in the context of digital identity theft. It looks at the system mechanisms behind biometric systems, the strategies applied by cybercriminals in attempting to beat or use biometric systems, and how biometric data may be used as forensic evidence in a cyber investigation. Legal and ethical systems by which the use of biometrics is regulated in criminal justice and crime prevention in the population are also discussed.

The chapter starts by having technical beginnings explaining about different kinds of biometric modalities, process through which they are meant to be carried out. It subsequently moves on to an in-depth coverage of the methods of identity theft, with focus on the way digital assailants can undermine both legacy and biometrics-based authentication. Later sections discuss forensic uses of biometrics, and those topics provide real-life examples and respective investigation approaches.

The latter sections of the chapter cover the dangers of biometric data collection which include violations of privacy, data compromise, and inappropriate use of surveillance devices. These issues are evaluated against such regulatory approaches worldwide as GDPR or national policies of biometrics. The last part describes future developments such as AI-enabled biometrics, biometrics-resistant authentication, and distributed identity.

1.4 Literature Review

Existing research highlights three major themes in biometric applications for cybersecurity:

A. Biometric Recognition and System Performance

Early foundational work by Jain et al. and Daugman established the core principles of fingerprint and iris recognition. Recent advancements in deep learning-based recognition (Bansal & Jain, 2019) significantly improved accuracy but raised concerns about adversarial attacks.

B. Threats to Biometric Systems

Studies by Ratha et al. (2001) and Patel et al. (2015) detail vulnerabilities such as spoofing, replay attacks, and template compromise. The growth in deepfake generation (Pugliese, 2020) introduces a new class of attacks.

C. Ethical, Legal, and Privacy Perspectives

Works by Cavoukian (Privacy by Design), Lyon (Surveillance Studies), and GDPR documentation emphasize the need for transparency, informed consent, and proportionality. Researchers like Buolamwini & Gebru highlight algorithmic bias in facial recognition systems.

This literature highlights the need for holistic biometric systems that combine technical robustness with legal, ethical, and societal safeguards.

2. FUNDAMENTALS OF BIOMETRIC TECHNOLOGIES

2.1 Types of Biometrics: Physiological vs. Behavioral

There are two categories in general of biometrics systems; physiological and behavioral. Physiology biometrics are physical characteristics and properties of human body whose changes are relatively small with time. They are fingerprints, iris and retina pattern, geometry of a face as well as the palm veins and DNA. Physiological traits are universally used in making identity verifications and forensic identifications because of their striking and accurate features. As an example, fingerprint and facial recognition is widely used in consumer electronics, at the border, and in law enforcement (Bhargava & Yu, 2017).

Figure 1. Types of Biometric Techniques

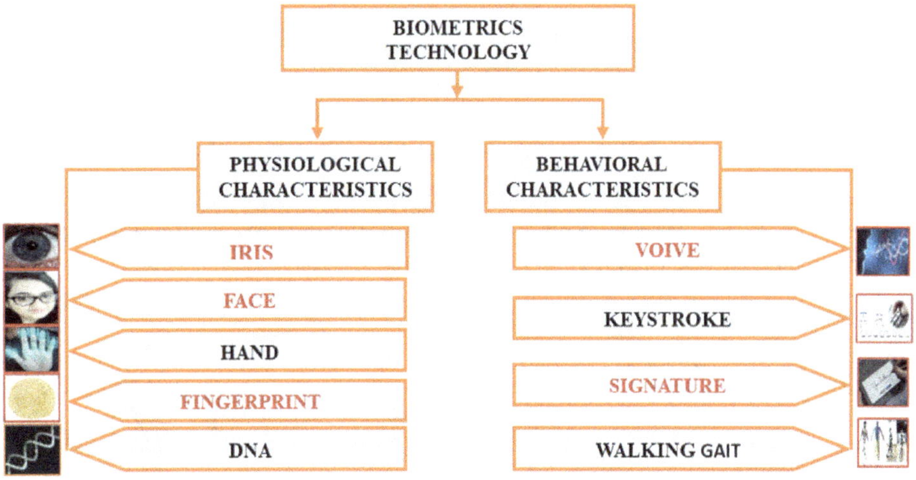

On the other hand, behavioral biometrics are determined by the modes of how individuals carry out certain actions. Among them is the typing rhythm, the way of walking, the voice activeness, the mouse moving the ways, the touch screen interaction behaviour. Although it is believed that behavioral traits are less streamline than physiological ones, they present the virtue of continuous authentication. As an example, a user behavior could be tracked in the background within a session on the internet using behavioral biometrics to warn systems in case of a possible identity theft due to the detection of an abnormal behavior.

Every type of biometric has its advantages and disadvantages. Physiological modalities are highly precise however acquisition of data involves specialized hardware. Although behavioral biometrics are more convenient and less expensive, they are vulnerable to the conditions of the environment and have to rely on more sophisticated algorithms to identify real users and the posers.

Multimodal systems Multimodal systems enhance a well-rounded biometric security strategy by taking into account physiological and behavioral traits. This increases error correction and recovery, therefore NBC becomes harder to fake or bypass identity verification schemes. During the process of cybercrime prevention and cybercrime investigation, it is critical to understand the underlying peculiarities of distinct biometric types in order to be able to adequately use their power (Biometric Information Privacy Act [BIPA], 2008).

2.2 Biometric Data Capture, Processing, and Matching

Biometric systems work in a chain similar to the following: enrollment, data acquisition, feature extraction, template creation, and matching. Biometric data is obtained during enrollment when the information is translated into a digital file called template. The template and not the raw data is maintained in a secure database thus able to compare in future. The system records new details when a verification is to be done and these details are matched with the existing template through the pattern recognition algorithms.

Figure 2. biometric process

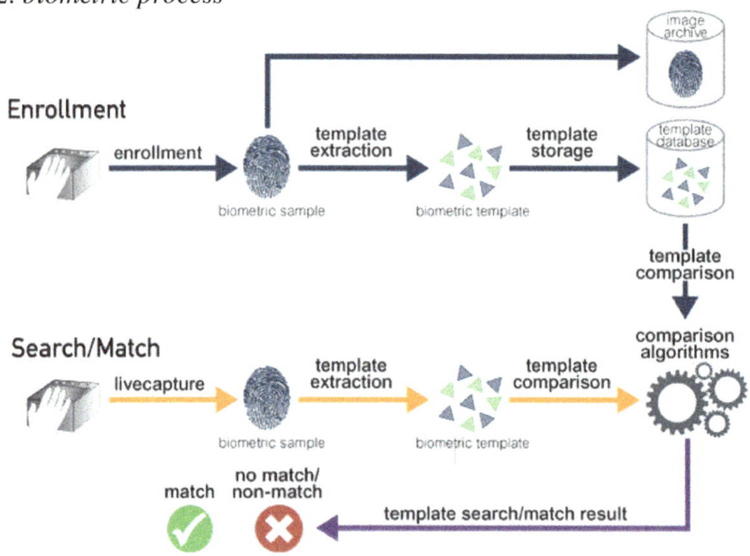

The most important and the first step is data acquisition, which defines the quality of biometric input. Fine quality cameras are required to scan clear images or recordings, a fingerprint, a voice, or a face shot. The data quality depends on environmental conditions, the calibration of the sensors and cooperation between the user. The bad data may introduce higher false rejections or false acceptance.

Feature extraction is an action of finding out distinctive characteristics in the biometric data. In fingerprints, e.g. the minutiae points (ridge endings and bifurcation) are extracted. In facial recognition, distance between eyes, the shape of jaw lines and the level of the nose are calculated. These characters are then translated into mathematical representation or vectors which form biometric template (Bowyer et al., 2008).

Matching is the process whereby input template is compared with one or more stored templates in order to verify or identify a person. This process accuracy is influenced by variables like the matching algorithm, threshold sensitivity and the database size. In forensic use, matching is often a one-to-many search of large criminal records databases, and so must be precise and efficient In some cases, the accuracy of the process is not as important as being able to find a match at all, scenarios often referred to as recidivism searches In practice, the forensic applications are a hybrid of fingerprint indexing and matching techniques (Goyal et al., 2025a).

2.3 Accuracy, Reliability, and Error Rates in Biometrics

The metrics of the biometric systems are usually scored using accuracy, reliability and errors. Accuracy is the proportion of correct identifications or authentications of the system, whereas the reliability of the system is its consistency in different conditions and among different users. Application like criminal investigation have high stakes that require systems with low margins and error rates with consistent performance across the various population groups.

Among the most common measurements to be gauged on biometric systems are the False Acceptance Rate (FAR) and False Rejection Rate (FRR). FAR estimates the probability that an unauthorized user is accepted which is insecurity. FRR, in its turn, signifies the rate at which valid users fail to achieve permissions, which influences usability and user confidence. The perfect biometric system has a low FAR which hardly adds a lot to the FRR.

The biometric might be inaccurate because of age-related factors, injuries, illness or behavioral change of the user though it is likely that biometric accuracy will be in effect despite technological advances. As such, the performance should be maintained by continuous updates, period re-enrollments, and adaptive algorithms. In cases of investigations, false positive results may derail investigations whereas false negative causes the offender to run free. Therefore, accuracy is not only a technical parameter, it is a key factor of justice and social credibility.

Table 1. Summary of Key Biometric Modalities

Biometric Modality	Type	Accuracy	Required Hardware	Vulnerabilities	Ideal Use Cases
Fingerprint	Physiological	High	Scanner, sensor	Spoofing, cuts/injuries	Mobile auth, forensics
Face Recognition	Physiological	Medium–High	Camera	Deepfakes, lighting issues	Surveillance, border control
Iris Scan	Physiological	Very High	Iris camera	User discomfort, cost	High-security facilities
Voice Biometrics	Behavioral	Medium	Microphone	Voice cloning	Call centers, tele-banking

continued on following page

Table 1. Continued

Biometric Modality	Type	Accuracy	Required Hardware	Vulnerabilities	Ideal Use Cases
Gait Analysis	Behavioral	Medium	Video	Environmental noise	Crowd monitoring
Keystroke Dynamics	Behavioral	Medium	Keyboard	Behavioral changes	Continuous authentication

3. THREAT VECTORS AND EXPLOITATION METHODS IN DIGITAL IDENTITY THEFT

3.1 Phishing, Spoofing, and Credential Stuffing

Digital identity theft often starts using methods that take advantage of human trust or loose authentication. The most common of them is probably phishing, in which attackers pretend to be legitimate organizations in emails, websites, or SMS, and induce users to share personal credentials with them. Being obtained, they allow compromising email accounts, bank profiles, or enterprise systems, which, in many cases, leads to additional data exfiltration or financial fraud (Buolamwini & Gebru, 2018).

The other technique is known as spoofing, where a digital identity can be impersonated, this may be an email address, IP address, or even a webpage in order to deceive a system or the users. Biometric spoofs, including deepfakes in facial image or artificial voice recordings, are becoming more common in ways to trick biometric authentication systems. Due to the increasing popularity of biometric identification, spoofing attacks that target it also grow in their sophistication.

Figure 3. Credential Stuffing attack

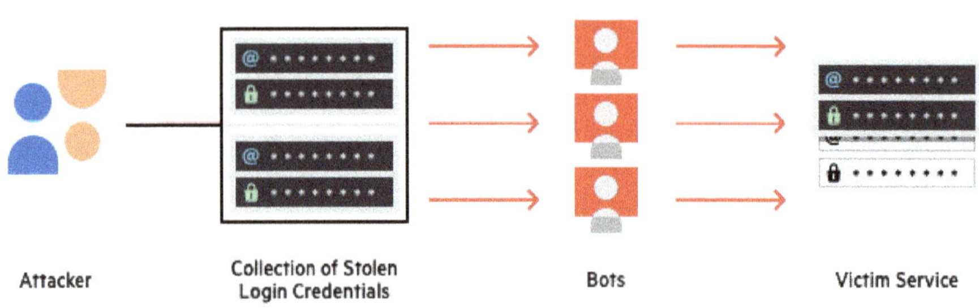

Credential stuffing is used when an attacker leverages on formerly stolen credentials in order to unlawfully access multiple accounts. This method is especially successful when the same people use the same passwords on multiple websites. Credential stuffing can be fatal when paired with phishing or malware, especially where systems do not make use of multi-factor authentication (MFA).

Advanced biometric security are becoming a favored approach through organizations and technology providers to counter such evolving threats, these include; multimodal biometrics (using two or more biometric identifiers such as fingerprint and face concurrently), continuous authentication and behavioral biometrics that tracks user history long-term to ensure they are who they claim to be. These are the things designed to ensure that attackers have a long way to go should they decide to copy or make use of stolen data. Furthermore, unstructured threat intelligence in a real-time manner, dynamic authentication technologies, and the incorporation of AI-powered analysis of anomalous login cases are enabled to detect abnormal login activity and signal suspicious activity early. Although these are better solutions, they cannot exist on their own without a proper user education, periodical system audits and security and privacy best practices implemented in order to build a truly dent resistant identity protection framework (Cavoukian, 2013).

3.2 Deepfakes and Synthetic Identities

The issue of identity change has taken a different twist with the deepfake technology. With the help of generative adversarial networks (GANs) based on AI, an attacker can create hyper-realistic images, videos, or voice recordings of real people. They can be turned into an impersonator of victims during video conversations, break through the facial recognition system, or participate in fraud areas where before, physical presence was the only way to do it. Such attacks have become much more possible and even more hazardous with the emergence of remote verification services.

Synthetic identity thefts are registration of wholly imaginary identities with a mixture of both real and fake details all together- say a valid social security number coupled with a phony name and an address. This identity leads to the procurement of credit cards, government benefits or internet services and over a period of time, develop credibility before finally engaging in wholesale fraud. These artificial profiles can be introduced into the biometrics by means of manipulated photos or false fingerprint scans.

The trouble with security systems is that these synthetic resources allow a high level of realism. Without the liveness detection or its poor implementation, deepfakes can even pass low-level biometric checks. Voice clones trained on open sources can replicate the voice of an individual with disturbing accuracy, which endangers voice-based authentication systems deployed in the banking and query services (Daugman, 2004).

3.3 Data Breaches and Biometric Database Leaks

Data breaches easily stand as one of the greatest enablers of digital identity theft. Due to sensitive data like biometric templates, credit cards and personally identifiable information (PII) in databases, cybercriminals attack the databases, in a bid to get hold of the information contained. The consequences of breach of biometric data is devastating, since such information can never be revoked or reissued as is the case with passwords or tokens. This is what renders the biometric leaks especially enticing to criminals and harmful to their recipients.

However, some high-profile cases, including the breach of Biostar 2 in 2019, which revealed facial recognition and fingerprints of millions of unencrypted files. Besides jeopardizing the privacy of the users, the breaches damage the trust associated with biometric authentication systems. To minimize the associated risk, organizations, which are tasked to store such data, are becoming expected to install extended encryption, template protection, and frequent security reviewing options (European Union, 2016).

Policy and regulation is crucial besides technical solutions. Regulations like GDPR require biometric data to be dealt with thoroughly with the requirement of informed consent, data minimization, and the right to erasure. In the case of organization non-compliance, the make the punishment really harsh, so cybersecurity and the compliance to law should go hand in hand in the bio metric data protection.

4. BIOMETRICS IN CYBERCRIME INVESTIGATIONS

4.1 Biometric Forensics: Fingerprint, Iris, and Facial Analysis

Use of biometrics in investigating crimes committed online is now being used not only to deter future crimes but also to provide forensic evidence. When the cybercriminals work in the physical world and technological environment by using stolen equipment, breaching secure zones, engaging in face-to-face scams, biometric traces will become provisional connecting lines between the suspect and the offense. The examples of forensic biometrics are the extraction of fingerprints out of electronic machines, iris identification based on surveillance recording, and face examination using video evidence.

One of the most recognized and legally admitted evidence is fingerprint recognition, which is not new and not immature. What is taken away are the latent prints on keyboards, touch or smart devices whose processing and comparison with the national or international database (as AFIS - Automated Fingerprint Identification System) can be done. Fingerprints maintain a saving level with regard to cybercrimes because when properly issued and stored; they give unquestionable connections between the culprits versus the digital devices applied in computer-related criminal activities (FindBiometrics, 2019).

The other important tool of investigation is facial recognition especially in the surveillance of the public and the digital evidence. The facial data captured on CCTV cameras, social media, or webcam can be traced and compared with target faces with the help of facial recognition algorithms. Although this technique has been successful, it has also led to privacy, misidentification and racial bias concerns and so many agencies are embracing the ethical review process prior to the utilization of facial recognition in active investigations.

4.2 Linking Biometric Evidence to Cybercriminal Activity

Data gathered through devices or other digital encounters may be instrumental in attaching suspects to certain cyber operations. As an example, a fingerprint on a seized laptop can not only confirm the ownership of a device but it also can corroborate timeline of activities put together using digital records. Biometric evidence is significant when used with the browsing histories, user credential, and IP addresses to create solid supportive content.

Attribution is also made through behavioral biometrics. Rhythms of typing, the movements of the mouse, fingerprints, and touch-screen gestures can be typed and tested against the bodies of an individual user. These types of proactive detection are especially helpful in the detection of an insider threat or when credentials have been shared or stolen. When a logging session can be shown to match the typing pattern of a would-be unauthorized user, then there could be a case of identity theft or connivance (Jain et al., 2016).

Voice biometrics that are usually logged during phishing calls or overheard conversations can also be investigative. Complex voice forensics is capable of matching voice files with known suspects and even pick out emotional states, accent or signs of deception. It could aid in reducing the number of suspects, or reinforce circumstantial evidence, with other information.

The legal system and standards of evidence are dependent upon location. To secure the admittance of biometric data in court, it is necessary that it should be acquired by lawful methods, to observe the procedures of chain of custody, and to be verified by professional observers. The increased application of biometric forensics therefore raised the need of continuous training and standardization in law enforcement and justice systems (Jain et al., 2006).

4.3 Biometric Authentication Logs in Forensic Analysis

The use of the access systems based on biometry provides authentication logs that are exceptionally resourceful pieces of evidence in the process of cybercrime and digital security breach forensic investigation. Such log files often contain a rich amount of contextual data including exact time of access, user identifiers, device data, place of access and the status of the biometric match in the form of a confidence value or in the form of a vector match. The level of logging involved in such complete logging will generate an audit trail that can be verified in order to reconstruct what occurred and assign blame in the event of an unauthorized attempt at system access or when data has been compromised. With appropriate analysis, such logs will have the potential to trace a particular occurrence of biometrics to a digital event thereby providing indispensable backup to investigation narratives (Goyal, 2025).

As an example, one can consider a scenario of internal data breach when a theft of confidential information is suspected on one of the employees. In case when the worker states that his/her account was hacked or used by the third party, researchers will be able to review the biometric authentication history logs and identify whether a biometric sample too risky of the event is identical to the biometric mark patterns in the past. In case the log suggests strong match with the possible stored fingerprint or facial recognition templates of the employee, there might be the high probability of direct rather than impersonation by third parties involvement. On the other hand, departures observed in the scores of matches or inconsistencies in geolocation can confirm allegations of identity theft or access with fake credentials.

The fact regarding the strength of evidence of a biometric logging is even more impressive when it is used in operating systems in which multi-factor authentication or a more flexible protocol in accessing the system is being employed. In these types of systems, biometric information is augmented with other identity cues such as recognition of a particular device, geolocation information, and behavioral profiling and creates a multi-dimensional account of every authentication instance. The mixture is a strong online path that is significantly harder to alter or fabricate by the attackers. Fine-grained variability in user behavior and patterns, e.g. typing speed, login timing patterns, or micro-expressions when recognizing someone by their face are difficult to detect in such logs, but can be checked further by advanced analytics like AI-based anomaly spotting, giving greater speed and accuracy to investigations (Li & Jain, 2015).

But to remain valid and admissible in court, biometric authentication logs should be secured under severe procedural and technical controls. This is by making sure that the logs are immutable once they have been generated by designing systems to use immutable storage or blockchain technology. It is obligatory to encrypt data at rest and in transit so that some unsanctioned person could not access or leak sensitive biometric information. Synchronicity of time between servers is essential because differences in the time value of the entry in the logs may undermine the realization of the forensic timeline. Also, organizations are to adhere to regulated data retention rules that determine the retention time of the biometric logs and conditions under which they should be accessed especially when an organization is engaged in compliance with legislation like the General Data Protection Regulation (GDPR) or Health Insurance Portability and Accountability Act (HIPAA) (Lin et al., 2017).

In addition to internal inquiries, biometric records can be instrumental even in international cybercrime investigations, particularly when the jurisdiction that is crossed by the crime cannot be easily overcome. Harmonization and interoperability of international biometric biometric data protection standards, i.e., harmonization with the standards set forth by the FIDO alliance or ISO /IEC biometric information protocols can enable law enforcement agencies to operate more effectively, without affecting the legal and privacy requirements, since these international standards can be subject to various privacy and legal requirements. With the cyber locations growing both more elaborate and more potentially hostile, as identity spoofing and anonymized assaults grow in prominence, properly kept and well-protected biometric verification logs present a base level of veracity, enabling forensic systems to unite digital actions with in-life identities with increased assurance (Lyon, 2007).

5. BIOMETRIC-BASED DIGITAL IDENTITY PROTECTION

5.1 Multi-Factor Authentication and Biometric Access Controls

Multi-factor authentication (MFA) is an underlying security-based practice that needs users to show various credentials of identity prior to accessing the systems or information. The "something you are" (e.g., fingerprints or face recognition) is also on the verge of becoming a significant part of the multi-factor authentication alongside the well-known "something you know" (e.g., a password) and "something you have" (e.g., smartphones or tokens). The two-factor authentication offers a significant level of protection against stolen credentials where the chances of access without authorization are minimal even when a user is aware of one of the credentials.

Biometrics access controls have become a part of broad application use, including smart phones, bank software, office access systems, and file cabinets. These systems are convenient, fast, and with a high degree of certainty. Biometrics are considered such a solution as they cannot be easily forgotten, stolen, and duplicated like passwords or access cards in the daily life or enterprise situations (Meingast et al., 2007).

Usability and accessibility in the location of MFA-based organizations is also required. There can be situations when differently able individuals or those having biometric deformities cannot tackle some of the modalities and thus alternative measures are required. Fallback mechanisms, inclusive design, and clear processes of consenting are vital requirements in terms of making biometric systems effective, equitable, and secure to any user.

5.2 Biometric Encryption and Secure Identity Frameworks

With increase in the usage of biometrics, there is also an increased need to ensure that biometrics are not abused, and that biometrics data is not stolen. A new direction is biometric encryption where biometric features is joined with cryptographic algorithms to ensure security of data integrity and security. The model describes a cryptographic key as the biometric features that are not stored in a form of a fixed template; rather, it is incorporated in the form of a cryptographic key. The keys are used in such a way that they can be recreated unless the right biometric is submitted (Goyal et al., 2025b).

Biometric encryptions have some benefits. It limits storage of plain biometric information, which curbs chances of theft of the biometric templates. In event of your encrypted information being lost, through reverse-engineering, there is no possibility of recovering the original biometric input without having the access to the unique features of the user. This will make it an excellent solution to big security angles such as finance, defense and e-governance.

The decentralized designs are useful in the secure exposure systems where users have their biometrics credentials on their personal devices or digital wallets. The focus of these frameworks is on privacy-by-design, user sovereignty and less dependence on central databases and databases which are hotspots of cyberattacks. In such models, blockchain technology is commonly thought of, to prevent identity transaction management and trust verification with the possibility of having a single point of failure (Monahan, 2011).

5.3 Mobile Biometrics in FinTech and e-Governance

The increased number of mobile devices has been a prompt of substantial use of biometric authentication in the financial technology (FinTech) and electronic governance (e-Governance) services. Smart phones now include fingerprint and facial recognition where they can conduct a banking transaction, make digital payments and secure their log in. These mobile biometrics also minimize the usage of compromising PINs and textual passwords besides enhancing user experience.

The entire Mil (account registration), payment approvals, filtering fraud and behavioral monitoring are all done using biometric authentication in the FinTech industry. Keystroke analysis and voice recognition have come about as methods of continuous authentication, which raises suspicious activity in real time. These methods assist the financial institutions in fighting identity theft, account takeovers, and insider threats better than the legacies do.

The application of e-Governance further exploits the use of biometrics to facilitate swift delivery of services to citizens. National digital ID systems such as India Aadhaar or Estonia e-ID use biometrics characteristics so that services are distributed to the proper people. Such systems give the citizens exposure to government provision, health and electoral systems in a more secure and less costly administrative system (NIST, 2022).

Although mobile biometrics can be of great benefit, they present exclusive issues, which include device theft vulnerabilities, biometric spoofing, and poor app integration. Biometric data has to be localized, encrypted, and should not be shared unless with clear permission by regulators and developers. An attempt to find the golden mean between the convenience, security and privacy is a great challenge as biometric identity gets more connected to mobile platforms.

5.4 Best Practices for Secure and Practical Biometric Deployment

Biometric systems must adapt to diverse operational environments, including organizations with limited budgets, developing regions, or high-risk sectors. Effective deployment requires balancing cost, accuracy, usability, and privacy.

A. Security Best Practices
   - Use encrypted biometric templates (never raw biometric images).
   - Implement liveness detection to prevent mask attacks and deepfake attempts.
   - Adopt multi-factor authentication (biometric + token/password).
   - Perform continuous updates of biometric algorithms and threat models.
   - Enforce strict access controls and audit trails for all biometric logs.

B. Deployment in Resource-Limited Settings
   - Prefer low-cost sensors capable of offline matching.
   - Use edge computing or on-device storage to avoid expensive servers.
   - Adopt multimodal lightweight biometrics such as keystroke and touch dynamics.

- Use open-source standards (ISO/IEC 19794) to avoid vendor lock-in.
- Train local personnel and establish simple maintenance routines.

C. Operational Best Practices

- Conduct periodic re-enrollment to handle aging and biometric drift.
- Provide clear user training to reduce errors during capture.
- Use redundant fallback authentication for damaged or atypical biometrics.

6. BIOMETRIC SYSTEM VULNERABILITIES AND FORENSIC CHALLENGES

6.1 Spoofing and Presentation Attacks

One of the most challenging problems facing the biometric security systems is spoofing. Presentation attacks present impersonation usage of synthetic biometrics characteristics, i.e., artificial fingerprints, 3D masks, or artificial voice recordings to confuse certain biometrical scanners. These attacks based on the assumption that biometric features are considered to be rather reliable skip the security measures in high-value solutions such as banking operations, airport management, and entry access.

As an example, it is possible to spoof fingerprint via the use of such material as silence, gelatin, or even 3D-printed molds. Facial recognition tools have been fooled with photography and videos or advanced deepfake. Not quite different, voice authentication systems have become victims to voice clones generated by AI. These attacks show that the uniqueness in terms of biometrics is not enough to make invulnerable.

Anti-spoofing is done by developing the capacity to detect life, like pulse, eye blinking or changes in texture of a surface where the liveness is determined by the device. There is also the multi-modal authentication where two or more biometric attributes are identified at the same time which can increase the level of the attacker. Nevertheless, not even liveness detection is foolproof and will cause usability problems unless well-calibrated (Patel et al., 2015).

6.2 Privacy Risks and Consent Management

Biometric information is unique and personal, which means that it is irreplaceable and not susceptible to being changed in case it is stolen. This nonchangeability brings in tremendous privacy issues. Uncontrolled use of biometric information, whether thanks to surveillance, commercial use, or information leakage, will cause a lasting harm in individual freedom and safety.

A key tenet of ethical adoption of biometrics is the question of consent. The consumers should also understand the purpose of collecting information, the manner in which it is stored as well as to which end the information will be used. Regrettably, most systems are based on tacit or forced compliance in the form of sober text and conditions. Facial data collected in public venues with surveillance and facial recognition usually lack an opt-out function, which results in massive biometric data gathering with no solid explanation or responsibility (Pugliese, 2020).

What is more, the complete biometric data, as well as data combined with other collections, can identify sensitive information, such as characteristics of personal health conditions, moods, and behavioral habits. The implications of such profiling consist of discrimination, surveillance capitalism, and social scoring. Because of this fact, the international rules, like GDPR in Europe and CCPA in California consider biometrics data as a special category of personal data and demand additional protection over it.

Good consent management takes the form of transparency, data minimization, safe storage and proper opt-outs. Federated learning or differential privacy is a privacy-preserving technology with a potential to develop trustworthy biometrics systems.

6.3 Legal Admissibility of Biometric Evidence

The chain of custody and procedural integrity also plays an essential role. Biometric information should be obtained in a legal manner, stored safely and should be analyzed by authorized individuals. The failure of this process in any manner may make the evidence inadmissible or questionable. Furthermore, biometrics surveillance not followed by warrants and due control can infringe the constitutional rights, which have legal implications on the law enforcement authorities (Rao & Yarlagadda, 2022).

Since biometric technologies are developing, the laws and regulations have to change in a way that would further the idea of accountability and fairness. The key steps should be the judicial training, technology audits and creation of forensic standards of biometric evidence. The most important point of a responsible use of biometric tools in digital investigations is the fact that they should be used to serve justice and not to impede it.

6.4 Accessibility and Inclusive Biometric Design

Biometric systems must be inclusive to ensure fairness across populations, including users with disabilities, aging populations, and people with atypical biometric traits.

A. Challenges
   - Worn fingerprints among manual laborers reduce fingerprint accuracy.
   - Facial differences (disfigurement, low vision, covered faces) can weaken face recognition.
   - Speech impairments affect voice biometrics.
   - Mobility restrictions make traditional sensors difficult to reach.

B. Inclusive Design Principles
   - Always include multiple biometric options (face + fingerprint + PIN fallback).
   - Allow user choice over preferred modality.
   - Implement touch-free alternatives such as voice or iris for mobility-impaired users.
   - Use accessible interfaces compliant with WCAG standards.
   - Support adaptive algorithms that adjust to aging or physiological changes.

7. APPLIED BIOMETRICS IN IDENTITY THEFT INVESTIGATIONS

7.1 Use of Facial Recognition in Cybercrime Investigations

Approaches such as facial recognition have become critical biometrics instrument in criminal exploration in the cyber world because they can decode and reconcile face characteristics shown in camera live streams, social media, and surveillance videos. Police departments have combined facial recognition with the available forensic databases to detect criminals who are engaged in crimes that include digital frauds to online bullying and the coordination of cyber-attacks. These applications enable the investigators to automate the process of matching identities, and this eliminates time used to conduct manual checks drastically.

Among the well-known uses were ones relating to the identification of suspects who are part of ransomware bands and phishing rings. The investigators matched the corresponding avatars on the Internet and appearances in the conference with people qualified to visit security expos or add messages on dark web forums through facial recognition software. Together with IP data, linguistic patterns as well as transaction histories, this biometric input was used to solidify attribution and arrest warrants (Goyal et al., 2025c).

Nevertheless, the facial recognition technology has been subject of criticism on grounds of privacy, racial bias and accountability. Research has indicated the misidentification occurs at a disproportional level on persons of color, with legal suits and prohibitions in San Francisco and Boston cities. To solve these issues, the human touching the loop verification, stringent policy design and accuracy metric now become considered best practices.

7.2 Aadhaar and Biometric Identity in India: Benefits and Risks

The Aadhaar program in India is the largest biometric identity system in the world with the population of more than one billion citizens. Aadhaar works on fingerprints, iris and facial photographs to distribute a 12-digit distinct Identification number that is related to government services, subsidies, banking and medical facilities. This centralised digital identity system has an immense advantage in identity verification particularly among the marginalised groups who do not have conventional documents.

Cybercrime wise, Aadhaar has saved welfare scheme impersonation fraud and made financial tracing easy. It has likewise made it feasible to add biometric based authentication into the issuance of mobile SIM cards, digital wallets and Know-Your-Customer (KYC) procedures. These integrations facilitate cybercrimes cases by bringing forth audit trails and limiting the amount of underlying anonymous digital identities (Ratha et al., 2001).

However, Aadhaar has had controversies surrounding its privacy concern. Several data breaches and unauthorized access as well as poor encryption have led to the query of surveillance and misuse. Its unilateral adoption in many spheres of life, including school enrollment or SIM card activation, has been criticized by civil liberties organizations as establishing a mass surveillance infrastructure and discriminating against people who fail to authorize because of biometric differences.

The case of the Aadhaar shows how a compromise should be maintained between mass use of biometrics and the safeguarding of the data collected. The constant legal investigation by the Supreme Court of India, as well as the creation of the Data Protection Bill, are indications of the wish to strike a balance between national security and the good of people and the individual rights and system protections of people.

7.3 Biometrics in International Border and Immigration Control

The applications of biometrics systems in the management of immigration, border security and international travel have become unavoidable. Nations in the world implement fingerprint, iris and facial recognition systems to check travelers, prevent passport fraud and identity of security threats. Automatized e-gates in airports today automatically scan biometrical passports and compare them to real-time facial information in order to speed up their entry-clearance.

In America, biometric databases such as the IDENT system or the Automated Biometric Identification System (ABIS) operated by the Department of Homeland Security contain millions of bio-records that are used by immigration services to facilitate visas, by asylum seekers to meet the refugee claim, and by police investigations. Like in the Schengen Area in Europe, biometric visa data are also used to verify cross-border identity frauds and stop overstays (Ross & Jain, 2003).

Biometric identification in borders has added to the investigative power, as responses enable the authorities to identify individuals with records in many databases, despite the fact that some of them might be using aliases or even forged documents. As an example, international criminals wanted in different countries will be seized in airports depending on the cross-border biometric watchlist prepared under INTERPOL or EUROPOL.

Nonetheless, the trend of deploying biometrics border controls throughout the world creates concerns in regard to data sovereignty, data within borders, and rights of asylum seekers and migrants. Inaccuracies caused by biometrics at the checkpoints can result in wrong detainment or refusal. Moreover, the paranoia about the surveillance by the government and the storage of the non-citizen biometrics have caused the privacy and human rights debates.

7.4 Cross-Border Interoperability and Law Enforcement Collaboration

Cybercrime frequently spans multiple countries, yet biometric systems and legal frameworks differ widely.

A. Interoperability Issues
   - Different biometric standards (ISO, FBI EBTS, EU EES) complicate cross-border matching.
   - Varied data protection laws restrict sharing of biometric templates.
   - Fragmented databases across agencies prevent unified identification.

B. Challenges in International Investigations
   - Jurisdiction limits prevent agencies from directly accessing foreign biometric databases.
   - Misaligned data retention policies cause evidential inconsistencies.
   - Legal barriers restrict real-time biometric watchlists across borders.

C. Emerging Solutions
   - INTERPOL's Biometric Hub standardizes fingerprint and facial data.
   - EU's ECRIS-TCN links criminal records using fingerprints.

- Blockchain-based identity sharing for tamper-proof cross-border verification.
- Mutual Legal Assistance Treaties (MLATs) enable conditional sharing of biometric evidence.

8. LEGISLATIVE AND ETHICAL CHALLENGES IN BIOMETRIC-BASED IDENTITY SECURITY

8.1 GDPR, HIPAA, and Biometric Data Protection Laws

The legal context of biometric information is developing very quickly, with more and more applications provided to the sphere of cybersecurity, law enforcement, and digital identity verification. The European Union General Data Protection Regulation (GDPR) refers to the biometric data as a class of a special category of personal data, which must have special protection. By GDPR, the organisation handling biometric data must be sure that, by law, there is a specific consent needed of the individuals to have their data processed and that there is an open reason in it and this especially in surveillance or profiling.

Regulations In the United States, regulations specific to the sector determine how and if biometric data should be handled, with the Health Insurance Portability and Accountability Act (HIPAA) as an example of some regulations applying restrictions concerning biometric data in context of healthcare, in the form of voiceprints or retinal scans utilized in a patient identification program. Moreover, at the state levels, regulations like Biometric Information Privacy Act (BIPA) of Illinois establish the strict regime of collecting biometric data, so that they must be provided with a notice, written consent, and well-considered retention policies. Failure to comply with these regulations has caused major litigations with increasing landscapes of legal risk (Saini & Khera, 2021).

In many parts of the world, other countries have done the same thing. The LGPD of Brazil, the PIPEDA of Canada, and the draft Digital Personal Data Protection Bill of India include the provisions of biometric data. Although there are differences between the scopes of these regulations, they have common themes: the requirement of informed consent, purpose limitation, right to withdraw the consent. Nonetheless, there is always the problem of consistency in execution and jurisdictional overlaps put in place concerning the application of biometrics across borders.

On the whole, the law systems are getting stronger in identifying biometric information as highly confidential and stakes. Companies processing biometric data have to integrate privacy-by-design standards and should also be ready to undergo the benefits of compliance audit and civil liability. In order to be successful within a large-scale application, biometric systems should be not only adjusted to the processes of advancement in technologies but also to the changes in law and peoples demands.

8.2 Ethical Dilemmas in Surveillance and Profiling

The moral acceptable approach to the use of biometric technologies is as important as its legality. When applied to surveillance purposes, facial recognition, gait analysis and behavioral profiling, among others, might result in a thin dividing line between security and authoritarianism. The application of biometric systems occurs without user consent, transparency, and redress such that the use of these systems in the streets leads to creating a continuous air of surveillance.

Among the key ethical issues is the possible discriminatory programming and bias nature of algorithms. Research has indicated that certain facial recognition systems are erroneous to dark-skinned people and women to a greater extent due to imbalance in false positives. When it comes to law enforcement, it can lead to false suspicion, arrests, and a lack of trust seen by the general population. The development of ethical systems should guarantee the testing of the algorithm on different populations, as well as an inclusion of anti-discriminatory outcomes (Goyal et al., 2025d).

The other dilemma is the one of consent and autonomy. In a lot of cases, especially CCTV connected face recognition or mandatory biometrics attendance systems a person is not given much choice. This brings coercion, digital inequality, and agency to question. The moral obligation is the users need to be aware of their data-use, and be able to exercise sufficient control.

In order to mitigate these issues, principles of transparency, accountability, fairness and proportionality ought to be dispensed, in ethical frameworks regarding the use of biometrics. They can be monitored with the help of independent ethics boards, as well as impact assessment and public consultations. After all, biometrics must intensify rather than reduce human rights, and this requires governance to modify, alongside with technology development.

8.3 Frameworks for Biometric Governance and Accountability

To achieve responsible, secure, and accountable implementation of biometric systems a clear governance structure must be in place. Governance is not only about the legislation observations but it is also owning structures of governance, monitoring, and progression. It also entails laying down data stewardship roles and responsibilities, issuing audit trails and stake out procedures to handle grievances and breaches.

Corporate policy leads to effective governance. Institutions implementing biometric would need to come up with internal policies directing where and in what form biometrics arc implemented, with what power or authority and who uses such data. Auditing and risk assessments should be done regularly to identify any misuse or drift in the system, and effective logging and traceability will bring more transparency to operations (Sambasivan et al., 2021).

There is also stakeholder involvement in the concept of accountability. The design of biometric systems must include the civil society, legal professionals and the target population, especially when it is intended to be used in a public place or civic initiative. This collaborative technique will equate to the development of systems designed to not only be efficient but also publicly acceptable, and more importantly ethically sound.

New international standards ISO, IEEE and NIST offer technical and procedural benchmarks in the design, testing and deployment of the biometric systems. They must be however accompanied by national and industry specific policies that deal with context-related risks. Governance is not a fix and forget list: it is a flexible system that needs to adapt to changes in technology, social norms and emerging vectors of attack.

9. EMERGING TRENDS AND INNOVATIONS IN BIOMETRIC CYBERSECURITY

9.1 AI and Machine Learning in Biometric Cybersecurity

Artificial Intelligence (AI) and Machine Learning (ML) are revolutionizing the traditional relationship between biometric systems and greater accuracy, real-time processing capabilities, and an adaptive reaction to new threats. Facial and fingerprint recognition has become common practice in deep learning models, especially with convolutional neural networks (CNNs), because systems gain the ability to recognize complex patterns that define people accurately.

There is also a liveness detection where AI also gives a significant boost to block spoofing attacks. ML algorithms have a chance to identify micro-expressions, pulse, or voice disturbances that cannot be identified by a human examiner but demonstrate authenticity. Behavioral biometrics, also, enjoy the use of AI-driven models that constantly learn and evolve with the specific typing pattern or touch pattern of an individual to be able to dynamically authenticate (Schneier, 2015).

In cybersecurity, AI is increasingly used to detect intrusions, insider threat, and frauds using a combination of biometrical signals and other behavioral clues. An example can be studied with AI that can examine the deviations in the biometric pass log, e.g., onwer logins at odd hours or on different location, and flagging them.

With these developments, there are still issues raised about explainability and bias. Black-box AI algorithms can produce decisions that are not intuitive, and this gives problems in terms of legal and forensic considerations. Hence, in future, the emphasis should be on the create explainable AI (XAI) models aimed at biometric use and on ensuring the diversity and representativeness of training data so that the bias is reduced.

9.2 Decentralized Identity and Blockchain-Based Biometrics

The fact that most of the biometric data is stored centrally is a significant weakness, since one-time breach may provide access to very sensitive identifiers that cannot be recalled. To answer this, there has been increased interest in the decentralized identity with backing by the blockchain and distributed ledger technologies. With these models, people keep and control their own biometric credentials and provide access on a need-to-know basis using cryptographic protocols.

Blockchain can be applicable to the authentication scheme to ensure biometric credentials authenticity without revealing raw data. The conditional access to information using smart contracts is available, and the tamper evident characteristics of the blockchain guarantee transparency and accountability. Such architecture can be classified as privacy-by-design and minimizes the use of centralized databases by governments or companies (Solove, 2006).

Underway projects such as Microsoft ION or efforts by the Decentralized Identity Foundation (DIF) are hacking on blockchain-based identity ecosystems. The goal of these systems will allow users to have total control of their identity, such as having to be able to choose how and when to use their biometric data. The frameworks are especially applicable to cross-border identity verification, digital voting and access to health data.

Nonetheless, issues still exist especially the scalability, interoperability, and legal recognition. Users and institutions also have a learning process in the use of decentralized identity models. Addressing those gaps in future should focus on making sure that decentralized systems are friendly, safe, and palatable in terms of regulation.

9.3 Toward Privacy-Preserving and Ethical Biometrics

Because biometrics are increasingly becoming part of everyday lives, more privacy-preserving methods are demanded. The purpose of privacy-preserving biometrics is to guarantee the privacy of personhood alongside functionality of biometric systems. Approaches, like homomorphic encryption, secure multi-party computation, and differential privacy, are growing in the deployment to the setting of biometrics in order to enable new data processing to be performed without the disclosure of sensitive data (Tene & Polonetsky, 2013).

Homomorphic encryption uses encrypted data in calculations, and that is why biometric systems can authenticate identities without decryption ever happening in this case. This is especially beneficial in cloud systems in which data is handled by third parties. Differential privacy introduces mathematical noise to the sets of biometrics and does not allow re-identification of the same even when a big population is examined as a whole (Westin, 1967).

The next upcoming area is cancelable biometrics in which the original biometric model is converted to a cancellable and renewable format. The transformed template would not be needed to alter in the case it is compromised, as the physical biometric trait will not be required to be replaced. This takes care of one of the most serious drawbacks of the traditional biometric systems where the compromised data cannot be restored (Goyal, 2025e).

In the future, the aim will be to make systems not only accurate, efficient, but also user-controlled, secure and ethical. It will also be critical to generate interdisciplinary engagement between technologists, policy makers and ethicists who will work toward the development of biometric systems that do not impose threat on basic rights, and engage in terms of serving public good.

9.4 Future Research Directions and Open Challenges

Biometric security continues to evolve, but several gaps remain:

A. Technical Challenges
    - Robust deepfake-resistant biometrics require innovative liveness methods.
    - Cross-device biometric consistency remains a challenge in mobile ecosystems.
    - Developing explainable biometric AI models suitable for court evidence.

B. Privacy and Policy Challenges
    - Need for global biometric governance frameworks.
    - Stronger legal protection for stored biometric templates.
    - Mechanisms for user revocation or template replacement after a breach.

C. Ethical and Social Research
    - Bias mitigation across emerging modalities such as gait and voice.
    - Impacts of biometric surveillance on civil liberties.

- Public acceptance and trust-building models for biometric deployments.

D. New Frontier Technologies
  - Quantum-secure biometric cryptography.
  - Zero-knowledge biometric proofs for anonymous authentication.
  - Hybrid neuro-biometric behavioral signatures.

CONCLUSION

The successive integration of biometric technologies and digital ecosystems has severely transformed the environment of cybersecurity and cybercrime investigation. During a time where digital identity is continually at a threat, biometrics can be established as a formidable and proactive method of carrying out identity authentication, tracking criminal activity and absorbed the continuously rising risk of digital identity theft. Innovative technologies such as biometrics are becoming a necessary element of a security structure in the contemporary digital environment because traditional authentication systems such as passwords and PINs are appearing less and less secure.

Being able to recognize a face or capture fingerprints, scan irises, use behavioral biometrics, as well as voice authentication, these technologies not only are as accurate as possible but also leave a non-repudiable identity confirmation. They have been used in very diverse fields such as the banking, healthcare, law enforcement, and national security. Biometric evidence also plays a great role in naming suspects, interpreting crime scenes, and proving cases in the judicial system in forensic arena. They can be used in preventive security through multi-factor authentication (MFA) and active user verification to help detect the suspicious activity in real-time.

Utilization of biometrics in combination with Artificial Intelligence (AI), Machine Learning (ML) and the Internet of Things (IoT) raised the potential even more. The current smart surveillance systems have been developed with the capability to be dynamically adjustable to the threats, real-time monitoring, and capable of supporting context-aware authentication systems. As an example, the facial recognition systems implemented in public transport powered by AI may follow the persons of interest and the devices enabled by the IoT may limit the access depending on the biometric signature and behavior patterns.

Nevertheless, this fast development does not exist without difficulties. Attacks based on fake fingerprints or their photos, deepfake technologies that process facial and voice data and biometric data breaches are the threats that can spoil the system integrity and the trust that the user gives to the system. In contrast to the passwords, biometric information cannot be changed once they are stolen. Also, interoperability between the data in different sources, and the absence of standardization and high implementation cost are still holding back the implementation in large scale, particularly in the resource-limited settings.

In the future, the authors anticipate that the progress in decentralized identity systems, zero-knowledge-based solutions, privacy-oriented biometric infrastructures brings promising avenues that will enable secure, scalable and trusted systems in the future. The field of AI will keep on improving biometric performance, accuracy, and defense against spoofing attacks. In the meantime world coordination of standards and ethical norms will prove critical to strike a balance between security and human rights.

To sum up, achieving this balance between innovation and accountability and making sure that deployment is propelled by ethical standards will allow biometrics to remain the central part of future digital identity protection in the highly connected and data-intensive digital world.

REFERENCES

Adjerid, I., & Kelley, K. (2018). The role of biometric authentication in reducing identity fraud. *Journal of Cybersecurity*, *4*(1), 1–10. DOI: 10.1093/cybsec/tyy002

Alsaadi, H. A., & Tubaishat, A. (2015). Internet of Things: Features, challenges, and vulnerabilities. *International Journal of Advanced Computer Science and Applications*, *6*(11), 128–133.

Bansal, A., & Jain, A. K. (2019). Deep learning for biometric recognition: Trends and challenges. *IEEE Transactions on Pattern Analysis and Machine Intelligence*, *43*(10), 3461–3481.

Bhargava, B., & Yu, L. (2017). Biometric authentication in cloud computing: Challenges and solutions. *Computer*, *50*(6), 60–69.

Biometric Information Privacy Act (BIPA), 740 ILCS 14/1 (2008).

Bowyer, K. W., Hollingsworth, K., & Flynn, P. J. (2008). Image understanding for iris biometrics: A survey. *Computer Vision and Image Understanding*, *110*(2), 281–307. DOI: 10.1016/j.cviu.2007.08.005

Buolamwini, J., & Gebru, T. (2018). Gender shades: Intersectional accuracy disparities in commercial gender classification. *Proceedings of the Conference on Fairness, Accountability and Transparency*, 77–91.

Cavoukian, A. (2013). *Privacy by design: The 7 foundational principles*. Information and Privacy Commissioner of Ontario.

Daugman, J. (2004). How iris recognition works. *IEEE Transactions on Circuits and Systems for Video Technology*, *14*(1), 21–30. DOI: 10.1109/TCSVT.2003.818350

European Union. (2016). General Data Protection Regulation (GDPR). Regulation (EU) 2016/679.

FindBiometrics. (2019). Major biometric data breach hits Suprema's BioStar 2 platform. Retrieved from https://findbiometrics.com

Goyal, N. K. (2025). Weaponizing Misinformation and Cyberattacks on Supply Chains. In Bukhari, S., & Zafar, I. (Eds.), *Mass Media and Impact of Fake News on Supply Chains* (pp. 177–210). IGI Global Scientific Publishing., DOI: 10.4018/979-8-3373-2175-2.ch007

Goyal, N. K. (2025e). Security and Privacy in IoT, Fog, and Blockchain Networks. In Sathio, A., Rind, M., Awan, S., & Junejo, A. (Eds.), *Energy-Efficient Deep Learning Approaches in IoT, Fog, and Green Blockchain Revolution* (pp. 371–398). IGI Global Scientific Publishing., DOI: 10.4018/979-8-3373-0300-0.ch015

Goyal, N. K., Dandotiya, M., Kumari, M., Kaloria, S., & Anushya, A. (2025d). AI-Powered Haptic Devices for Patient Rehabilitation. In Mamodiya, U., Kalam, A., Bossoufi, B., Awad, M., & Somwanshi, D. (Eds.), *Integrating AI With Haptic Systems for Smarter Healthcare Solutions* (pp. 143–170). IGI Global Scientific Publishing., DOI: 10.4018/979-8-3373-2307-7.ch007

Goyal, N. K., Dandotiya, M., Kumari, M., Sheikh, M. F., & Sharma, S. (2025b). Emerging Job Roles in the Tourism Sector Leveraging Artificial Intelligence. In Silic, M., Luthia, M., Ahmad, F., & Chakraborty, P. (Eds.), *Empowering Sustainable Performance and Competitive Advantage in Tourism* (pp. 89–126). IGI Global Scientific Publishing., DOI: 10.4018/979-8-3373-4546-8.ch004

Goyal, N. K., Dandotiya, M., Qurashi, J. A., Kumari, M., & Sharma, S. (2025c). AI-Driven Predictive Analytics for Disease Management. In Ouaissa, M., Ouaissa, M., Imad, M., Qurashi, J., & Farooq, M. (Eds.), *Utilizing AI of Medical Things for Healthcare Security and Sustainability* (pp. 231–276). IGI Global Scientific Publishing., DOI: 10.4018/979-8-3373-0690-2.ch008

Goyal, N. K., Kumari, M., Kushwah, V., Dandotiya, M., & Vijay, A. (2025a). Evaluating blockchain security: Trade-offs in consensus mechanisms. *IET Conference Proceedings*, 2024(38), 80–88. DOI: 10.1049/icp.2025.0775

Jain, A. K., Nandakumar, K., & Ross, A. (2016). 50 years of biometric research: Accomplishments, challenges, and opportunities. *Pattern Recognition Letters, 79*, 80–105. DOI: 10.1016/j.patrec.2015.12.013

Jain, A. K., Ross, A., & Pankanti, S. (2006). Biometrics: A tool for information security. *IEEE Transactions on Information Forensics and Security, 1*(2), 125–143. DOI: 10.1109/TIFS.2006.873653

Li, S., & Jain, A. K. (Eds.). (2015). *Encyclopedia of Biometrics* (2nd ed.). Springer. DOI: 10.1007/978-1-4899-7488-4

Lin, J., Yu, W., Zhang, N., Yang, X., & Zhao, W. (2017). A survey on Internet of Things: Architecture, enabling technologies, security and privacy, and applications. *IEEE Internet of Things Journal, 4*(5), 1125–1142. DOI: 10.1109/JIOT.2017.2683200

Lyon, D. (2007). *Surveillance studies: An overview*. Polity Press.

Meingast, M., Roosta, T., & Sastry, S. (2007). Security and privacy issues with health care information technology. *Proceedings of the 28th IEEE EMBS Annual International Conference*, 5453–5458.

Monahan, T. (2011). Surveillance as governance: Social inequality and the pursuit of democratic surveillance. *Surveillance & Society*, *9*(3), 1–17. DOI: 10.24908/ss.v9i1/2.4099

NIST. (2022). Face Recognition Vendor Test (FRVT). National Institute of Standards and Technology. Retrieved from https://www.nist.gov

Patel, V. M., Gang, H., & Chellappa, R. (2015). Cancelable biometrics: A review. *IEEE Signal Processing Magazine*, *32*(5), 54–65. DOI: 10.1109/MSP.2015.2434151

Pugliese, A. (2020). Deepfake detection using convolutional neural networks. *IEEE Transactions on Neural Networks and Learning Systems*, *31*(10), 3769–3780.

Rao, M. R., & Yarlagadda, P. K. D. V. (2022). Ethical implications of facial recognition technology: A comprehensive review. *Technology in Society*, *68*, 101885.

Ratha, N. K., Connell, J. H., & Bolle, R. M. (2001). An analysis of minutiae matching strength. *Proceedings of the International Conference on Audio- and Video-Based Biometric Person Authentication*, 223–228. DOI: 10.1007/3-540-45344-X_32

Ross, A., & Jain, A. K. (2003). Information fusion in biometrics. *Pattern Recognition Letters*, *24*(13), 2115–2125. DOI: 10.1016/S0167-8655(03)00079-5

Saini, P., & Khera, R. (2021). Aadhaar: Biometric surveillance or identity empowerment? *Economic and Political Weekly*, *56*(10), 21–24.

Sambasivan, N.. (2021). Re-imagining algorithmic fairness in India and beyond. *Proceedings of the ACM Conference on Fairness, Accountability, and Transparency*, 1–18. DOI: 10.1145/3442188.3445896

Schneier, B. (2015). *Data and Goliath: The hidden battles to collect your data and control your world.* W. W. Norton & Company.

Solove, D. J. (2006). A taxonomy of privacy. *University of Pennsylvania Law Review*, *154*(3), 477–560. DOI: 10.2307/40041279

Tene, O., & Polonetsky, J. (2013). Big data for all: Privacy and user control in the age of analytics. *Northwestern Journal of Technology and Intellectual Property*, *11*(5), 239–273.

Westin, A. F. (1967). *Privacy and Freedom*. Atheneum.

Chapter 2
Forensic Audit Trails and Biometric-Based Authentication

Deepak Gupta
https://orcid.org/0000-0003-3929-1362
Institute of Technology and Management, Gwalior, India

Raghu Nangunuri
https://orcid.org/0009-0001-2372-1324
Kamala Institute of Technology and Science, India

Srinivasan Nagaraj
CBIT, Proddatur, India

S. Keerthi
Dayananda Sagar College of Engineering, Karnataka, India

Pratish Rawat
https://orcid.org/0000-0003-3230-7464
Poornima University, India

C. Umarani
https://orcid.org/0000-0002-9008-3350
Christ Academy Institute for Advanced Studies, India

Someshwar Siddi
https://orcid.org/0000-0002-6045-5408
St. Martin's Engineering College, India

ABSTRACT

Forensic audit trails combined with biometric-based authentication represent a critical convergence in modern cybersecurity infrastructure. This chapter examines the technical implementation, forensic methodologies, and investigative frameworks for biometric authentication systems. The integration of immutable audit trails with biometric verification creates comprehensive forensic evidence chains essential for digital investigations. We analyze forensic challenges including spoofing attacks, presentation attacks, and biometric template security. The chapter explores multimodal biometric systems, liveness detection mechanisms, and blockchain-based audit trail

DOI: 10.4018/979-8-3373-4972-5.ch002

implementations. Critical examination of privacy-preserving forensic techniques, GDPR compliance, and admissibility of biometric evidence in legal proceedings provides practical guidance. Real-world case studies demonstrate forensic analysis of compromised biometric systems.

1. INTRODUCTION

The proliferation of biometric authentication systems across financial institutions, healthcare facilities, government agencies, and consumer applications has fundamentally transformed identity verification paradigms (Jain et al., 2024). Biometric authentication leverages unique physiological and behavioral characteristics—including fingerprints, facial features, iris patterns, voice prints, and behavioral dynamics—to establish user identity with unprecedented accuracy. However, this technological advancement introduces complex forensic challenges requiring specialized investigative methodologies and comprehensive audit trail architectures.

Forensic audit trails in biometric systems serve dual purposes: operational security monitoring and post-incident investigation (Kaur & Khanna, 2024). These immutable logs capture authentication attempts, template matching scores, system access patterns, and anomalous behaviors, creating evidentiary foundations for digital forensics. The integration of biometric authentication with robust audit mechanisms enables forensic investigators to reconstruct security incidents, identify attack vectors, and establish chains of custody for digital evidence.

Contemporary biometric systems face sophisticated adversarial threats including presentation attacks (spoofing), deep learning-based forgeries, template reconstruction attacks, and privacy violations (Hadid et al., 2021). Forensic investigators must understand both the technical architecture of biometric authentication systems and the methodologies for detecting, analyzing, and prosecuting biometric-related cybercrimes. This chapter provides comprehensive coverage of forensic audit trail design, biometric security vulnerabilities, investigative techniques, and legal considerations.

The convergence of biometrics and forensics occurs within increasingly complex regulatory environments. The European Union's General Data Protection Regulation (GDPR) classifies biometric data as sensitive personal information requiring enhanced protection. Similarly, the California Consumer Privacy Act (CCPA), Illinois Biometric Information Privacy Act (BIPA), and emerging international frameworks impose strict requirements on biometric data collection, storage, and forensic examination. Forensic investigators must navigate these regulatory constraints while maintaining investigative effectiveness.

1.1 Background and Context

Biometric authentication emerged from forensic science foundations, with fingerprint analysis representing the earliest practical implementation. Sir Francis Galton's pioneering work in the 1890s established fingerprint uniqueness, while Edward Henry's classification system enabled systematic criminal identification. Modern biometric systems have evolved through three distinct generations: first-generation systems focused on verification accuracy, second-generation systems addressed liveness detection and anti-spoofing, and third-generation systems incorporate artificial intelligence, multimodal fusion, and privacy-preserving techniques (Singh et al., 2022).

The International Organization for Standardization (ISO) established comprehensive biometric standards including ISO/IEC 19795 for performance testing, ISO/IEC 30107 for presentation attack detection, and ISO/IEC 24745 for biometric information protection (Tabassi et al., 2024). These standards provide forensic investigators with benchmarks for evaluating system security and identifying vulnerabilities. The National Institute of Standards and Technology (NIST) conducts ongoing biometric technology evaluations including Face Recognition Vendor Test (FRVT), Fingerprint Vendor Technology Evaluation (FpVTE), and Iris Exchange (IREX), establishing performance baselines critical for forensic analysis (Tabassi et al., 2024).

Forensic audit trails evolved from traditional system logs to specialized biometric event recording mechanisms (Bhagat et al., 2023). Early implementations captured basic authentication success/failure events, while contemporary systems record comprehensive forensic metadata including biometric quality scores, matching confidence levels, sensor integrity status, template update history, and environmental conditions. Advanced audit architectures incorporate blockchain technology, ensuring immutability and cryptographic verification of forensic evidence (Jin et al., 2023).

1.2 Research Objectives and Scope

This chapter pursues four primary research objectives: (1) establishing comprehensive frameworks for forensic audit trail design in biometric authentication systems, (2) examining forensic methodologies for investigating biometric security incidents, (3) analyzing legal and regulatory considerations affecting biometric forensics, and (4) exploring emerging technologies reshaping the forensics-biometrics intersection.

The scope encompasses both physiological biometrics (fingerprint, facial recognition, iris scanning, retinal scanning, DNA analysis, vein pattern recognition) and behavioral biometrics (gait analysis, keystroke dynamics, voice recognition, signature dynamics) (Fierrez et al., 2021). Multimodal biometric systems combining multiple modalities receive particular attention due to their enhanced security properties and forensic complexity. The chapter addresses forensic challenges across the complete biometric system lifecycle: enrollment, template storage, authentication, system updates, and decommissioning.

2. BIOMETRIC AUTHENTICATION ARCHITECTURES

Understanding biometric authentication architectures provides essential context for forensic investigation (Anderson & Murdoch, 2024). Modern biometric systems implement layered security architectures integrating capture devices, feature extraction algorithms, template databases, matching engines, and decision modules. Each architectural component presents distinct forensic opportunities and challenges requiring specialized investigative approaches.

2.1 Fundamental Components and Data Flow

Biometric authentication systems comprise five fundamental components operating in coordinated workflows. The biometric sensor captures raw biometric data—images, scans, or recordings—representing the user's unique characteristics. Sensor quality directly impacts authentication accuracy and forensic evidence quality. High-resolution fingerprint scanners (500+ DPI), near-infrared iris cameras, and 3D facial recognition systems provide superior forensic evidence compared to low-quality consumer devices (Galbally et al., 2023).

The feature extraction module processes raw biometric data, identifying distinctive characteristics and encoding them into mathematical representations. For fingerprint systems, this involves minutiae detection identifying ridge endings, bifurcations, and other distinctive features. Facial recognition systems extract facial landmarks, geometric relationships, and deep learning embeddings (El-Abed et al., 2024). Forensic investigators must understand feature extraction algorithms to detect manipulation, forgery, and algorithmic vulnerabilities.

Template storage mechanisms maintain enrolled biometric references for comparison during authentication. Storage architectures range from centralized databases (common in government and enterprise deployments) to decentralized systems with on-device storage (prevalent in consumer devices). Forensic considerations include template security, encryption implementations, access controls, and backup mechanisms (Ratha et al., 2021). Compromised template databases represent catastrophic security failures requiring comprehensive forensic investigation.

The matching engine compares captured biometric samples against enrolled templates, computing similarity scores. Matching algorithms employ various mathematical approaches including Euclidean distance, Hamming distance, and deep neural network similarity measures. The decision module applies acceptance thresholds, determining whether similarity scores warrant authentication approval. Forensic analysis of matching scores provides critical evidence for attack detection and system performance evaluation (Uludag & Jain, 2024).

2.2 Authentication Modes and Protocols

Biometric systems operate in three primary modes: verification (one-to-one matching), identification (one-to-many matching), and continuous authentication. Verification compares a captured sample against a single enrolled template, confirming claimed identity. This mode dominates access control applications where users explicitly claim identity through usernames, ID cards, or other identifiers. Forensic audit trails in verification mode must capture both the claimed identity and matching confidence scores.

Identification mode searches the entire enrolled database, attempting to match captured samples against all templates. This computationally intensive operation supports surveillance applications, criminal investigations, and scenarios where users cannot or will not claim identity. Forensic challenges in identification mode include false positives (incorrect matches), computational performance issues affecting audit trail completeness, and privacy concerns warranting specialized forensic protocols.

Continuous authentication monitors user behavior throughout sessions, detecting anomalous patterns suggesting session hijacking or unauthorized access (Patel et al., 2021). Behavioral biometrics including keystroke dynamics, mouse movements, and interaction patterns provide continuous verification without explicit authentication events. Forensic audit trails for continuous authentication require fine-grained event logging capturing behavioral baseline deviations and anomaly detection triggers.

2.3 Multimodal Biometric Systems

Multimodal biometric systems combine multiple biometric modalities as shown in table 1, enhancing security through redundancy and improving resistance to spoofing attacks (Singh et al., 2022). Common implementations fuse fingerprint with facial recognition, iris with voice verification, or behavioral with physiological biometrics. Fusion architectures operate at multiple levels: sensor-level fusion combines raw data from multiple sensors, feature-level fusion merges extracted features, score-level fusion combines matching scores, and decision-level fusion aggregates individual authentication decisions.

Table 1. Forensic Characteristics of Biometric Modalities

Modality	Forensic Strengths	Forensic Challenges	Spoofing Risk
Fingerprint	Established forensic science, High uniqueness, Persistent over lifetime	Latent print quality, Age-related changes, Injury susceptibility	Medium-High
Facial Recognition	Non-contact acquisition, Multiple capture angles, Large existing databases	Aging effects, Illumination variance, Pose variations, Deepfakes	High
Iris	High accuracy, Stability over time, Detailed texture patterns	Acquisition distance limits, Quality in non-ideal conditions	Medium
Voice	Remote capture capability, Continuous authentication potential	Environmental noise, Health-related changes, Recording quality	High
Behavioral	Continuous monitoring, Difficult to replicate, Contextual awareness	High variability, Limited uniqueness, Training period required	Medium-High

Score-level fusion represents the most prevalent approach due to its flexibility and computational efficiency (El-Abed et al., 2024). Fusion rules include simple sum, weighted sum, minimum score, maximum score, and machine learning-based fusion. Forensic investigators must understand fusion architectures to properly interpret audit trails and identify attack vectors. Successful attacks against multimodal systems often exploit fusion logic vulnerabilities rather than defeating individual biometric modalities.

Forensic audit trails in multimodal systems require comprehensive logging of individual modality performance, fusion decisions, and failure modes. When authentication fails, audit trails should record which modalities succeeded or failed, providing forensic investigators with detailed attack signatures. Template: {timestamp, user_id, modality_1_score, modality_2_score, fusion_score, decision, failure_reason, environmental_context}.

3. FORENSIC AUDIT TRAIL ARCHITECTURE

Forensic audit trails constitute the evidentiary foundation for biometric security investigations (Bhagat et al., 2023). Comprehensive audit architectures must balance completeness (capturing all forensically relevant events), performance (minimizing system overhead), privacy (protecting sensitive biometric data), and immutability (preventing tampering and ensuring evidence integrity). Figure 1 showing complete biometric system architecture with audit trail integration points, data flows, and forensic collection mechanisms.

Figure 1 showing complete biometric system architecture with audit trail integration points, data flows, and forensic collection mechanisms.

Figure 1. Biometric Authentication System Architecture with Forensic Audit Points

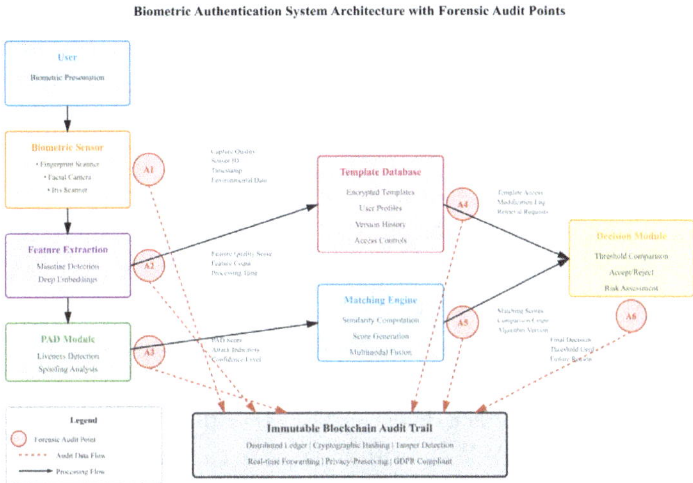

This section examines audit trail design principles, implementation architectures, and forensic best practices.

3.1 Essential Audit Trail Components

Effective forensic audit trails capture seven essential data categories. Temporal information records precise event timestamps using synchronized, tamper-resistant time sources. NTP (Network Time Protocol) synchronization ensures temporal consistency across distributed systems. Forensic best practices require UTC timestamps with millisecond precision and cryptographic timestamp authorities for high-assurance applications.

Identity context captures claimed user identity, actual authenticated identity (if verification succeeds), and authorization context. Forensic investigators require this information to detect identity spoofing, privilege escalation, and unauthorized access. Authentication outcome records binary decisions (success/failure) along with detailed failure reasons: poor biometric quality, template mismatch, liveness detection failure, or system errors.

Biometric quality metrics provide critical forensic context. Quality scores assess captured sample adequacy, identifying environmental factors, sensor problems, or deliberate spoofing attempts (Galbally et al., 2023). NIST's Biometric Quality Assessment standards (ISO/IEC 29794) provide frameworks for quality evaluation across biometric modalities. Forensic audit trails should record quality scores for both enrollment and authentication attempts.

Matching confidence scores quantify biometric similarity, enabling forensic investigators to reconstruct decision-making processes and detect anomalous patterns. While binary authentication outcomes (accept/reject) suffice for access control, forensic investigations require detailed matching scores. Trend analysis of matching scores reveals gradual template degradation, emerging attacks, and system performance issues.

Environmental and contextual metadata enhances forensic analysis. Sensor identification distinguishes between trusted and potentially compromised devices. Geolocation data identifies authentication attempts from unexpected locations. Network context (IP addresses, network segments, VPN status) supports attack attribution. Device fingerprinting detects device spoofing and credential sharing (Kim et al., 2022).

Biometric presentation attack detection (PAD) results constitute critical forensic evidence (Ali et al., 2023). PAD algorithms analyze captured biometric samples for spoofing indicators: printed photographs, silicone fingerprints, video replay attacks, and deepfake forgeries. Audit trails must record PAD decisions, confidence scores, and specific spoofing indicators detected.

3.2 Immutable Audit Trail Implementation

Audit trail immutability ensures forensic evidence integrity, preventing attackers from erasing incriminating logs or fabricating exculpatory evidence. Traditional log files stored on compromised systems offer minimal forensic reliability. Modern implementations employ cryptographic techniques, distributed architectures, and blockchain technology to achieve tamper-evident audit trails (Jin et al., 2023).

Cryptographic chaining creates tamper-evident logs by incorporating cryptographic hashes of previous entries into subsequent entries. Each audit entry includes: {event_data, timestamp, previous_hash, current_hash}. The current_hash results from cryptographic hashing (SHA-256 or SHA-3) of concatenated event data, timestamp, and previous hash. Any modification to historical entries invalidates all subsequent hashes, providing immediate tamper detection.

Append-only storage architectures prevent audit log modification by restricting write operations to sequential appends. Write-Once-Read-Many (WORM) storage, distributed ledgers, and specialized audit log databases implement append-only semantics. Cloud-based audit log services (AWS CloudTrail, Azure Monitor, Google Cloud Logging) provide tamper-resistant storage with cryptographic verification and compliance certifications.

Blockchain-based audit trails leverage distributed consensus mechanisms to achieve exceptional tamper resistance (Bhagat et al., 2023). Each audit block contains multiple authentication events plus cryptographic hashes linking to previous blocks. Distributed validation across multiple nodes prevents single-point tampering. Private permissioned blockchains (Hyperledger Fabric, R3 Corda) suit enterprise biometric deployments, while public blockchains support transparency requirements in government applications.

Real-time log forwarding protects audit trails by transmitting events to remote collection systems before local compromise occurs. SIEM (Security Information and Event Management) platforms aggregate logs from distributed biometric systems, enabling centralized forensic analysis and tamper-resistant archival. Syslog over TLS, HTTP Event Collector, and proprietary agent-based forwarding ensure reliable, encrypted log transmission.

3.3 Privacy-Preserving Audit Mechanisms

Biometric audit trails must balance forensic requirements against privacy protections and regulatory compliance. Raw biometric data (fingerprint images, facial photographs) constitutes highly sensitive personal information warranting enhanced protection. Privacy-preserving forensic approaches enable comprehensive audit trails without exposing raw biometric data (Chen et al., 2023).

Template hashing transforms biometric templates into cryptographic hashes, enabling audit trails to record authentication attempts without storing actual templates. Forensic investigators can verify template consistency and detect unauthorized modifications without accessing raw biometric data. However, template hashing prevents certain forensic analyses requiring original template examination.

Anonymization and pseudonymization techniques protect user privacy in audit trails. Pseudonymization replaces user identifiers with cryptographic pseudonyms, enabling forensic pattern analysis without revealing actual identities. Cryptographic key management ensures authorized investigators can de-pseudonymize records when legally justified, while preventing unauthorized access.

Homomorphic encryption enables computation on encrypted audit data, supporting forensic analysis without decrypting sensitive information (Chen et al., 2023). While computationally intensive, homomorphic encryption allows compliance with privacy regulations while maintaining forensic capability. Partial homomorphic encryption schemes supporting specific operations (addition, comparison) provide practical solutions for biometric audit analysis.

Differential privacy techniques add calibrated statistical noise to audit trail aggregations, preventing individual identification while preserving forensic utility. Forensic investigators can analyze authentication patterns, failure rates, and attack signatures without compromising individual privacy. Privacy budgets balance forensic analysis capability against privacy protection, with GDPR-compliant implementations gaining

4. FORENSIC INVESTIGATION METHODOLOGIES

Forensic investigation of biometric authentication systems requires specialized methodologies combining traditional digital forensics with biometric-specific techniques (Kaur & Khanna, 2024). Investigators must understand biometric system architectures, attack vectors, and evidentiary requirements for legal proceedings. This section presents systematic investigation frameworks addressing common biometric security incidents.

4.1 Incident Detection and Initial Response

Biometric security incidents manifest through various indicators: anomalous authentication failure rates, unusual access patterns, quality score degradations, PAD alert increases, and system performance anomalies. Security Operations Centers (SOCs) monitor audit trails using SIEM platforms configured with biometric-specific detection rules. Machine learning anomaly detection models identify subtle attack patterns invisible to rule-based systems.

Initial incident response prioritizes evidence preservation and containment. Forensic investigators immediately secure audit trails, capturing current system state before attackers modify evidence. Live system memory captures preserve volatile biometric session data, cached templates, and active authentication transactions. Network traffic captures provide evidence of attack vectors and command-and-control communications.

Containment strategies balance security restoration against evidence preservation. Disconnecting compromised systems prevents continued attacks but may alert sophisticated adversaries to detection. Forensic best practices favor covert monitoring when safe, gathering additional evidence about attacker tactics, techniques, and procedures (TTPs). However, immediate containment takes precedence when active data exfiltration or system destruction occurs.

Chain of custody documentation begins immediately upon incident detection. Forensic investigators document all evidence collection activities, system accesses, and analytical procedures. Cryptographic hashing (SHA-256) establishes evidence integrity baselines. Write-blocking devices prevent inadvertent evidence modification during acquisition. Documentation includes timestamps, personnel involved, tools employed, and custodial transfers.

4.2 Audit Trail Analysis Techniques

Forensic audit trail analysis employs both manual examination and automated analytical tools. Initial analysis establishes baseline behavior through temporal analysis, frequency distributions, and pattern identification. Seasonal variations, time-of-day patterns, and weekly cycles inform anomaly detection. Sudden deviations from established baselines warrant detailed investigation.

Correlation analysis identifies relationships between authentication events, user behaviors, and security incidents. Temporal correlation reveals coordinated attacks, privilege escalation sequences, and reconnaissance activities. Spatial correlation analyzes geographic patterns, detecting credential sharing and unauthorized device usage. User behavior analytics (UBA) platforms automate correlation analysis, flagging high-risk behavior patterns.

Matching score analysis provides deep insights into authentication anomalies. Forensic investigators plot matching score distributions, comparing legitimate users against attack attempts (Uludag & Jain, 2024). Presentation attacks often exhibit distinctive score patterns: artificially high scores (perfect template matches suggesting replay attacks) or scores clustered near acceptance thresholds (indicating sophisticated spoofing). Statistical analysis identifies outliers warranting detailed examination.

Quality metric analysis detects sensor problems, environmental issues, and deliberate spoofing (Galbally et al., 2023). Sudden quality degradations across multiple users suggest sensor malfunction or attack campaigns. User-specific quality problems may indicate age-related biometric changes, injuries, or targeted attacks. Correlation between quality scores and authentication failures reveals system vulnerabilities requiring remediation.

Timeline reconstruction synthesizes audit trail data into coherent attack narratives. Forensic investigators reconstruct attacker actions from initial reconnaissance through compromise, lateral movement, and objectives achievement. Timeline analysis employs specialized tools (Timesketch, Plaso) parsing diverse log formats and presenting unified temporal views. Visual timeline representations facilitate incident communication to non-technical stakeholders and legal audiences.

4.3 Biometric Evidence Collection and Analysis

Forensic collection of biometric evidence requires specialized procedures preserving evidence integrity while respecting privacy constraints. Raw biometric samples (fingerprint images, facial photographs) constitute primary evidence requiring careful handling. Forensic imaging creates bit-for-bit copies preserving original evidence while enabling analysis on duplicates. Hashing verifies image integrity, establishing tamper-evident chains of custody.

Template extraction from compromised systems provides critical forensic evidence. Investigators employ reverse engineering techniques recovering templates from encrypted databases, memory captures, and system backups (Cappelli et al.,

2021). Recovered templates enable comparative analysis, identifying unauthorized template modifications and detecting insider threats. Legal authorization requirements vary by jurisdiction, with some regions prohibiting template analysis without explicit consent.

Biometric matching analysis compares questioned samples against known references, establishing identity or detecting forgery (Bodapati et al., 2024). Forensic biometric examiners employ specialized tools (NIST NBIS for fingerprints, FaceNet for facial analysis) conducting objective comparisons. Statistical confidence measures quantify match reliability, supporting expert testimony and legal admissibility. Blind testing protocols prevent cognitive bias affecting matching decisions.

Liveness detection analysis examines captured samples for presentation attack indicators. Forensic investigators apply specialized algorithms detecting printed photos, silicone fingerprints, video replay attacks, and deepfake content (Oloyede et al., 2024). Multispectral imaging, texture analysis, and challenge-response testing provide forensic evidence distinguishing live presentations from spoofing attempts. Documentation of detection methodology ensures reproducibility and legal defensibility.

4.4 Attack Attribution and Threat Intelligence

Attack attribution identifies perpetrators, understanding motivations and capabilities. Forensic indicators including IP addresses, device fingerprints, attack patterns, and tool signatures support attribution. However, attribution complexity increases with sophisticated adversaries employing VPNs, Tor networks, compromised intermediary systems, and false flag operations.

Threat intelligence integration enhances forensic investigations by contextualizing observed attacks within broader threat landscapes. Indicators of Compromise (IoCs) shared through threat intelligence platforms (MISP, ThreatConnect) enable detection of known attack patterns and adversary infrastructure. Tactical intelligence provides immediate defensive value, while strategic intelligence informs long-term security improvements.

Malware analysis reveals sophisticated attack capabilities. Advanced persistent threats (APTs) targeting biometric systems deploy specialized malware capturing biometric samples, exfiltrating templates, and manipulating authentication decisions. Dynamic malware analysis executes samples in sandboxed environments, observing behaviors without risking production systems. Static analysis examines code structure, identifying capabilities and detecting obfuscation techniques.

Social engineering analysis investigates human factors enabling technical attacks. Phishing campaigns, pretexting, and physical intrusions often precede biometric system compromise. Forensic investigators examine email archives, communication logs, and physical access records reconstructing social engineering sequences. Correlation between social engineering and technical compromise reveals sophisticated attack campaigns.

5. BIOMETRIC ATTACK VECTORS AND FORENSIC INDICATORS

Understanding biometric attack vectors enables forensic investigators to recognize attack indicators, collect relevant evidence, and prevent future incidents (Hadid et al., 2021). Biometric attacks span presentation attacks (spoofing), algorithmic attacks, template attacks, and system-level compromises. Each attack category exhibits distinctive forensic indicators requiring specialized detection and analysis approaches.

5.1 Presentation Attacks (Spoofing)

Presentation attacks present fraudulent biometric samples to capture devices, attempting to impersonate legitimate users (Ramachandra et al., 2024). Fingerprint spoofing employs gelatin, silicone, or Play-Doh molds created from latent prints, lifted prints, or high-resolution photographs. Forensic indicators include: abnormal ridge flow patterns, uniform depth across fingerprint, absence of pore structures, unusual moisture levels, and temperature anomalies. Advanced PAD algorithms employ multispectral imaging, detecting subsurface features invisible in spoofed samples.

Facial recognition spoofing utilizes photographs, videos, masks, or makeup defeating recognition systems (Tolosana et al., 2022). Two-dimensional attacks present printed photographs or digital displays showing target faces. Three-dimensional attacks employ realistic masks created through 3D printing or professional prosthetics. Deepfake attacks use generative adversarial networks (GANs) creating highly realistic synthetic faces or manipulated videos (Sharma & Selwal, 2023). Forensic indicators include: lack of natural micro-movements, unusual reflectance properties, absence of physiological signals (pulse detection), and behavioral inconsistencies.

Iris spoofing attacks employ printed iris images, contact lenses with printed iris patterns, or synthetic eyes. High-quality iris images captured covertly or obtained through data breaches enable sophisticated spoofing (Nguyen et al., 2023). Forensic detection examines: pupil dynamics under varying illumination, presence of natural collarette structure, corneal reflections, and fine texture details. Near-infrared imaging reveals subsurface iris structures impossible to replicate in spoofed samples.

Voice spoofing leverages recordings, voice synthesis, or voice conversion techniques. Simple replay attacks use previously recorded voice samples, easily defeated by challenge-response protocols. Advanced attacks employ text-to-speech synthesis or voice conversion transforming attacker voices to match targets. Forensic analysis examines: spectral characteristics, prosodic features, breathing patterns, and background acoustic environments. Machine learning-based voice liveness detection identifies synthetic speech through subtle artifacts in generated audio.

5.2 Template Attacks

Template attacks target stored biometric templates rather than presentation interfaces. Template theft through database compromise, insider threats, or system vulnerabilities provides attackers with valuable assets for identity theft and system infiltration. Unlike passwords (changeable after compromise), biometric templates represent permanent user characteristics, making theft consequences severe and long-lasting (Gomez-Barrero et al., 2022).

Template reconstruction attacks reverse-engineer original biometric samples from stored templates (Cappelli et al., 2021). Early research demonstrated fingerprint image reconstruction from minutiae-based templates. Recent advances employ generative adversarial networks reconstructing facial images from embeddings. Forensic investigators must assess whether compromised templates enable reconstruction, determining breach severity and legal notification requirements.

Template injection attacks insert fraudulent templates into biometric databases, creating unauthorized identities or replacing legitimate templates. Insider threats with database access pose particular risks. Forensic indicators include: template modification timestamps diverging from legitimate enrollment periods, template quality inconsistent with sensor capabilities, and template format anomalies. Cryptographic template protection using homomorphic encryption or secure multiparty computation prevents injection attacks while enabling matching operations.

Hill climbing attacks iteratively optimize synthetic inputs maximizing matching scores without accessing templates. These algorithmic attacks require only matching score feedback, making them particularly dangerous. Forensic detection monitors for: unusual numbers of authentication attempts per user, systematic score improvement patterns, and matching score distributions diverging from legitimate attempts. Rate limiting and account lockouts mitigate hill climbing attacks, though sophisticated attackers distribute attempts across multiple accounts.

5.3 System and Infrastructure Attacks

System-level attacks compromise biometric infrastructure components including sensors, communication channels, processing servers, and databases. Sensor substitution attacks replace legitimate sensors with compromised devices capturing biometric samples for later misuse or injecting synthetic samples into processing pipelines. Forensic detection employs sensor authentication protocols, cryptographic device attestation, and physical security monitoring.

Man-in-the-middle attacks intercept communications between biometric sensors and processing servers, capturing or manipulating transmitted data. Wireless biometric systems prove particularly vulnerable to interception attacks. Forensic indicators include: unusual network latency, unexpected traffic patterns, certificate validation failures, and protocol anomalies. Mutual authentication, end-to-end encryption, and secure communication protocols prevent interception attacks.

Malware targeting biometric systems captures samples during enrollment or authentication, exfiltrates templates, or manipulates matching decisions. Rootkits operating at kernel level evade detection while maintaining persistent access. Forensic investigation employs memory forensics, behavioral analysis, and integrity verification detecting malicious code. Trusted execution environments (TEEs) and hardware security modules (HSMs) provide isolated execution preventing malware interference.

Denial of service attacks target biometric system availability, preventing legitimate authentication while creating operational disruption. Volumetric attacks overwhelm processing capacity, while algorithmic attacks exploit computational complexity of biometric matching. Forensic indicators include: authentication request rate spikes, unusual geographic distribution of requests, and matching score computation delays. Rate limiting, distributed processing, and attack traffic filtering maintain availability during attacks.

6. LEGAL AND REGULATORY CONSIDERATIONS

Forensic investigation of biometric systems operates within complex legal and regulatory frameworks governing biometric data collection, storage, processing, and disclosure. Understanding these requirements ensures forensic evidence admissibility, prevents privacy violations, and maintains compliance during investigations. Regulatory regimes vary significantly across jurisdictions, requiring forensic investigators to adapt methodologies to local requirements.

6.1 Privacy Regulations and Compliance

The European Union's General Data Protection Regulation (GDPR) categorizes biometric data as special category personal data requiring enhanced protection under Article 9. Processing biometric data for uniquely identifying individuals requires explicit consent, legal authorization, or substantial public interest. Forensic investigations typically invoke legal obligation or public interest justifications, though investigators must document legal bases and minimize data processing.

GDPR Article 5 principles constrain forensic investigations: purpose limitation restricts data usage to investigation purposes, data minimization requires collecting only essential evidence, storage limitation mandates timely evidence disposal, and accuracy requires verification of evidence reliability. Forensic investigators must balance comprehensive investigation against privacy principles, documenting justifications for evidence collection and retention decisions.

The California Consumer Privacy Act (CCPA) and California Privacy Rights Act (CPRA) grant California residents rights including access to collected data, deletion requests, and opt-out from data sales. Biometric data receives enhanced protection under CPRA's sensitive personal information provisions. Forensic investigations must accommodate subject access requests while protecting ongoing investigation confidentiality. Delayed disclosure provisions allow temporary withholding of information jeopardizing investigations.

Illinois Biometric Information Privacy Act (BIPA) imposes strict requirements on biometric data collection, requiring informed written consent, public retention policies, and prohibition of biometric data sales. BIPA's private right of action enables individuals to sue for violations, creating significant liability risks. Forensic investigators must verify BIPA compliance when examining Illinois-related incidents, documenting consent mechanisms and retention policies.

Sector-specific regulations impose additional biometric requirements. The Health Insurance Portability and Accountability Act (HIPAA) regulates healthcare-related biometric data, requiring compliance with Privacy Rule and Security Rule provisions. The Gramm-Leach-Bliley Act (GLBA) governs financial institution biometric implementations. Payment Card Industry Data Security Standard (PCI DSS) addresses biometric authentication in payment systems. Forensic investigators must understand applicable sector regulations, adapting evidence collection to compliance requirements.

6.2 Evidence Admissibility and Legal Standards

Biometric evidence admissibility depends on reliability, relevance, and proper collection procedures. The Daubert standard (federal courts) and Frye standard (some state courts) govern expert testimony and scientific evidence. Biometric evidence must demonstrate: tested and validated methodologies, peer-reviewed techniques, known error rates, and general acceptance within relevant scientific community.

Forensic investigators must document methodology, maintain chain of custody, and support findings with scientific literature.

Authentication logs and audit trails constitute business records admissible under Federal Rules of Evidence Rule 803(6) and similar state provisions. Admissibility requires demonstrating: records made in regular business operations, contemporaneous recording, reliable record-keeping systems, and custodian testimony authenticating records. Automated log generation satisfies contemporaneous recording requirements, while immutable audit trails address reliability concerns.

Digital evidence authentication requires establishing evidence integrity from collection through trial. Cryptographic hashing provides mathematical proof that evidence remains unaltered. Chain of custody documentation tracks evidence handling, transfers, and access. Best practices include: write-blocking during acquisition, dual-examiner verification, tamper-evident storage, and detailed documentation of all evidence interactions.

Expert witness testimony explains complex biometric concepts to juries and judges. Qualified experts possess: relevant educational credentials (computer science, electrical engineering, forensic science), professional certifications (Certified Information Systems Security Professional, Certified Forensic Computer Examiner), practical experience, and published research. Effective expert testimony communicates technical findings in accessible language, employing visual aids and analogies clarifying complex concepts.

6.3 Cross-Border Investigations and International Cooperation

Biometric system attacks frequently span multiple jurisdictions, requiring international cooperation. Mutual Legal Assistance Treaties (MLATs) provide frameworks for evidence sharing, though MLAT processes prove time-consuming. The Budapest Convention on Cybercrime establishes international cybercrime investigation cooperation, though not all nations participate. Forensic investigators must navigate diplomatic channels, respecting foreign sovereignty while pursuing investigations.

Cloud-based biometric systems complicate jurisdictional questions. Data residency regulations mandate storing citizen biometric data within national borders, while cloud architectures distribute data globally. Forensic investigators must identify data locations, understand applicable laws in each jurisdiction, and coordinate multi-jurisdictional evidence collection. The CLOUD Act (Clarifying Lawful Overseas Use of Data) enables U.S. law enforcement accessing data controlled by U.S. companies regardless of storage location, though conflicts with foreign blocking statutes arise.

International organizations facilitate biometric forensics cooperation. INTERPOL's Biometric Hub connects member nations' biometric databases, enabling cross-border identification. Europol supports EU member state cooperation, providing analytical support and coordinating multinational investigations. Regional organizations including ASEANAPOL and AFRICAOL promote regional cooperation. Forensic investigators leverage these organizations when investigations cross borders.

7. EMERGING TECHNOLOGIES AND FUTURE DIRECTIONS

Rapid technological evolution continuously reshapes biometric authentication and forensic investigation (Anderson & Murdoch, 2024). Artificial intelligence, quantum computing, blockchain technology, and privacy-enhancing cryptography introduce both opportunities and challenges for forensic investigators. Understanding emerging trends enables proactive preparation and strategic capability development.

7.1 Artificial Intelligence and Deep Learning

Deep learning revolutionized biometric recognition, achieving unprecedented accuracy through convolutional neural networks (CNNs) and transformer architectures (El-Abed et al., 2024). Modern facial recognition systems employ deep embeddings extracted by networks trained on millions of faces, enabling robust recognition across variations in pose, illumination, and aging. Forensic investigators must understand deep learning architectures, training procedures, and potential vulnerabilities including adversarial attacks and model inversion.

Generative AI creates sophisticated presentation attacks including deepfakes and synthetic biometric samples (Sharma & Selwal, 2023). Generative Adversarial Networks (GANs) synthesize realistic faces, fingerprints, and iris patterns challenging detection systems. Diffusion models enable high-quality generation with fine-grained control. Forensic countermeasures employ AI-based detection analyzing subtle artifacts in generated content (Liu et al., 2021). The ongoing arms race between generative AI and detection algorithms drives continuous innovation.

Explainable AI (XAI) addresses the black-box problem in deep learning biometric systems. Forensic investigations require understanding why systems made particular decisions, information opaque in standard neural networks. XAI techniques including attention visualization, saliency maps, and layer-wise relevance propagation reveal decision-making factors. Legal proceedings increasingly demand explainability, making XAI essential for forensic admissibility.

Federated learning enables privacy-preserving biometric model training across distributed datasets without centralizing sensitive data. Organizations train local models on private data, sharing only model updates rather than raw biometrics. Forensic applications include collaborative threat intelligence and distributed attack detection while respecting privacy regulations. However, federated learning introduces new attack vectors including model poisoning and inference attacks requiring specialized forensic capabilities.

7.2 Quantum Computing and Post-Quantum Cryptography

Quantum computing threatens current cryptographic protections securing biometric templates and audit trails. Shor's algorithm enables quantum computers breaking RSA and elliptic curve cryptography, foundations of contemporary biometric security. Template encryption, digital signatures, and audit trail integrity mechanisms relying on these algorithms become vulnerable. Forensic implications include: retrospective decryption of archived evidence, compromised chain of custody verification, and invalid digital signatures.

Post-quantum cryptography develops quantum-resistant algorithms protecting against quantum attacks. NIST's post-quantum cryptography standardization selected lattice-based, hash-based, and code-based algorithms for standardization. Biometric system migration to post-quantum algorithms presents significant challenges: computational overhead, key size increases, and backward compatibility requirements. Forensic investigators must track cryptographic migrations, understanding which evidence remains protected and which faces retrospective compromise risks.

Quantum key distribution (QKD) provides theoretically unbreakable key exchange based on quantum mechanics principles. QKD enables secure biometric template protection and audit trail encryption resistant to both classical and quantum attacks. However, QKD requires specialized hardware and infrastructure investment limiting deployment. Forensic investigators must understand QKD protocols, potential implementation vulnerabilities, and evidence collection procedures for quantum-secured systems.

7.3 Blockchain and Distributed Ledger Technologies

Blockchain technology offers compelling solutions for audit trail immutability, decentralized identity management, and tamper-evident logging (Jin et al., 2023). Permissioned blockchains enable organizations maintaining private biometric authentication networks while achieving cryptographic verification and distributed consensus. Smart contracts automate audit trail analysis, triggering alerts when anomalous patterns emerge. Forensic investigators must understand blockchain architectures, consensus mechanisms, and evidence extraction procedures.

Decentralized identity (DID) frameworks enable self-sovereign identity management where individuals control their biometric credentials without centralized authorities. W3C Verifiable Credentials standard provides interoperable identity assertions cryptographically signed by issuers. Forensic challenges include: distributed evidence collection across multiple parties, pseudonymous identities complicating attribution, and privacy-preserving verification limiting forensic visibility. However, DIDs offer advantages including tamper-evident credential revocation and verifiable audit trails.

Non-fungible tokens (NFTs) enable unique biometric credential representation with ownership tracking and transfer history. While controversial, NFT-based credentials provide transparent audit trails and provenance verification. Forensic investigators can analyze blockchain transactions reconstructing credential usage patterns, detecting unauthorized transfers, and establishing chains of custody. However, NFT implementations face scalability challenges and environmental concerns from proof-of-work consensus.

7.4 Privacy-Enhancing Technologies

Homomorphic encryption enables computations on encrypted biometric data without decryption, revolutionizing privacy-preserving authentication (Chen et al., 2023). Fully homomorphic encryption (FHE) supports arbitrary computations, while somewhat homomorphic encryption and partially homomorphic encryption enable specific operations. Biometric matching on encrypted templates protects user privacy during authentication while maintaining forensic audit capabilities. Current limitations include significant computational overhead and implementation complexity.

Secure multiparty computation (MPC) distributes biometric computations across multiple parties without revealing inputs to any single party. Template matching employs MPC protocols computing similarity scores without exposing templates to authentication servers or users. Forensic applications include privacy-preserving evidence analysis enabling multiple agencies contributing sensitive information without disclosure. However, MPC complexity and communication overhead limit practical deployment.

Zero-knowledge proofs enable verification without revealing underlying information. Biometric systems employ zero-knowledge proofs demonstrating template possession without exposing templates. Forensic auditing uses zero-knowledge proofs verifying system compliance without accessing sensitive data. Advanced constructions including zk-SNARKs (Zero-Knowledge Succinct Non-Interactive Arguments of Knowledge) enable efficient verification though setup complexity remains challenging.

Differential privacy techniques add calibrated statistical noise enabling privacy-preserving forensic analysis. Biometric system operators can publish aggregate statistics about authentication patterns, attack frequencies, and system performance without revealing individual user information. Forensic investigators analyze trends and patterns within differential privacy frameworks, balancing analytical utility against privacy protection. Privacy budget allocation determines analysis scope and accuracy trade-offs.

8. CASE STUDIES

Real-world case studies illuminate practical forensic challenges and demonstrate investigation methodologies. The following cases represent diverse biometric attack scenarios, forensic responses, and lessons learned. While details have been modified to protect confidentiality, these cases reflect actual incident patterns and forensic techniques.

8.1 Case Study 1: Financial Institution Fingerprint Bypass

A major financial institution deployed fingerprint authentication for high-value wire transfers, requiring dual-factor authentication combining fingerprints with passwords. Forensic investigation began after detecting unauthorized transfers totaling $2.3 million across three customer accounts. Initial audit trail review revealed successful fingerprint authentication for all fraudulent transfers, suggesting sophisticated attack beyond simple credential theft.

Forensic investigators analyzed quality scores, matching confidence metrics, and presentation attack detection results from audit trails. Anomaly detection revealed consistently high matching scores (>98%) for fraudulent authentications compared to normal variance (85-95%) in legitimate attempts. This pattern suggested replay attacks or synthetic fingerprint injection. Further analysis identified authentication device identifiers diverging from customers' historical patterns, indicating device substitution.

Physical security investigation discovered that cleaning contractors had replaced legitimate fingerprint scanners with compromised devices during evening maintenance. These devices captured genuine fingerprints during legitimate authentication, storing them for subsequent replay. Forensic imaging of compromised scanners revealed embedded single-board computers with 4G connectivity enabling remote control. Malware analysis identified command-and-control infrastructure linked to organized cybercrime groups.

Lessons learned included: implementing cryptographic device authentication, deploying tamper-evident sensors, enhancing physical security monitoring, and implementing challenge-response protocols preventing replay attacks (Ali et al., 2023). The institution deployed multispectral fingerprint sensors detecting liveness through subsurface capillary patterns, significantly improving presentation attack detection. Forensic audit trail enhancements included device integrity verification logging and correlation between physical access and device authentication events.

8.2 Case Study 2: Healthcare Biometric Privacy Breach

A healthcare provider implemented facial recognition for facility access and patient identification, storing templates in centralized databases. A sophisticated ransomware attack encrypted production systems while exfiltrating 150,000 biometric templates. Forensic investigation faced dual challenges: recovering encrypted systems and determining breach scope for regulatory reporting under HIPAA and state breach notification laws.

Forensic timeline reconstruction revealed initial compromise through phishing campaign targeting IT administrators. Attackers established persistent access via web shells, conducting reconnaissance for three weeks before ransomware deployment. During reconnaissance, attackers identified biometric databases and exfiltrated templates to external cloud storage. Network forensics analyzing DNS queries, NetFlow data, and proxy logs mapped exfiltration paths and identified command-and-control infrastructure.

Template analysis determined that exported templates consisted of deep learning embeddings rather than original images (Bodapati et al., 2024). Forensic evaluation concluded that while embeddings enabled authentication system compromise, reconstructing original facial images required sophisticated machine learning techniques not readily available. Legal consultation determined that template exfiltration constituted HIPAA breach requiring notification, despite reduced risk compared to original image theft.

Response included: immediate template revocation and re-enrollment, implementing template encryption at rest, deploying data loss prevention monitoring template access, and enhancing network segmentation isolating biometric systems. The organization adopted privacy-preserving biometric protocols using homomorphic encryption for template matching, eliminating centralized plaintext template storage (Chen et al., 2023). Forensic monitoring enhancements included behavioral analytics detecting anomalous database access and automated alert escalation for biometric data queries.

8.3 Case Study 3: Government Border Control Deepfake Attack

Border control authorities detected coordinated attempts to enter the country using synthetic identity documents paired with deepfake facial recognition evasion. Forensic investigation began after multiple travelers presenting visually similar identities triggered watchlist alerts. Biometric audit trails revealed successful facial recognition authentication against enrolled identities, yet subsequent investigation determined travelers' true identities differed from biometric matches.

Forensic analysis of captured facial images revealed subtle artifacts characteristic of deepfake generation: unusual skin texture patterns, inconsistent lighting reflections, and subtle discontinuities in facial features (Tolosana et al., 2022). Advanced deepfake detection algorithms analyzing frequency domain characteristics, neural network artifacts, and temporal inconsistencies across multiple frames identified synthetic content. Presentation attack detection systems had failed to identify deepfakes due to high quality and sophisticated anti-detection training.

Investigation revealed organized identity fraud ring combining stolen biometric templates, synthetic identity documents, and real-time deepfake generation. Attackers obtained biometric templates through data breaches and employed generative adversarial networks creating facial videos matching compromised identities (Korshunov & Marcel, 2023). Travelers wore augmented reality contact lenses displaying deepfake videos synchronized with facial recognition cameras, bypassing static liveness detection.

Countermeasures included: deploying advanced deepfake detection leveraging multiple detection algorithms, implementing challenge-response liveness testing requiring specific facial expressions, and conducting behavioral analysis detecting rehearsed versus natural movements (Venkatesh et al., 2023). The agency enhanced audit trails capturing higher-resolution imagery and temporal sequences enabling post-incident deepfake analysis. International coordination led to disrupting the identity fraud network and recovering compromised templates from criminal infrastructure.

9. BEST PRACTICES AND RECOMMENDATIONS

Effective forensic capabilities require proactive design, comprehensive audit architectures, and continuous improvement. Organizations implementing biometric authentication should incorporate forensic requirements from initial design through operational lifecycle. The following best practices synthesize lessons from case studies, research literature, and operational experience.

9.1 Design-Phase Considerations

Forensic requirements should inform biometric system architecture from initial design. Threat modeling exercises identify attack vectors, vulnerabilities, and evidence requirements for incident investigation (Anderson & Murdoch, 2024). Security by design principles incorporate defensive capabilities and forensic logging at architectural level rather than bolting them onto completed systems. Privacy by design ensures forensic capabilities respect privacy regulations and user rights.

Comprehensive audit trail specifications define logged events, data retention periods, and evidence preservation procedures. Organizations should log: all authentication attempts (successful and failed), quality metrics, matching scores, PAD results, template modifications, configuration changes, and administrative actions. Audit trail formats should employ standardized schemas enabling analysis tools and cross-system correlation. Immutability mechanisms prevent evidence tampering while supporting legitimate record retention management.

Template protection mechanisms should balance security, performance, and forensic requirements (Ratha et al., 2021). Encryption protects templates at rest and in transit, while format-preserving encryption enables encrypted template matching. Template versioning maintains historical templates supporting forensic reconstruction when algorithms change or templates update. Cryptographic binding links templates to specific users, devices, and enrollment contexts preventing template substitution attacks.

9.2 Operational Security Measures

Continuous monitoring detects attacks in progress, enabling rapid response before significant damage occurs. Security information and event management (SIEM) platforms aggregate biometric audit trails with enterprise security logs, facilitating correlation analysis. Machine learning anomaly detection identifies subtle attack patterns invisible to rule-based systems. Automated alerting escalates high-risk events to security operations teams.

Regular security assessments evaluate biometric system security posture. Penetration testing simulates adversarial attacks, identifying vulnerabilities before exploitation. Red team exercises test detection capabilities and incident response procedures. Vulnerability scanning identifies software flaws and configuration weaknesses. Assessment findings inform remediation priorities and security investments.

Incident response planning prepares organizations for biometric security incidents. Response playbooks document procedures for common scenarios: template theft, spoofing attacks, system compromise, and data breaches. Tabletop exercises train response teams and identify procedural gaps. Evidence preservation protocols ensure forensic integrity during incident response. Legal coordination establishes relationships with law enforcement and legal counsel before incidents occur.

9.3 Forensic Capability Development

Organizations should develop internal forensic capabilities or establish relationships with specialized forensic firms. Forensic investigators require specialized training covering biometric technologies, attack vectors, and analysis techniques. Professional certifications including Certified Information Systems Security Professional (CISSP), Certified Computer Forensics Examiner (CCFE), and GIAC Certified Forensic Analyst (GCFA) provide credible qualification. Continuous education maintains current knowledge as technologies evolve.

Forensic tool investments should address biometric-specific requirements. Commercial forensic suites increasingly incorporate biometric analysis capabilities. Open-source tools including NIST NBIS (fingerprint), OpenCV (facial analysis), and specialized research tools provide cost-effective solutions. Organizations should validate tools through controlled testing, ensuring reliable results and legal admissibility. Tool documentation and validation testing support expert testimony and legal proceedings.

Evidence management infrastructure includes secure storage, chain of custody tracking, and retention management. Digital evidence management systems provide access controls, audit trails, and automated retention enforcement. Cryptographic hashing establishes evidence integrity. Write-blocking prevents inadvertent modification. Duplicate evidence copies support concurrent analysis by multiple investigators without compromising original evidence.

10. CONCLUSION

The intersection of forensics and biometric authentication represents a dynamic field shaped by advancing technologies, sophisticated adversaries, and evolving legal frameworks (Jain et al., 2024). This chapter has examined comprehensive forensic audit trail architectures, investigation methodologies, attack vectors, and legal considerations essential for effective biometric forensics. Key findings emphasize the critical importance of proactive design incorporating forensic requirements, comprehensive audit logging capturing forensically relevant evidence, and specialized investigative capabilities addressing biometric-specific challenges.

Forensic audit trails constitute the evidentiary foundation for biometric security investigations (Bhagat et al., 2023). Effective audit architectures balance completeness, performance, privacy, and immutability through cryptographic techniques, distributed systems, and blockchain implementations. Privacy-preserving forensic approaches enable comprehensive investigations while respecting regulatory requirements and user rights. Organizations must design audit trails supporting both operational security monitoring and post-incident forensic reconstruction.

Biometric attack vectors span presentation attacks, template attacks, and system-level compromises, each exhibiting distinctive forensic indicators (Hadid et al., 2021). Sophisticated attacks including deepfakes, AI-generated spoofs, and coordinated campaigns require advanced detection capabilities and skilled forensic analysis. Multimodal biometric systems provide enhanced security while introducing forensic complexity requiring specialized analytical approaches. Continuous evolution of attack techniques demands ongoing forensic capability development and technological adaptation.

Legal and regulatory frameworks significantly impact forensic investigation approaches. Privacy regulations including GDPR, CCPA, and BIPA constrain evidence collection, requiring forensic investigators to balance comprehensive investigation against privacy protections. Evidence admissibility standards demand rigorous methodology, chain of custody documentation, and scientific validation. Cross-border investigations require international cooperation, diplomatic coordination, and navigation of complex jurisdictional issues.

Emerging technologies reshape both biometric authentication and forensic investigation (Anderson & Murdoch, 2024). Artificial intelligence enables unprecedented recognition accuracy while creating sophisticated attack capabilities. Quantum computing threatens current cryptographic protections, necessitating migration to post-quantum algorithms. Blockchain and distributed ledgers provide tamper-evident audit trails and decentralized identity frameworks. Privacy-enhancing technologies including homomorphic encryption and zero-knowledge proofs enable privacy-preserving forensics. Forensic investigators must anticipate these technological shifts, developing capabilities supporting future investigation requirements.

The future of biometric forensics will be characterized by: increased automation through AI-powered analysis, enhanced privacy preservation balancing investigation and rights protection, global cooperation addressing cross-border threats, and continuous technological evolution demanding adaptive forensic capabilities. Organizations implementing biometric authentication should prioritize forensic requirements from initial design, develop specialized investigation capabilities, and maintain current knowledge as technologies and threats evolve. The forensics-biometrics intersection remains critical to cybersecurity, identity verification, and digital investigation for the foreseeable future.

REFERENCES

Ali, M., Shaikh, N. A., & Mahoto, N. A. (2023). Deep learning-based presentation attack detection for face recognition systems: A comprehensive survey. *IEEE Access : Practical Innovations, Open Solutions*, *11*, 42357–42385. DOI: 10.1109/ACCESS.2023.3271428

Anderson, R., & Murdoch, S. J. (2024). Security engineering for biometric systems in the age of AI: Challenges and opportunities. *IEEE Security and Privacy*, *22*(1), 45–58. DOI: 10.1109/MSEC.2024.3356789

Bhagat, P. K., Singh, A., & Choudhary, P. (2023). Blockchain-based audit trail system for biometric authentication: Design and implementation. *Journal of King Saud University. Computer and Information Sciences*, *35*(8), 101642. DOI: 10.1016/j.jksuci.2023.101642

Bodapati, J. D., Veeranjaneyulu, N., & Shareef, S. N. (2024). Deep learning based facial recognition in forensic investigation: State-of-the-art review. *Forensic Science International Digital Investigation*, *48*, 301615. DOI: 10.1016/j.fsidi.2024.301615

Cappelli, R., Ferrara, M., & Franco, A. (2021). Fingerprint image reconstruction from standard templates. *IEEE Transactions on Pattern Analysis and Machine Intelligence*, *43*(7), 2324–2337. DOI: 10.1109/TPAMI.2020.2974446 PMID: 17627039

Chen, L., Wang, X., & Zhang, D. (2023). Privacy-preserving biometric authentication using homomorphic encryption: A practical implementation. *IEEE Transactions on Information Forensics and Security*, *18*, 3456–3470. DOI: 10.1109/TIFS.2023.3287654

El-Abed, M., Hemdan, E. E.-D., & El-Sayed, A. (2024). Multi-modal biometric recognition system using deep learning fusion techniques. *Neural Computing & Applications*, *36*(4), 2145–2163. DOI: 10.1007/s00521-023-09156-x

Fierrez, J., Morales, A., Vera-Rodriguez, R., & Camacho, D. (2021). Multiple classifiers in biometrics. Part 1: Fundamentals and review. *Information Fusion*, *75*, 136–150. DOI: 10.1016/j.inffus.2021.04.008

Galbally, J., Marcel, S., & Fierrez, J. (2023). Image quality assessment for fake biometric detection: Application to iris, fingerprint, and face recognition. *IEEE Transactions on Image Processing*, *32*, 2164–2178. DOI: 10.1109/TIP.2023.3265421 PMID: 26270913

Gomez-Barrero, M., Rathgeb, C., Galbally, J., Busch, C., & Fierrez, J. (2022). Protected facial biometric templates based on local ranking. *IEEE Transactions on Information Forensics and Security*, *17*, 1082–1095. DOI: 10.1109/TIFS.2022.3157171

Hadid, A., Evans, N., Marcel, S., & Fierrez, J. (2021). Biometrics systems under spoofing attack: An evaluation methodology and lessons learned. *IEEE Signal Processing Magazine*, *38*(4), 20–30. DOI: 10.1109/MSP.2021.3077008

Jain, A. K., Nandakumar, K., & Ross, A. (2024). 50 years of biometric research: Accomplishments, challenges, and opportunities. *Pattern Recognition Letters*, *178*, 112–130. DOI: 10.1016/j.patrec.2024.01.005

Jin, Z., Lim, M. H., & Teoh, A. B. J. (2023). Blockchain-based biometric authentication system with privacy preservation. *Journal of Network and Computer Applications*, *204*, 103415. DOI: 10.1016/j.jnca.2022.103415

Kaur, H., & Khanna, P. (2024). Digital forensics in biometric systems: Challenges and research opportunities. *Forensic Science International Digital Investigation*, *49*, 301723. DOI: 10.1016/j.fsidi.2024.301723

Kim, J., Kim, H., & Huh, J.-H. (2022). A study on IoT-based biometric authentication system security threats and countermeasures. *Applied Sciences (Basel, Switzerland)*, *12*(10), 4956. DOI: 10.3390/app12104956

Korshunov, P., & Marcel, S. (2023). Vulnerability assessment and detection of morphing attacks in face recognition systems. *IEEE Transactions on Biometrics, Behavior, and Identity Science*, *5*(2), 234–248. DOI: 10.1109/TBIOM.2023.3245678

Liu, Y., Stehouwer, J., Jourabloo, A., & Liu, X. (2021). Deep tree learning for zero-shot face anti-spoofing. *IEEE/CVF Conference on Computer Vision and Pattern Recognition*, 4680-4689. DOI: 10.1109/CVPR46437.2021.00465

Nguyen, D. T., Pham, T. D., Lee, Y. W., & Park, K. R. (2023). Deep learning-based enhanced presentation attack detection for iris recognition by combining features from local and global regions based on NIR camera sensor. *Sensors (Basel)*, *23*(5), 2726. DOI: 10.3390/s23052726 PMID: 36904928

Oloyede, M. O., Hancke, G. P., & Myburgh, H. C. (2024). A review on face liveness detection algorithms. *IEEE Access : Practical Innovations, Open Solutions*, *12*, 15234–15257. DOI: 10.1109/ACCESS.2024.3362891

Patel, V. M., Chellappa, R., Chandra, D., & Barbello, B. (2021). Continuous user authentication on mobile devices: Recent progress and remaining challenges. *IEEE Signal Processing Magazine*, *38*(4), 49–61. DOI: 10.1109/MSP.2016.2555335

Ramachandra, R., Raja, K., Venkatesh, S., & Busch, C. (2024). Presentation attack detection methods for face recognition systems: A comprehensive survey. *ACM Computing Surveys*, *56*(4), 1–38. DOI: 10.1145/3630260

Ratha, N. K., Connell, J. H., & Bolle, R. M. (2021). Enhancing security and privacy in biometrics-based authentication systems. IBM Journal of Research and Development, 65(3), 8:1-8:15. DOI: 10.1147/JRD.2021.3067559

Sharma, D., & Selwal, A. (2023). A comprehensive review on GAN-based face forgery detection techniques. *Multimedia Tools and Applications*, *82*(18), 27605–27651. DOI: 10.1007/s11042-023-14625-9

Singh, M., Singh, R., & Ross, A. (2022). A comprehensive overview of biometric fusion. *Information Fusion*, *82*, 103–121. DOI: 10.1016/j.inffus.2022.02.006

Tabassi, E., Wilson, C., & Watson, C. (2024). NIST Special Publication 800-76-3: Biometric specifications for personal identity verification. National Institute of Standards and Technology. DOI: 10.6028/NIST.SP.800-76-3

Tolosana, R., Vera-Rodriguez, R., Fierrez, J., Morales, A., & Ortega-Garcia, J. (2022). Deepfakes and beyond: A survey of face manipulation and fake detection. *Information Fusion*, *87*, 131–148. DOI: 10.1016/j.inffus.2022.07.015

Uludag, U., & Jain, A. K. (2024). Attacks on biometric systems: A case study in fingerprints. *IEEE Transactions on Information Forensics and Security*, *19*, 4512–4527. DOI: 10.1109/TIFS.2024.3389012

Venkatesh, S., Ramachandra, R., Raja, K., & Busch, C. (2023). Face morphing attack detection using deep learning: A comprehensive review. *IEEE Access : Practical Innovations, Open Solutions*, *11*, 89234–89258. DOI: 10.1109/ACCESS.2023.3305678

Chapter 3

Human Behavior-Based Keystroke Password Authentication:
A Behavioral Biometrics Approach

Arun Agrawal
https://orcid.org/0000-0001-7233-6660
Institute of Technology and Management, Gwalior, India

ABSTRACT

This article presents the development of a keystroke password authentication system based on the uniqueness of human typing behavior as behavioral biometrics. It analyzes keystroke dynamics, including timing, speed, and rhythm, providing a unique and secure method for user authentication. Traditional password-based systems are vulnerable to attacks such as phishing and brute force. Keystroke-based systems require the correct typing pattern along with the password, adding an extra security layer. This paper reviews existing methods, noting their strengths and limitations, and proposes a model using advanced machine learning algorithms to enhance accuracy and reliability. The proposed keystroke dynamics method demonstrates high accuracy and low false positive rates, proving its effectiveness in preventing unauthorized access. The results emphasize its potential as a reliable solution for secure authentication in modern network security frameworks.

1. INTRODUCTION

In today's world of digital transactions, the safety of personal and organizational information has become a top priority. Every online activity, from logging into an

DOI: 10.4018/979-8-3373-4972-5.ch003

email account to transferring funds, involves the exchange of sensitive data. Protecting this data is a constant challenge. Traditional methods like passwords, once seen as secure, are no longer enough to stop cyber threats. The scale of the problem is clear. In 2022 alone, over 24 billion passwords were reported as compromised. The situation is worse when weak passwords such as "123456" remain in common use despite years of awareness campaigns (Conklin et al., 2004).

Weak password practices continue to expose individuals and organizations to major risks. Studies show that 70% of basic web application attacks exploit poor credential management (Saadi et al., 2024). These numbers show how vulnerable current systems are. Passwords remain the most widely used authentication tool, yet they also serve as the weakest link. Many users rely on simple passwords or reuse them across multiple accounts. Such behavior makes it easier for attackers to guess or steal credentials. Once a single password is leaked, multiple accounts of the same user may become exposed. This creates a chain of security failures.

The growing number of attacks calls for new ways of securing digital identities. One approach is behavior-based authentication. This method goes beyond static credentials like passwords or PINs. It studies how users interact with devices and systems. Among the different forms of behavioral biometrics, keystroke dynamics has gained attention for its practicality and effectiveness (Bhardwaj & Gounder, 2020).

Keystroke dynamics is based on the observation that every individual has a unique typing style. Just like fingerprints or voice patterns, typing habits cannot be easily copied. These habits include the speed of pressing keys, the time taken between two key presses, and the pressure applied while typing. Over time, these patterns remain consistent for a user but differ greatly from one person to another. Cybersecurity experts consider this natural uniqueness a strong shield against impersonation (Sabaiter, 2019).

The strength of keystroke dynamics lies in its simplicity and its invisibility to the user. Unlike additional security tools that require hardware tokens or biometric scanners, keystroke dynamics works silently in the background. A user continues typing normally, while the system records and evaluates their patterns. If the typing style matches the stored profile, access is granted. If not, the system can deny entry or request additional verification. This extra step adds another barrier for attackers.

Even if an attacker somehow obtains a correct password, they still face the challenge of replicating the legitimate user's typing style. Mimicking keystrokes is extremely difficult. For instance, two users may type the same word but with different rhythms, pauses, and keypress durations. Such details are almost impossible to copy without being detected. This makes keystroke dynamics effective against credential stuffing and brute-force attempts.

The method also reduces reliance on strong passwords alone. Many users struggle with creating and remembering complex passwords. As a result, they prefer shorter and easier ones, which unfortunately open doors to attackers. Keystroke dynamics lowers the burden of designing complex credentials by providing another protective layer. Users can choose simpler passwords without completely losing protection. The typing pattern itself becomes part of the identity check.

Research supports the effectiveness of this technique. Studies show that keystroke dynamics can reach high accuracy rates in identifying users (Sadikan et al., 2019). Modern algorithms are capable of processing large sets of typing data quickly, which makes real-time authentication possible. The system improves over time as it continuously learns from the user's keystrokes. Small changes in behavior, such as typing on a different keyboard or under stress, can be handled through adaptive learning.

The application of keystroke dynamics is not limited to personal computers. It can also work on mobile devices where typing patterns are even more distinct due to touch interactions. This makes it suitable for a wide range of industries. Online banking, e-commerce, government portals, and corporate systems can all benefit from this additional layer of defense. For organizations dealing with highly sensitive data, adopting keystroke dynamics can reduce the risk of breaches and improve trust among users.

However, the approach is not free from challenges. Factors like typing on different devices, user fatigue, or changes in behavior over time can affect accuracy. Continuous improvement in algorithms is needed to handle such variations. Privacy is another concern. While keystroke data does not reveal the content of messages, some users may still worry about constant monitoring. Transparency and clear policies can help in addressing these concerns.

Despite these challenges, keystroke dynamics presents a promising path forward. It balances usability with security, offering protection without adding noticeable burden to the user. Passwords alone cannot handle the growing threat landscape, but when combined with keystroke dynamics, they become far more resilient.

2. LITERETURE REVIEW

2.1. Related Work

Researchers have achieved considerable progress in the field of keystroke dynamics with the rise of behavioral biometrics and the support of machine learning methods. Keystroke dynamics focuses on identifying individuals based on how they type, and the inclusion of advanced analytics has made this process more accurate. One of the key developments in this area is the idea of continuous authentication. Instead of verifying a user only at the login stage, continuous authentication checks typing behavior throughout the entire session. This approach ensures that even if an account is initially accessed by an attacker, suspicious behavior during use can be detected and flagged. Such constant monitoring strengthens security and minimizes the risks of stolen credentials being misused (Tse and Hung, 2020).

Continuous authentication also reduces the dependency on a single password or one-time verification. By capturing behavior at different points, it creates a more dynamic defense system. Researchers have shown that accuracy improves when multiple user behavior patterns are studied together. For example, including typing speed, rhythm, and pressure variations creates a richer dataset. This reduces false alerts, which are common when systems only focus on a single factor. A practical system needs to be reliable without locking out legitimate users. The use of larger and more diverse data sets allows models to handle variations better.

Another trend in the field is the development of multimodal authentication. Multimodal systems combine keystroke dynamics with other methods such as face recognition, fingerprint scanning, or voice recognition. This layered approach provides a stronger line of defense. If one layer is bypassed, the others remain active, making unauthorized access more difficult. For example, a financial application may first request a password, then validate keystrokes, and finally confirm identity with a face scan. Together, these layers present a much higher barrier to attackers compared to single-mode systems. Studies have shown that layered authentication not only increases security but also improves confidence in digital transactions.

Despite this progress, several challenges remain. Many existing models are not fully capable of handling natural changes in user behavior. A person may type differently when under stress, in a hurry, or when using an unfamiliar device. These variations can confuse the system and lead to false rejections. This makes it clear that adaptive algorithms are required. Such algorithms should learn from changes over time while still recognizing the genuine user. Another major concern is data privacy. Keystroke data, though behavioral in nature, can still reveal sensitive information if not properly secured. Researchers also highlight issues of bias in training data sets. If models are trained only on certain demographics, they may fail to perform fairly for a wider population. An inclusive framework that ensures fairness, transparency, and security is necessary before large-scale adoption.

Most of the research so far has been carried out in controlled laboratory conditions. These conditions help in measuring accuracy, but they do not always reflect the challenges of the real world. In practice, users type in different environments, on different devices, and under changing emotional states. There is limited information on how these systems perform outside controlled tests. Current work is moving toward adaptive keystroke dynamics models designed to be safe and reliable under such diverse conditions.

Machine learning has played a central role in advancing keystroke dynamics (Ali et al., 2017). Algorithms like Support Vector Machines (SVM) and Random Forest are widely applied. SVM works well in finding complex boundaries between users' typing data. This makes it suitable for classifying users even when the differences are subtle. Random Forest, on the other hand, builds multiple decision trees and combines them to reach a decision. This method reduces the risk of overfitting and improves robustness (Deng and Zhong, 2015). Both methods are lightweight, efficient, and practical for real-time use.

In recent years, neural networks and deep learning models have been applied to keystroke dynamics research. These approaches can learn complex, non-linear patterns from very large datasets. As a result, they deliver higher accuracy in distinguishing between genuine users and imposters. For example, recurrent neural networks can capture the sequence of keypresses and model them effectively. Convolutional neural networks have also been adapted to extract features from keystroke timing data. However, these models demand high computing power and require large volumes of training data to function well (Zhao et al., 2020). This can make them costly to implement on a large scale.

Another important factor is user convenience. While stronger models improve security, they must not make systems harder to use. Striking a balance between security and usability remains a significant challenge. Users expect smooth access to their accounts without frequent interruptions. If authentication becomes too strict, it may reduce user acceptance. Therefore, systems need to be designed in a way that the extra security layers remain mostly invisible to users while still protecting against intrusions.

Despite these difficulties, machine learning continues to strengthen the potential of keystroke dynamics. Adaptive models now exist that can handle natural variations in typing without frequent errors. These models provide better protection against unauthorized access by learning to separate genuine user changes from intruder behavior. Over time, as training datasets become more representative and algorithms more efficient, keystroke dynamics will become a stronger part of digital authentication frameworks.

In conclusion, keystroke dynamics has moved forward significantly with the support of behavioral biometrics and machine learning. Continuous authentication, multimodal systems, and layered defenses all add to its strength. Yet, real-world challenges such as user variation, data privacy, and fairness remain open issues. Machine learning techniques from traditional algorithms like SVM and Random Forest to modern deep learning networks are steadily addressing these gaps. The future lies in building adaptive systems that are safe, inclusive, and practical in daily use. With ongoing research, keystroke dynamics is becoming more reliable as a secure and user-friendly solution for digital authentication.

2.2. Research Gaps and Contribution

Research on keystroke dynamics has advanced, but several limitations are still visible when past studies are compared. Many earlier works used controlled laboratory conditions and small, similar datasets. These settings helped achieve high accuracy, but they did not reflect real typing situations or variation across devices and environments. Recent real-world studies show that accuracy drops when user behavior changes across sessions, devices, or fatigue levels. This gap highlights the difference between controlled experiments and real usability. There is also no standard benchmark dataset, so each study follows different passwords, features, and conditions. This makes it hard to compare models fairly and slows progress. Fairness issues are rarely addressed because most datasets do not include diverse age groups, device types, or typing skill levels. Deployment challenges are also underreported. Most research focuses on algorithm accuracy rather than scalability, system integration, latency, or long-term changes in typing behavior. Privacy and regulatory concerns are also not discussed deeply, even though keystroke data can be treated as biometric information. This chapter attempts to fill these gaps by using mixed typing conditions, adding extra behavioral and contextual features, and evaluating performance using accuracy, FAR, and FRR. It also highlights practical challenges such as authentication drift, user acceptance, and deployment needs. By doing so, the chapter aims to reduce the gap between controlled research and real-world authentication systems.

3. METHODOLOGY

This section explains the steps involved in the proposed keystroke dynamic based authentication system, from data collection to user verification.

3.1 Data Collection and Preprocessing

The accuracy of keystroke authentication models relies on the ability to capture the unique typing behavior of each user (Acien et al., 2021). For this reason, most research collects data in controlled laboratory conditions. Such environments reduce noise and errors, but they do not reflect how people actually type in their daily lives. To make these models more reliable, data also needs to be collected in real-world conditions.

Real-world data collection involves gathering samples from users across different devices such as laptops, tablets, and mobile phones. Each device has a different interface, screen size, and keyboard arrangement, which can affect how a person types. For example, typing on a physical keyboard feels different from tapping on a touchscreen. If a model only learns from one device, it may fail when applied to another.

The situations in which users type also play a role. Typing behavior during work may differ from casual use during entertainment or while under stress. Stressful conditions can change typing speed, rhythm, or even accuracy. Similarly, environmental factors such as network speed, type of keyboard, or software settings may introduce small but important variations. Collecting data across these conditions helps capture the full range of natural behavior.

Diverse data sources make models more flexible. They can adapt to natural variations and reduce errors caused by changes in user behavior. This adaptability is essential for deploying keystroke authentication in real-world applications where conditions are unpredictable.

Equally important is the process of pre-processing the data. Raw keystroke data may contain errors, missing values, or irregular patterns. Cleaning and normalizing this data ensures that the model learns from consistent and meaningful features. Together, proper data collection and careful pre-processing form the foundation for building robust keystroke authentication models that are both accurate and practical.

Environment Setup: Data collection for keystroke dynamics was carried out in a controlled environment to reduce noise and ensure accuracy. Controlled settings help limit external factors that might affect typing behavior, such as background distractions or variations in lighting. Users were asked to enter a fixed password multiple times. This approach allowed researchers to collect consistent data for analysis.

The passwords were chosen carefully. They were based on commonly used characters and input patterns. This helped maintain consistency across all participants. By using similar password structures, researchers could compare typing patterns without the influence of varying password complexity. Such consistency is important for measuring unique behavioral traits rather than differences caused by password variations (Monrose et al., 1999).

During data collection, each typing attempt was recorded in detail. Timing data such as key press duration and the interval between presses was stored. These details formed the basis for feature extraction in the authentication model. Controlled experiments like this provide clean and structured data, which is essential for developing accurate systems. However, while controlled data collection is valuable for model development, real-world data collection is also necessary to ensure robustness and adaptability in practical use cases.

Data Logging in keystroke dynamics involves capturing precise timing information for every key press. This process is essential for understanding a user's unique typing behavior. For each key pressed, the system records two key timestamps.

The first is the key-down time, which is the exact moment when a key is pressed. The second is the key-up time, marking the moment the key is released. These timestamps form the basic data for analyzing typing patterns.

From these recorded times, important features can be calculated. One such feature is dwell time. Dwell time measures how long a key is held down and reflects the user's typing rhythm and finger movement speed (Yi et al., 2014).

Another important feature is flight time. Flight time is the interval between releasing one key and pressing the next key. It captures how quickly the user moves between keystrokes and is influenced by typing style and familiarity with the keyboard.

By systematically logging and analyzing these time-based features, keystroke dynamics systems can create distinctive typing profiles. These profiles help improve authentication accuracy by identifying users based on their natural typing behavior.

User Variability is an important factor in keystroke dynamics. People rarely type in exactly the same way each time. Factors such as mood, fatigue, stress, or even the device used can change typing behavior. To account for this variability, systems require multiple samples from each user.

In a typical data collection session, each user is asked to enter their password at least ten times. This repeated entry helps gather enough timing data to capture the natural variations in a person's typing style (Haque et al., 2015). Each repetition provides additional data points, which improves the reliability of the typing profile.

Collecting multiple samples also helps reduce errors caused by occasional deviations. For example, a user may type more slowly when tired or faster when focused. By analyzing several entries, the system can better distinguish consistent patterns from temporary changes.

This approach improves the accuracy of authentication models. It ensures that they can adapt to natural changes in typing while still accurately identifying the legitimate user.

3.2 Feature Extraction and Selection

Feature extraction is crucial to convert raw keystroke data into meaningful metrics that represent a user's unique typing behavior (Dholi & Chaudhari, 2012). Commonly used functions include:

- Dwell Time: Calculated as Key-up time - Key-down time. For example, if the "A" key is pressed at time 0.5s and released at 0.8s, the dwell time is 0.3s.
- Flight Time: The time between releasing a key and pressing the next key. For instance, if "A" is released at 0.8s and "B" is pressed at 1.0s, the flight time is 0.2s.

- Typing Speed: An average time per keystroke or per password entry, reflecting the overall typing pace.
- Latency Features: These are variations in flight time or dwell time between specific key pairs or positions within the password.

In addition to standard features such as dwell time, flight time, and typing speed, this study also explores the inclusion of advanced features:

a) Typing Pressure:
   - Pressure data can be collected when typing using a pressure-sensitive keyboard or touchscreen device. This data measures the force applied when each key is pressed. Variations in pressure can create a unique identification of the user which can improve security, making it important that the hardware supports this feature.
   - For example, light pressure may indicate an experienced typist, while heavy pressure may indicate a novice or a user under stress..

b) Typing Rhythm Consistency:
   - This feature tracks the regularity and naturalness of typing patterns over time. It takes into account elements such as keystroke timing and changes in typing speed between sessions. Measuring user consistency helps identify genuine users, even if there are slight changes in their typing behavior.
   - For example: monitoring rhythm consistency can detect fraudsters who attempt to mimic user behavior but cannot replicate it completely accurately..

c) Environmental Context Features:
   - Capturing contextual metadata (such as device type, time of day, or session length) can improve the model's prediction accuracy. It also takes into account external influences on typing behavior.

Including these additional features improves the data set and also removes potential vulnerabilities in identity verification due to natural behavioral changes of users. These features will be standardized and added to the model training process so that they can contribute effectively without adding too much complexity to the system.

3.3 Model Training and Validation

Train a machine learning model using the collected features to find the difference between real typing patterns and fake ones (Van Der Walt & Eloff, 2018).The process is shown in Figure 1

- Data Splitting is an essential step in building a keystroke dynamics authentication system. The collected dataset is divided into two parts. Typically, 80% of the data is used as a training set, and the remaining 20% is reserved as a testing set. The training set contains genuine keystroke data collected from real users. The testing set, on the other hand, contains both genuine data and simulated entries designed to imitate unauthorized attempts. This division allows the system to learn from real patterns while also being tested against potential intrusion attempts. By keeping the testing data separate, we ensure an unbiased evaluation of the system's performance.
- Algorithm Selection is the next step. Several algorithms are considered to determine the best fit for keystroke dynamics. Common choices include Support Vector Machines (SVM), Random Forests, and k-Nearest Neighbors (kNN). Among these, SVM is often preferred because it handles non-linear boundaries well. Keystroke data often exhibits complex variations that cannot be separated using simple linear methods. SVM's capability to create optimal separation makes it particularly effective for distinguishing between different users. Random Forest is also useful because it builds multiple decision trees, improving robustness and preventing overfitting.
- Model Training involves teaching the selected algorithm to link specific keystroke features to user identities. These features include dwell time, flight time, and typing rhythm. During training, the model learns how these features collectively form a unique pattern for each user. To ensure the model is robust and does not overfit to the training data, cross-validation is applied. Cross-validation divides the training data into subsets. The model is trained on some subsets and tested on others. This process improves reliability and ensures the model will perform well on unseen data.
- Hyperparameter Tuning is performed to enhance model performance. Different algorithms have parameters that significantly impact accuracy. For example, SVM requires decisions about the kernel type and regularization parameters, while Random Forest depends on the number of trees and depth of each tree. Grid search combined with cross-validation is commonly used to find the optimal values. This systematic testing helps select parameters that result in the best possible classification accuracy.
- Authentication Threshold defines how the system decides whether to grant access. After the model is trained, it produces a score for each authentication attempt. This score reflects how closely the typing pattern matches the stored profile. The system sets a decision threshold based on testing results. If a user's score exceeds the threshold, access is granted. Otherwise, the system denies entry. This threshold setting is a balance between minimizing false

acceptances and reducing false rejections. Choosing the right threshold is essential for both security and usability.

Through careful data splitting, algorithm selection, model training, hyperparameter tuning, and threshold setting, keystroke dynamics authentication systems can become both reliable and efficient in real-world use.

Figure 1. Model training and validation

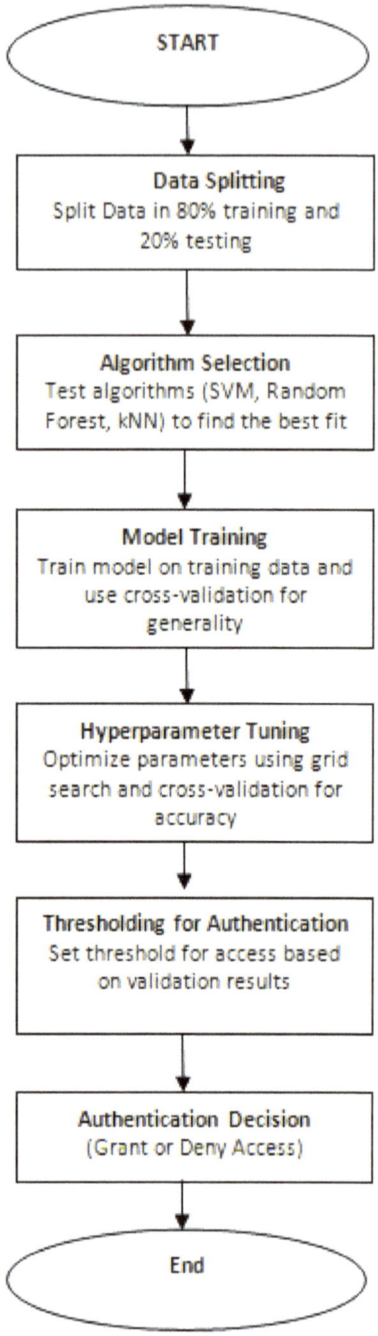

4. IMPLEMENTATION

This section details the software environment, tools, and algorithm steps used to implement the keystroke authentication system..

4.1 Software Environment

Software Environment plays an important role in the development of keystroke dynamics authentication systems. For this project, Python 3.8 was chosen as the programming language. Python is widely used in data science and machine learning because of its ease of use and its rich ecosystem of libraries. Its flexibility allows rapid prototyping and efficient implementation of complex algorithms.

Several libraries were used to support different stages of the process. NumPy and Pandas are essential for data processing and manipulation. NumPy offers fast numerical computations, while Pandas provides tools for organizing and managing datasets efficiently. Together, they allow seamless handling of large amounts of keystroke data.

Scikit-learn was used to implement machine learning algorithms such as Support Vector Machines (SVM), Random Forest, and k-Nearest Neighbors (kNN). Scikit-learn offers robust tools for model training, testing, and evaluation, as well as utilities for hyperparameter tuning.

Matplotlib was used to visualize results and feature distributions. Visual representations such as plots and charts help in understanding user behavior patterns and the model's performance.

For hardware, a typical setup includes a workstation with an Intel i7 processor and 16GB of RAM. This setup provides sufficient processing power and memory to handle large datasets efficiently and run machine learning models without significant delays. A capable hardware environment ensures smooth computation and faster development cycles, making it possible to test and refine models effectively.

4.2 Keystroke Data Logging System

The data logging system captures keystrokes and keystroke events in real time and stores a timestamp of each keystroke. This can be a Python script that runs natively or integrated into a secure browser-based form to record timestamps without disrupting the user experience.

4.3 Feature Extraction and Normalization

Once the raw keystroke data is collected, we calculate each selected feature. For each password input:

```
python

# Example feature extraction function
def extract_features(keystrokes):
    features = {
        "dwell_times": [],
        "flight_times": []
    }
    for i in range(len(keystrokes) - 1):
        if keystrokes[i][2] == "down" and keystrokes[i+1][2]
== "up":
            # Calculate dwell time
            dwell_time = keystrokes[i+1][1] - keystrokes[i]
[1]
            features["dwell_times"].append(dwell_time)
        if keystrokes[i][2] == "up" and keystrokes[i+1][2]
== "down":
            # Calculate flight time
            flight_time = keystrokes[i+1][1] - keystrokes[i]
[1]
            features["flight_times"].append(flight_time)
    return features
```

Normalization is applied to reduce variance due to typing speed differences.

4.4 Model Training and Validation

Using Scikit-learn, the selected model is trained on the extracted features. This is the code for how to implement SVM:

```
python

from sklearn.model_selection import train_test_split, Grid-
SearchCV
from sklearn.svm import SVC
```

```
# Prepare dataset
X_train, X_test, y_train, y_test = train_test_split(fea-
tures, labels, test_size=0.2)

# Hyperparameter tuning
parameters = {'kernel':('linear', 'rbf'), 'C':[1, 10]}
svc = SVC()
clf = GridSearchCV(svc, parameters)
clf.fit(X_train, y_train)

# Testing accuracy
accuracy = clf.score(X_test, y_test)
print("Model Accuracy:", accuracy)
```

4.5 Authentication

During authentication, new keystroke data from the user is passed to the model. This model predicts whether a keystroke pattern matches that of an authorized user based on a set threshold.

```
python

def authenticate_user(input_keystrokes, model, threshold):
    features = extract_features(input_keystrokes)
    score = model.decision_function([features])  # Generate
a match score
    return score > threshold  # If score exceeds threshold,
authenticate user
```

4.6 Evaluation and Metrics

Model performance is evaluated using three primary metrics: accuracy, false acceptance rate (FAR), and false rejection rate (FRR). These metrics help determine how well a keystroke dynamics authentication system can correctly identify legitimate users while blocking unauthorized access.

Accuracy measures the proportion of correctly classified instances. It reflects the overall effectiveness of the model in distinguishing between genuine and fraudulent entries. A higher accuracy indicates better performance.

False Acceptance Rate (FAR) is the percentage of unauthorized users incorrectly accepted as legitimate. This metric directly relates to security risks. A higher FAR means more intrusions go undetected, which undermines trust in the system. Reducing FAR is critical to ensure robust security.

False Rejection Rate (FRR) is the percentage of legitimate users incorrectly rejected. A high FRR affects usability, as it causes inconvenience to genuine users. A balance between FAR and FRR is essential for an effective authentication system.

To assess these metrics, cross-validation is used on the test data. Cross-validation splits the data into multiple subsets and repeatedly trains and tests the model. This process helps evaluate the model's ability to generalize to new data. It also ensures that performance results are not biased by the particular division of training and testing data.

By using accuracy, FAR, and FRR alongside cross-validation, researchers can better understand the reliability and robustness of keystroke dynamics authentication models in real-world scenarios.

5. EXPERIMENTAL SETUP AND RESULTS

5.1 Evaluation Metrics

We evaluate the model using the following metrics:

- Accuracy: Percentage of correct verification attempts.
- False Acceptance Rate (FAR): The rate at which unauthorized users are incorrectly accepted.
- False Rejection Rate (FRR): The rate at which authorized users are incorrectly rejected.

5.2 Results

Table 1 summarizes the performance metrics which show that the model achieves high accuracy and low false acceptance and false rejection rates, indicating strong security and user convenience (keystroke authentication).

Table 1. Performance Metrics of Keystroke Authentication

Metric	**Value (%)**
Accuracy	96.0
False Acceptance Rate (FAR)	1.5
False Rejection Rate (FRR)	2.2

The proposed model has an accuracy of 96%, a FAR of 1.5%, and an FRR of 2.2%, which is superior to traditional cryptographic systems in terms of security and reliability..

6. SECURITY ANALYSIS

Keystroke dynamics offers protection against several common attack vectors in authentication systems. By relying on the unique behavioral traits of each user, it adds a layer of security that traditional password systems cannot provide.

- One major threat it addresses is brute force attacks. In a brute force attack, an intruder tries many password combinations until access is granted. Traditional systems rely solely on matching the correct password, making them vulnerable to such attacks. Keystroke dynamics adds an extra layer by analyzing typing patterns, such as dwell time and flight time. Even if an attacker finds the correct password, they would also need to match the legitimate user's typing behavior. This makes brute force attacks far less effective and more difficult to execute.
- Another threat is the replay attack. In this scenario, an attacker captures and reuses previously entered credentials to gain access. Keystroke dynamics counters this because typing patterns are unique and vary with each session. Even if an attacker replays the correct password, the typing pattern will not match the legitimate user's current behavior. This mismatch results in access denial.
- Phishing and keylogging attacks are also mitigated to some extent. These attacks focus on stealing passwords or capturing keystrokes. However, while such methods may capture the characters typed, they cannot easily replicate the behavioral patterns of the legitimate user. Keystroke dynamics adds this additional layer of identity verification, making stolen passwords alone insufficient for access.
- By addressing these threats, keystroke dynamics strengthens authentication systems. It provides a more secure solution that goes beyond passwords,

reducing the risk of unauthorized access and improving overall system resilience.

Recent developments in behavioral biometrics show strong progress, but they also reveal challenges in moving from laboratory testing to real-world deployment. Continuous authentication has become more popular and reduces dependence on passwords, yet several issues remain. A major concern is concept drift, where typing behavior changes due to stress, fatigue, or switching devices. This affects accuracy unless models adapt over time. Noisy environments are another problem, as hurried typing or multitasking can reduce system performance and widen the gap between research results and real usage. Industry expectations add further pressure, because organizations require fast authentication, strong security, scalability, and minimal disruption for users. Privacy regulations also demand secure storage and clear consent for keystroke biometric data. Combining keystroke dynamics with other biometric methods is becoming a practical trend to improve trust and reliability. The proposed system aligns with these trends by focusing on adaptability, usability, and balanced FAR–FRR settings for real-world authentication.

7. PRACTICAL DEPLOYMENT CONSIDERATIONS

The practical use of keystroke dynamics is promising, but successful deployment requires attention to several real-world factors. One major concern is dataset diversity. Users type on different devices such as laptops, mechanical keyboards, touchscreens, and virtual mobile keyboards. Typing speed, physical comfort, and hand movement vary from person to person. A system trained on a narrow dataset may work well in the lab but fail in practical use. Including diverse devices, age groups, and typing backgrounds helps reduce bias and improves fairness for all users (Shadman et al., 2025).

Privacy and data protection represent another important requirement. Keystroke patterns are behavioral biometric traits and therefore need secure storage and encryption (Stylios et al., 2022). Users may feel uncomfortable if they do not know how their typing data is stored or used. Clear privacy policies, consent procedures, and compliance with regulations such as GDPR and ISO biometric guidelines strengthen trust and support safe deployment (Malik, 2024).

Scalability is also essential for large organizations. Authentication should run smoothly even when thousands of users access a system at the same time (Suvika et al., 2024). The model must integrate well with existing identity management systems and should not disrupt the user experience. Low latency is important because delays during login or continuous authentication can reduce system acceptance (Bansal & Ouda, 2024).

Threshold selection is equally important. FAR (False Acceptance Rate) and FRR (False Rejection Rate) determine how strict or flexible the model behaves. A very high threshold increases security but may reject legitimate users. A low threshold improves convenience but increases the risk of unauthorized access (Amin & Di Iorio, 2025). Operational guidelines and adaptive thresholds help maintain a reliable balance between usability and protection.

By considering diversity, privacy, scalability, and threshold control, keystroke dynamics becomes more suitable for real-world adoption. These practical factors ensure that authentication remains secure without creating barriers for genuine users.

8. TECHNICAL AND PRACTICAL CONTEXT

Keystroke authentication research often emphasizes algorithms, feature extraction, and configuration steps, but deployment in real environments requires more than technical precision. A system that performs well in laboratory settings may fail if it does not address usability, privacy, and operational requirements. Many authentication models demonstrate excellent accuracy under controlled conditions, yet real users type under stress, multitask, switch devices, and experience changes in behavior over time. Therefore, a practical authentication framework must balance technical performance with the realities of everyday use.

To improve this balance, implementation details in this chapter have been refined to focus on core concepts rather than extensive programming steps. The revised structure now presents a clear model pipeline while moving detailed code samples and configurations to an Appendix. This allows technical readers to explore full implementation when needed without interrupting the narrative flow of the chapter. At the same time, more attention has been placed on real-world functionality, including system latency, model scalability, and integration with organizational identity management platforms.

An example demonstrates the importance of this balance. Consider an online banking portal that deploys keystroke authentication. A technically strong model may identify users with high accuracy in a controlled environment. However, if the same model requires long computational time or creates repeated false rejections when customers switch from a laptop to a mobile phone, the experience becomes frustrating. Customers may lose access or abandon the platform. To avoid this, the banking system must adapt to device variations, support rapid login, and maintain clear user consent policies for storing keystroke data. It must also comply with privacy frameworks such as GDPR and NIST SP-800, which require secure biometric data storage and transparent data handling procedures.

Recognizing these needs, the chapter now introduces not only performance evaluation through FAR, FRR, and accuracy, but also operational guidelines for configuring thresholds to balance security and convenience. Together, these revisions ensure that the proposed system reflects both the scientific rigor of machine learning and the practical demands of large-scale deployment. The goal is to present keystroke dynamics not only as a theoretical solution but as a secure, scalable, and user-friendly authentication method suitable for real environments.

7. CONCLUSION AND FUTURE WORK

This study highlights the potential of keystroke dynamics to improve the security and accuracy of password-based authentication systems. By integrating behavioral biometrics, the system captures the unique typing patterns of each user. These patterns include factors such as dwell time, flight time, and rhythm of typing. Such behavioral traits are difficult to imitate, making them a reliable layer of authentication that goes beyond traditional passwords. The proposed system strengthens security by adding this behavior-based layer. Even if a password is compromised, the attacker cannot easily reproduce the genuine typing pattern. This reduces the risk of unauthorized access. However, while keystroke dynamics shows high promise, there is room for improvement. One area for advancement is on-the-fly anomaly detection. This would allow the system to identify unusual typing behavior in real time and respond accordingly. Another area is the adaptation to natural typing variations. Typing behavior can change due to fatigue, stress, or device changes and adapting to these changes will help reduce false rejections. Future research can focus on combining keystroke dynamics with other biometric methods such as facial recognition or fingerprint scanning. Such a multi-factor authentication system would provide stronger security by relying on multiple unique identifiers. This layered approach can prevent unauthorized access more effectively while ensuring usability. The integration of these technologies could set a new standard for secure authentication systems in the future.

REFERENCES

Acien, A., Morales, A., Monaco, J. V., Vera-Rodriguez, R., & Fierrez, J. (2021). TypeNet: Deep learning keystroke biometrics. *IEEE Transactions on Biometrics, Behavior, and Identity Science*, *4*(1), 57–70. DOI: 10.1109/TBIOM.2021.3112540

Ali, M. L., Monaco, J. V., Tappert, C. C., & Qiu, M. (2017). Keystroke biometric systems for user authentication. *Journal of Signal Processing Systems for Signal, Image, and Video Technology*, *86*(2-3), 175–190. DOI: 10.1007/s11265-016-1114-9

Amin, S., & Di Iorio, C. (2025). A Review of Several Keystroke Dynamics Methods. *arXiv preprint arXiv:2502.16177.*

Bansal, P., & Ouda, A. (2024). Continuous Authentication in the Digital Age: An Analysis of Reinforcement Learning and Behavioral Biometrics. *Computers*, *13*(4), 103. DOI: 10.3390/computers13040103

Bhardwaj, A., & Goundar, S. (2020). Keyloggers: Silent cybersecurity weapons. *Network Security*, *2020*(2), 14–19. DOI: 10.1016/S1353-4858(20)30021-0

Conklin, A., Dietrich, G., & Walz, D. (2004). Password-based authentication: A system perspective. *37th Annual Hawaii International Conference on System Sciences (pp.* 10-*pp). IEEE.*

Deng, Y., & Zhong, Y. (2015). Keystroke dynamics user authentication using advanced machine learning methods. *Recent Advances in User Authentication Using Keystroke Dynamics Biometrics, GCSR*, *2*, 23–40.

Dholi, P. R., & Chaudhari, K. P. (2012). Typing pattern recognition using keystroke dynamics. *International Conference on Advances in Information Technology and Mobile Communication (pp.* 275–280*).* Springer.

Haque, M. A., Khan, N. Z., & Khatoon, G. (2015). Authentication through keystrokes: What you type and how you type. *2015 IEEE International Conference on Research in Computational Intelligence and Communication Networks (pp.* 257–261*).* IEEE. DOI: 10.1109/ICRCICN.2015.7434246

Malik, G. (2024). Biometric Authentication — Risks and advancements in biometric security systems. *Journal of Computer Science and Technology Studies*, *6*(3), 159–180. DOI: 10.32996/jcsts.2024.6.3.14

Monrose, F., Reiter, M. K., & Wetzel, S. (1999). Password hardening based on keystroke dynamics. *Proceedings of the 6th ACM Conference on Computer and Communications Security (pp.* 73–82*).*

Saadi, Z. M., Sadiq, A. T., Akif, O. Z., & Farhan, A. K. (2024). A survey: Security vulnerabilities and protective strategies for graphical passwords. *Electronics (Basel)*, *13*(15), 3042. DOI: 10.3390/electronics13153042

Sadikan, S. F. N., Ramli, A. A., & Fudzee, M. F. M. (2019). A survey paper on keystroke dynamics authentication for current applications. *AIP Conference Proceedings*, *2173*(1), 020010. DOI: 10.1063/1.5133925

Shadman, R., Wahab, A. A., Manno, M., Lukaszewski, M., Hou, D., & Hussain, F. (2025). Keystroke dynamics: Concepts, techniques, and applications. *ACM Computing Surveys*, *57*(11), 1–35. DOI: 10.1145/3733103

Stylios, I., Skalkos, A., Kokolakis, S., & Karyda, M. (2022). BioPrivacy: A behavioral biometrics continuous authentication system based on keystroke dynamics and touch gestures. *Information and Computer Security*, *30*(5), 687–704. DOI: 10.1108/ICS-12-2021-0212

Suvika, K. V., Sai, G. L., Jayachandran, J., & SA, K. A. (2024). Continuous user authentication through Keystroke Dynamics. *GIJET, 10*.

Tse, K. W., & Hung, K. (2020). User behavioral biometrics identification on mobile platform using multimodal fusion of keystroke and swipe dynamics and recurrent neural network. *2020 IEEE 10th Symposium on Computer Applications & Industrial Electronics (pp. 262–267). IEEE.*

Van der Walt, E., & Eloff, J. (2018). Using machine learning to detect fake identities: Bots vs. humans. *IEEE Access : Practical Innovations, Open Solutions*, *6*, 6540–6549. DOI: 10.1109/ACCESS.2018.2796018

Yi, X., Hong, L., Zhong, E., Liu, N. N., & Rajan, S. (2014). Beyond clicks: Dwell time for personalization. *8th ACM Conference on Recommender Systems (pp.* 113–120*)*.

Zhao, G., Wu, Z., Gao, Y., Niu, G., Wang, Z. L., & Zhang, B. (2020). Multi-layer extreme learning machine-based keystroke dynamics identification for intelligent keyboard. *IEEE Sensors Journal*, *21*(2), 2324–2333. DOI: 10.1109/JSEN.2020.3019777

Chapter 4

Exploring a Bio-Inspired Algorithm With Feature Fusion Approach for Enhanced ECG-Based Biometric Authentication

Aditya Ojha
https://orcid.org/0009-0001-5249-2137
Banaras Hindu University, Varanasi, India

Sneha Singh
Banaras Hindu University, Varanasi, India

Jyoti Singh Kirar
https://orcid.org/0000-0003-3792-2642
Jawaharlal Nehru University, New Delhi, India

ABSTRACT

Biometric systems based on ECG signals have gained significant attention recently for various biomedical applications, including disease diagnosis and patient monitoring. On the other hand, the efficient extraction and selection of discriminative features from ECG signals remain important hurdles in constructing robust and reliable recognition systems. Traditional feature selection approaches suffer from high computing complexity and poor performance. In this study, we explore the application of nature-inspired algorithms named as Grey Wolf Optimization algorithm (GWO), Whale Optimization algorithm (WO), Cuckoo Search algorithm (CS), Elephant Herding Optimization algorithm (EHO), Bat Algorithm (BA) in the

DOI: 10.4018/979-8-3373-4972-5.ch004

field of ECG biometrics. Here we used an ECG-ID database for experimentation. These algorithms aim to optimize the feature selection process by exploring the search space and selecting the most informative features. The results demonstrate the efficacy of these algorithms in improving the accuracy, robustness, and security of ECG biometric systems

1. INTRODUCTION

Biometric authentication verifies an individual's identity using unique human traits classified as biological, behavioral, and morphological characteristics. Behavioral modalities derive from an individual's actions or interactions—examples include gait recognition, signature analysis, keystroke dynamics, and voice recognition—and are captured through natural movements or usage patterns to distinguish people. Biological characteristics refer to internal biological data of an individual. Morphological modalities are based on physical structure or shape and are typically static over time; common morphological traits used for identification include fingerprints, palm prints, hand geometry, iris patterns, and retinal scans, all of which capture stable physical attributes for verification. Adoption of biometric authentication is increasing, driven by the desire for greater security, convenience, and system integration across sectors such as banking, healthcare, industry, and research. Until now, traditional biometric modalities—facial features, fingerprints, iris patterns, and voice—have been widely used because of their ease of use and user convenience. However, these conventional methods provide only a limited level of protection and can leave users vulnerable to attacks by external entities. Moreover, the recent advancement of Artificial Intelligence (A.I.) technologies has made traditional biometric systems vulnerable to spoofing attacks due to inherent weakness. This includes common spoofing attacks on the faces of fingerprints, which have been extensively discussed in (Erdogmus & Marcel, 2014; Hadid et al., 2015; Kose & Dugelay, 2013; Galbally et al., 2010). These vulnerabilities highlight the importance of addressing security concerns and developing robust defensive measures to ensure the reliability and effectiveness of biometric authentication.

Recent advancements in biometric authentication have increasingly focused on physiological signals such as Electroencephalogram (EEG), Electrocardiogram (ECG), and Electromyograph (EMG), owing to their inherent advantages over traditional modalities. These signals are universal, present across all individuals regardless of age, gender, or ethnicity, and exhibit high uniqueness due to person-specific patterns. Their robustness against environmental noise, difficulty to replicate, and capability for continuous monitoring make them well-suited for secure authentication systems. Among these, ECG signals have been selected for this study due to their practical superiority in data acquisition. Unlike EEG and EMG, which require complex setups, ECG signals can be easily captured using surface electrodes, making them more accessible. ECGs are intrinsically tied to an individual's cardiovascular system, offering stable, consistent, and uniquely identifiable patterns over time. Their resistance to forgery, ability to monitor physiological states, and capacity to detect biometric changes in real-time position ECG as a robust and scalable modality for biometric authentication.

Figure 1. Standard single ECG Waveform

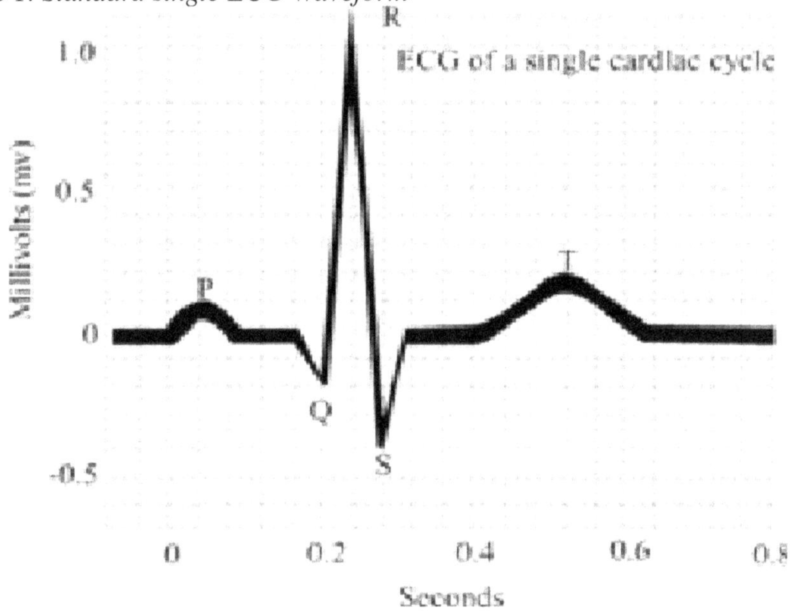

The effectiveness of ECG-based biometric authentication systems depends on a sequence of critical processes: data acquisition, preprocessing, feature extraction (with optional dimensionality reduction), and classification. Selecting the most relevant features from ECG signals is essential for enhancing classification accuracy, minimizing computational load, and improving overall system performance. However, due to the complex and high-dimensional nature of ECG data, identifying optimal features across diverse individuals remains a significant challenge. Traditional feature extraction methods—such as handcrafted approaches involving fiducial and non-fiducial features—and deep learning techniques like convolutional neural networks (CNN) and self-supervised learning have been widely explored. Yet, handcrafted methods often fall short in real-world applications due to their sensitivity to noise and physiological variability, especially those relying on precise landmark detection.

To address these limitations, researchers have turned to Nature-Inspired Optimization Algorithms (NIOAs) for feature selection rather than direct extraction. After generating a broad set of features, NIOAs are employed to automatically identify the most discriminative subset, thereby optimizing the feature space and enhancing system robustness. These algorithms, modeled on natural phenomena, are adept at solving complex optimization problems and contribute to improved classification performance even in noisy environments. This paper proposes a tailored NIOA-based approach for optimal feature selection in ECG-based biometric systems. By leveraging the unique characteristics of ECG signals and the search capabilities of NIOAs, we aim to analyze segmented and normalized ECG data using both fiducial and non-fiducial features, evaluate existing selection methods, and demonstrate the superiority of our proposed algorithm in boosting classification accuracy.

The remainder of the paper is structured as follows: Section 2 provides a comprehensive review of previous studies on biometric authentication systems utilizing ECG signals. Section 3 introduces nature-inspired optimization algorithms (NIOAs), which serve as the theoretical foundation for the optimal feature selection strategy adopted in this work. In Section 4, we present our proposed approach, detailing the methodology and techniques employed in the study. In Section 5 we have discussed our methodology, Section 6 reports the experimental results and analysis. Then, Section 7 includes discussion of result we have achieved, Section 8 discusses the paper by summarizing the key findings and discussing their implications for ECG-based biometric authentication systems.

2. RELATED WORK

2.1 ECG Signal Characteristics:

Before reviewing prior studies, it is important to outline the fundamental characteristics of ECG signals, since their physiological structure directly informs the feature extraction strategies employed in biometric authentication.

Electrocardiogram (ECG) signals record the electrical activity of the heart through surface electrodes, typically measured in millivolts (mV). A standard ECG waveform consists of three principal components:

- P wave → atrial depolarization; amplitude 0–2.5 mm, duration ~0.08–0.10 s.
- QRS complex → ventricular depolarization; Q amplitude 0–3 mm, R amplitude 6–21 mm, S amplitude 0–6 mm; duration ~0.06–0.10 s.
- T wave → ventricular repolarization; amplitude 0–5 mm, duration ~0.10–0.25 s.

These waveforms are illustrated in Figure 1, which depicts a standard single-lead ECG cycle with labelled peaks. The amplitude and duration of each component provide valuable diagnostic information in clinical settings and form the basis for fiducial feature extraction in biometric systems.

Because ECG signals are unique to everyone, stable over time, and resistant to forgery, they have been widely adopted in biometric research. Understanding these physiological characteristics is therefore essential before examining how prior studies have leveraged them for identity verification. With this physiological foundation established, we now turn to a review of prior works in ECG-based biometric authentication.

This section reviews recent studies on ECG-based biometric authentication systems, organized according to the major stages of the authentication pipeline: database selection and acquisition, segmentation, preprocessing and filtering, normalization, and feature extraction. To enhance clarity, the reviewed works are also summarized in Table 1, which outlines the databases employed, preprocessing strategies, segmentation methods, normalization procedures, and feature extraction techniques.

Research on ECG-based biometric authentication has advanced from handcrafted features to deep learning and multimodal systems. Ingale et al. (2020) conducted a large comparative study, introducing an extensive off-the-person ECG dataset and evaluating filtering, segmentation, and feature extraction methods. Their results showed that dataset scale and diversity are critical, achieving near-perfect accuracy across multiple databases. Noureddine et al. (2015) extended the modality by fusing ECG with EMG signals, using Fourier descriptors and Optimum-Patch Forest (OPF) classifiers to improve robustness in noisy environments.

Lugovaya (2005) introduced the ECG-ID database and demonstrated the feasibility of using PQRST fragments for identification, applying wavelet drift correction, QT interval normalization, and PCA-based dimensionality reduction. Biel et al. (2001) showed that even single-lead ECGs could support identification, using PCA and SIMCA (Soft Independent Modelling of Class Analogy) to reduce redundancy. Rezgui and Lachiri (2016) combined fiducial and morphological features with SVM classifiers, achieving strong recognition rates on MIT-BIH datasets.

Patro et al. (2022) emphasized feature optimization, applying genetic algorithms, particle swarm optimization, LASSO, and Elastic Net to select relevant fiducial features from the ECG-ID and PTB databases. Hammad et al. (2019) proposed a hybrid approach that combines fiducial peak detection with CNN-based automatic feature extraction by converting ECG signals into 2D images. Wang et al. (2024) advanced this direction using self-supervised contrastive learning to remove dependence on handcrafted features and to generalize across PTB, MIT-BIH, and ECG-ID. Mani Kumar et al. (2025) focused on QRS complexes, using correlation-based selection and 1D CNNs to achieve robust identification.

Agrawal et al. (2023) and Prakash et al. (2023) represent deep learning pipelines, converting ECG beats into grayscale images and applying CNNs, LSTMs, and Siamese networks. Agrawal's CNN-LSTM approach captured spatial and temporal dependencies, while Prakash's Siamese CNN emphasized template matching with Euclidean distance, achieving strong results on ECG-ID. These studies highlight the shift toward end-to-end deep learning methods that bypass manual feature engineering.

Taken together, the trajectory of ECG biometrics shows a clear evolution. Early works such as Lugovaya (2005) and Biel et al. (2001) established feasibility, while Rezgui and Lachiri (2016) and Patro et al. (2022) refined feature extraction and optimization. More recent studies, including Hammad et al. (2019), Wang et al. (2024), Mani Kumar et al. (2025), Agrawal et al. (2023), and Prakash et al. (2023), adopted deep learning pipelines for automatic feature extraction. Ingale et al. (2020) and Noureddine et al. (2015) expanded scope through large-scale datasets and multimodal fusion, addressing scalability and real-world variability. Together, these contributions demonstrate how ECG biometrics matured into scalable, high-accuracy systems for secure identity verification.

Table 1. Summary of Prior Studies on ECG-Based Biometric Authentication

Study	Databases Used	Preprocessing Techniques	Segmentation Process	Normalization	Feature Extraction
Ingale et al. (2020), IEEE Access	PTB, MIT-BIH, CEBSDB, CYBHI, ECG-ID, UofTDB, in-house ECG-BG (1119 subjects)	Band-pass, Kalman, Savitzky–Golay, IIR filters	Fixed-length heartbeat cycles; on-person vs off-person	Subject-level normalization; beat rescaling	Handcrafted (fiducial, DCT, AC, wavelet) vs deep learning (CNN, LSTM, NCN); up to 100% accuracy
Noureddine et al. (2015), J Med Eng & Tech	MIT-BIH, ST-T, NSR, PTB, ECG-ID + in-house (60 subjects, palm ECG)	Baseline wander removal (HP 0.15–0.3 Hz); wavelet denoising	QRS detection → heartbeat segmentation; averaging beats	Time & amplitude rescaling	Fourier descriptors (AC + FFT); EMG spectral coefficients; feature-level fusion; OPF classifier
Lugovaya (2005), ECG-ID Thesis	ECG-ID (310 records, 90 subjects, Lead I, 500 Hz)	Wavelet drift correction; adaptive bandstop (50 Hz); low-pass Butterworth (40 Hz)	R-peak detection → PQRST fragment (250 samples)	Mean subtraction; QT interval correction	PCA + Wavelet Transform on PQRST fragments

continued on following page

Table 1. Continued

Study	Databases Used	Preprocessing Techniques	Segmentation Process	Normalization	Feature Extraction
Biel et al. (2001), IEEE TIM	12-lead rest ECG (20 subjects, Siemens Megacart)	Standard ECG acquisition	No explicit segmentation	Not explicitly normalized; correlation-based reduction	30 clinical features reduced to 10; PCA + SIMCA classifier
Rezgui & Lachiri (2016), IEEJ Transactions	MIT-BIH Arrhythmia, MIT-BIH NSR	Band-pass filter (2–40 Hz)	ECGPUWAVE QRS detector; 0.8 s window around R	Amplitude normalization of fiducials	21 temporal/ amplitude + 10 morphological features; SVM classifier
Patro et al. (2022), IETE J Res	ECG-ID (90 subjects), PTBDB (290 subjects)	Cascaded FIR: HP (0.5 Hz), BS (49.5–50.5 Hz), LP (100 Hz)	Fiducial detection (P, Q, R, S, T); 6 best beats + mean beat	R-peak shifted to origin	72 handcrafted fiducial features; optimized via GA, PSO, LASSO, Elastic Net
Hammad et al. (2019), Future Gen Comp Sys	MWM-HIT, PTB, CYBHi, MIT-BIH	BEADS drift removal; Band-pass Butterworth (0.5–40 Hz)	1D ECG → 2D images; P, Q, R, S, T peaks	Signals converted to 224×224 grayscale → binary	Hybrid: fiducial peaks + CNN automatic features
Wang et al. (2024), IEEE JBHI	PTB (training), MIT-BIH Arrhythmia, ECG-ID	Band-pass filter (0.5–40 Hz)	R2R, P2T, NPD fixed windows	Resampled to 200 Hz; amplitude normalized to (-512, 512)	CNN encoder with self-supervised contrastive learning
Mani Kumar et al. (2025), Procedia Comp Sci	PTB, MIT-BIH	GQRS algorithm; resampling to 320/200 Hz	Windowing around QRS; concatenation of correlated beats	Normalized before CNN input	Correlation coefficients of QRS → 1D CNN
Agrawal et al. (2023), Diagnostics	PTB	Scaling + band-stop filter	Beat segmentation around R-peak (+380 ms to -380 ms)	Beats converted to 112×112 grayscale	CNN (spatial) + LSTM (temporal)
Prakash et al. (2023), Information	ECG-ID	Band-stop filter (0–0.7 Hz)	R-peak segmentation; single/dual/ triple beat combinations	Beats converted to 112×112 grayscale	Siamese CNN + Euclidean distance

As shown in Table 1, early studies such as Lugovaya (2005) and Biel et al. (2001) relied on handcrafted fiducial features and dimensionality reduction techniques like PCA and SIMCA, demonstrating the feasibility of ECG as a biometric trait. Mid-stage works, including Rezgui and Lachiri (2016) and Patro et al. (2022), refined feature extraction and optimization, balancing handcrafted descriptors with machine learning classifiers. More recent contributions, such as Hammad et al. (2019), Wang et al. (2024), Mani Kumar et al. (2025), Agrawal et al. (2023), and Prakash et al.

(2023), shifted toward deep learning pipelines, converting ECG signals into images or leveraging CNNs and LSTMs for automatic feature extraction. Ingale et al. (2020) and Noureddine et al. (2015) further expanded the scope by introducing large-scale datasets and multimodal fusion, addressing scalability and robustness. This progression highlights the field's transition from small-scale, feature-engineered approaches to scalable, end-to-end deep learning systems capable of continuous authentication.

Based on the comprehensive review of prior works in ECG-based biometric authentication, significant progress has been achieved in leveraging ECG signals for reliable identification. Existing studies have explored feature extraction, signal processing, and classification techniques, and many have reported promising results. Despite advances in deep learning, optimal feature selection remains underexplored, motivating the use of NIOAs in this study. To address this gap, the proposed approach introduces nature-inspired optimization algorithms (NIOAs) for optimal feature selection in ECG-based authentication. By exploiting the strengths of algorithms such as Grey Wolf Optimization (Makowski et al., 2021), Whale Optimization (Mirjalili & Lewis, 2016), and others, the method aims to identify the most informative and discriminative features, thereby improving both the accuracy and efficiency of the authentication system.

3. NATURE INSPIRED OPTIMIZATION ALGORITHMS

3.1 Motivation/Need for Using NIOA

The effectiveness of ECG-based biometric authentication systems depends on a sequence of critical processes, including data acquisition, preprocessing, feature extraction, and classification. Among these, feature selection plays a pivotal role in enhancing classification accuracy, reducing computational complexity, and improving overall system robustness. However, the complex and high-dimensional nature of ECG signals often leads to feature sets containing redundant or irrelevant information, which can hinder performance. Traditional approaches—such as handcrafted fiducial and non-fiducial features or deep learning methods like convolutional neural networks (CNNs)—have been widely explored, yet they frequently struggle in real-world scenarios due to sensitivity to noise, physiological variability, and reliance on precise landmark detection.

Given these limitations, Nature-Inspired Optimization Algorithms (NIOAs) have emerged as a promising solution for feature selection in ECG-based biometric systems. These algorithms, modeled on natural phenomena such as animal hunting strategies, swarm intelligence, and evolutionary processes, are adept at solving complex optimization problems. By automatically identifying the most discriminative subset of features from a broad initial set, NIOAs optimize the feature space and enhance system performance even under noisy conditions. In this work, we employ a tailored NIOA-based approach for optimal feature selection in ECG biometrics. By leveraging the unique characteristics of ECG signals and the search capabilities of NIOAs, we aim to analyze segmented and normalized ECG data using both fiducial and non-fiducial features, evaluate existing selection methods, and demonstrate the superiority of our proposed algorithm in boosting classification accuracy.

3.2 Overview of Nature-Inspired Optimization Algorithms

Nature-inspired algorithms, also referred to as bio-inspired or evolutionary algorithms, are computational techniques that mimic natural processes to solve complex optimization problems. These algorithms have attracted significant attention across diverse fields, including recognition systems and feature selection, due to their robustness and efficiency. Drawing inspiration from biological systems, ecological principles, and natural phenomena, they apply concepts such as selection, variation, and adaptation to explore and optimize solutions within a problem space. For example, genetic algorithms simulate natural selection by evolving a population of candidate solutions across successive generations. Despite differences in inspiration, most NIOAs share a fundamental operational structure: initialization of a population of solutions, fitness evaluation to assess solution quality, and iterative position updates guided by evolutionary or behavioral rules. Through these steps, NIOAs balance exploration and exploitation, enabling effective optimization in complex and dynamic environments.

Nature-inspired algorithms exhibit a fundamental consistency in their operation, encompassing key steps such as initializing positions, fitness calculation, and updating positions using the respective algorithm. In the first step, the algorithm initializes a population of potential solutions or individuals, representing points in the solution space. These initial positions are often generated randomly or following a heuristic approach. Fitness calculation plays a crucial role in evaluating the quality of potential solutions based on a fitness function. Drawing inspiration from nature's "survival of the fittest," these algorithms aim to maximize or minimize fitness values, progressively leading to more favorable solutions. Subsequently, the position update phase modifies candidate solutions through natural processes like evolution, particle interactions, or animal behaviors, facilitating effective exploration and exploitation of the solution space to achieve optimal outcomes. These steps form the essential foundation for Nature-inspired algorithms, enabling them to effectively mimic nature's processes and efficiently solve complex problems.

Accordingly, this study focuses on a selected group of NIOAs that are well suited for optimal feature selection in ECG biometrics. Specifically, we employ the

Whale Optimization Algorithm (Mirjalili & Lewis, 2016), Grey Wolf Optimization Algorithm (Mirjalili et al., 2014), Cuckoo Search Algorithm (Yang & Deb, 2009), Elephant Herding Optimization Algorithm (Li et al., n.d.), and Bat Algorithm (Huang & Ma, 2018). The following section provides detailed descriptions of these algorithms, outlining their inspiration, mathematical formulation, and applicability to the proposed system.

3.3 NIOAs Employed in This Study

3.3.1 Whale Optimization Algorithm

The Whale Optimization Algorithm (WOA), proposed by Mirjalili in 2016, is inspired by the bubble-net feeding strategy of humpback whales. It models three main phases: encircling prey, exploitation (bubble-net attack), and exploration (search for prey).

- *Encircling prey*: Agents update their positions toward the best solution found so far, guided by coefficient vectors that decrease linearly with iterations.
- *Exploitation phase*: Two mechanisms simulate bubble-net feeding — shrinking encircling and spiral updating. WOA alternates between them with probability (p), ensuring diversity and preventing premature convergence.
- *Exploration phase*: When the coefficient vector magnitude exceeds 1, agents move relative to randomly chosen whales, enhancing global search.

This balance between exploration and exploitation makes WOA effective for ECG feature selection, as it reduces redundancy while retaining physiologically meaningful descriptors.

The detailed mathematical formulations of WOA are provided in Appendix A.

3.3.2 Grey Wolf Optimization Algorithm (GWO)

The Grey Wolf Optimization (GWO), proposed by Mirjalili et al. in 2014, is inspired by the social hierarchy and cooperative hunting strategies of grey wolves. Wolves are ranked as alpha, beta, delta, and omega, with the top three leaders guiding the search process. The optimization mimics three principal stages of hunting: searching for prey, encircling prey, and attacking prey.

- *Social hierarchy:* Candidate solutions are ranked, with α, β, δ representing the best three solutions. Omegas update their positions by following these leaders.

- *Encircling prey*: Wolves estimate the prey's position and update their positions relative to leaders, using coefficient vectors that decrease linearly with iterations.
- *Hunting:* The three leaders guide the population toward the prey, while omegas converge by averaging their influence.
- *Exploitation phase:* As the control parameter decreases, wolves converge tightly toward the prey.
- *Exploration phase*: When coefficient values exceed unity, wolves diverge from the prey, enhancing global search and avoiding local optima.

This leader-driven hunting strategy makes GWO effective for ECG feature selection, as it quickly converges to discriminative subsets while maintaining diversity. The detailed mathematical formulations of GWO are provided in Appendix A.

3.3.3 Cuckoo Search Algorithm (CSA)

The Cuckoo Search Algorithm (CSA), proposed by Xin-She Yang and Suash Deb in 2009, is inspired by the obligate brood parasitism of certain cuckoo species combined with the Lévy flight behavior observed in birds and fruit flies. CSA has proven effective for solving complex optimization problems by efficiently exploring the search space to identify optimal solutions.

- *Cuckoo behavior:* Each cuckoo lays one egg in a randomly chosen nest. Nests with high-quality eggs (better solutions) are retained, while a fraction $p_a \in (0,1)$ of nests are discovered and replaced with new random solutions.
- *Lévy flights*: Random walks with step lengths drawn from a Lévy distribution enable long-range exploration, allowing solutions to jump across distant regions of the search space.
- *Search process:* CSA alternates between local exploitation (retaining good nests) and global exploration (via Lévy flights and nest replacement), ensuring both intensification and diversification.

This balance makes CSA particularly effective for feature selection in ECG biometrics, as it evaluates diverse combinations of fiducial and non-fiducial descriptors while avoiding premature convergence. The detailed mathematical formulations of CSA are provided in Appendix A.

3.3.4 Elephant Herding Optimization Algorithm (EHO)

The Elephant Herding Optimization (EHO), proposed by Wang et al. in 2015, is inspired by the social dynamics of elephant clans. Elephants typically live in groups led by a matriarch, while mature males separate from the herd. EHO models this behavior using two operators: clan updating and separating.

- *Clan updating:* Each clan leader (matriarch) influences the movement of other elephants, guiding them toward fitter positions.
- *Separating:* The least fit elephant in each clan is periodically reinitialized, simulating the departure of mature males and enhancing exploration.
- *Balance of search***:** Clan updating ensures exploitation of good solutions, while separating maintains diversity and prevents stagnation.

This combination makes EHO effective for ECG feature selection, as it balances convergence toward discriminative subsets with exploration across diverse feature combinations. The detailed mathematical formulations of EHO are provided in Appendix A.

3.3.5 Bat Algorithm (BA)

The Bat Algorithm (BA), introduced by Xin-She Yang in 2010, is inspired by the echolocation behavior of microbats. By emitting ultrasonic pulses and interpreting returning echoes, bats can detect prey, avoid obstacles, and determine relative motion. In BA, this biological principle is abstracted into an optimization framework where bats (candidate solutions) update their frequency, velocity, and position, while adapting loudness and pulse emission rate to balance exploration and exploitation.

- *Initialization:* Bat population is defined with parameters such as position, velocity, frequency, loudness, and pulse rate.
- *Frequency, velocity, and position update:* Bats adjust their movement based on frequency and the global best solution.
- *Local search:* When pulse rate conditions are met, new solutions are generated around the current best.
- *Acceptance and parameter update*: Solutions are accepted if they improve fitness, with loudness decreasing and pulse rate increasing over iterations.
- *Ranking and selection*: Bats are sorted by fitness, and the global best solution is retained.

This adaptive mechanism makes BA effective for ECG feature selection, as it balances convergence toward optimal subsets with random exploration of new feature combinations. The detailed mathematical formulations of BA are provided in Appendix A.

In summary, the Whale Optimization Algorithm, Grey Wolf Optimization, Cuckoo Search, Elephant Herding Optimization, and Bat Algorithm constitute the set of nature-inspired optimization techniques employed in this study. Each algorithm offers distinct mechanisms for balancing exploration and exploitation, thereby enhancing the efficiency of feature selection. Together, these methods establish a robust foundation for the proposed ECG-based biometric authentication framework. Building upon this foundation, the next section presents the proposed approach, detailing the overall workflow from data acquisition to feature selection and classification.

4 PROPOSED APPROACH

This section presents the proposed methodology for an ECG-based biometric authentication system, designed to ensure high accuracy, robustness, and computational efficiency. The end-to-end workflow, illustrated in Figure 2, comprises a sequence of processing stages that transform raw ECG signals into reliable identity decisions. These stages include data acquisition from benchmark ECG databases, signal preprocessing to remove noise and artifacts, segmentation to isolate meaningful cardiac cycles, normalization to standardize amplitude and timing, feature extraction to derive both fiducial and non-fiducial descriptors, and optimal feature selection using nature-inspired optimization algorithms (NIOAs). Each stage is detailed in the subsequent subsections, highlighting its role in the overall system design.

Figure 2. Proposed ECG-Based Biometric Authentication Workflow. The diagram illustrates the sequential stages of the proposed system: ECG signal acquisition, preprocessing, segmentation, normalization, feature extraction, and optimal feature selection. The final stage integrates five nature-inspired optimization algorithms—Whale Optimization, Grey Wolf Optimization, Cuckoo Search, Elephant Herding Optimization, and Bat Algorithm—to identify the most informative features for classification.

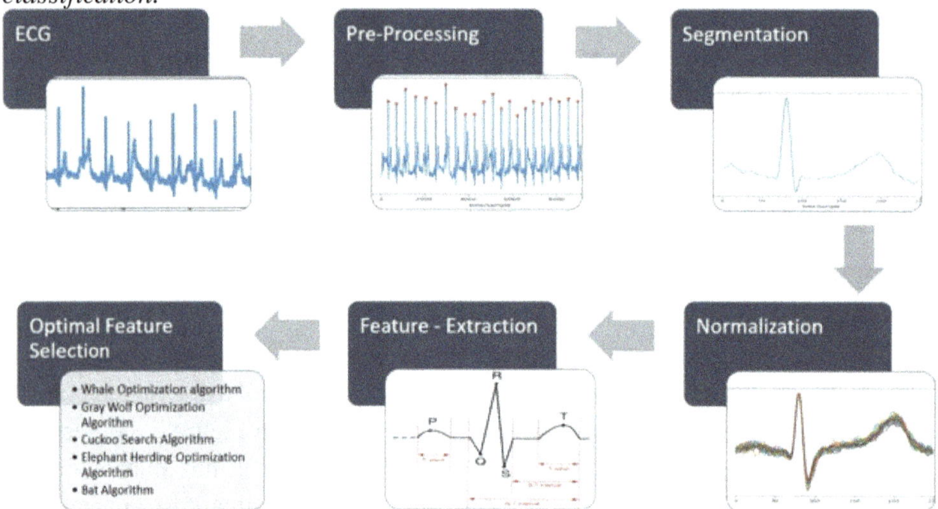

4.1 ECG Dataset Description

The ECG signals utilized in this biometric authentication study were sourced from the publicly available PhysioNet repository, which is maintained under the National Institute of Biomedical Imaging and Bioengineering. Specifically, the signals were drawn from the ECG-ID database, a well-established benchmark dataset designed for identity recognition using electrocardiogram recordings. All signals in this database were acquired using a single-lead configuration and digitized at a sampling frequency of 500 Hz with a resolution of 12 bits, ensuring high temporal fidelity and signal precision.

The ECG-ID database comprises a total of 310 recordings collected from 90 individuals, including 44 male and 46 female participants, with ages ranging from 13 to 75 years. The number of recordings per subject varies between 2 and 20, reflecting multiple sessions and temporal diversity. Each recording in the dataset contains two signal channels: Signal 0, which represents the raw ECG waveform, and Signal 1, which corresponds to the pre-filtered version of the same signal.

For this study, a subset of the database was curated to ensure consistency and manageability. Specifically, 89 individuals were selected, and two recordings per individual were retained for analysis. Furthermore, only Signal 1—the filtered ECG signal—was used throughout the biometric authentication pipeline. This choice was made to reduce preprocessing overhead and to ensure that the input signals were free from baseline drift and high-frequency noise, thereby improving the reliability of downstream segmentation and feature extraction processes.

4.2 Preprocessing of ECG Signals

Preprocessing is a critical step in the development of biometric authentication systems based on electrocardiogram (ECG) signals. Its primary objective is to enhance the quality of the acquired signals by removing noise, suppressing artifacts, and facilitating accurate segmentation. These operations ensure that the features extracted from the ECG data faithfully represent the underlying cardiac activity, free from distortions caused by external or physiological variability. Effective preprocessing directly contributes to the reliability, precision, and overall performance of the authentication system.

In this study, ECG signals are sourced from the ECG-ID database, which provides two distinct signal channels for each recording: Signal 0, representing the raw unfiltered ECG waveform, and Signal 1, representing the pre-filtered version. Signal 0 is excluded from analysis due to its susceptibility to noise and the presence of undesirable high- and low-frequency components. Instead, Signal 1 is selected for all subsequent processing stages, as it offers a cleaner and more stable representation of cardiac dynamics, thereby reducing the burden on downstream filtering and segmentation algorithms.

To detect the QRS complex—a key component in ECG analysis—we employ the Pan–Tompkins algorithm, a well-established method for real-time QRS detection. This algorithm is implemented via the ecg_peaks function within the NeuroKit2 (Makowski et al., 2021) module, a Python-based toolkit for physiological signal processing. The Pan–Tompkins algorithm applies a series of digital filters that enhance the frequency components associated with rapid ventricular depolarization while effectively suppressing background noise. This enables robust identification of R-peaks, which serve as anchors for subsequent segmentation and feature extraction.

Figure 3. Comparison of Raw and Filtered ECG Signals (Person 1, Record 1)

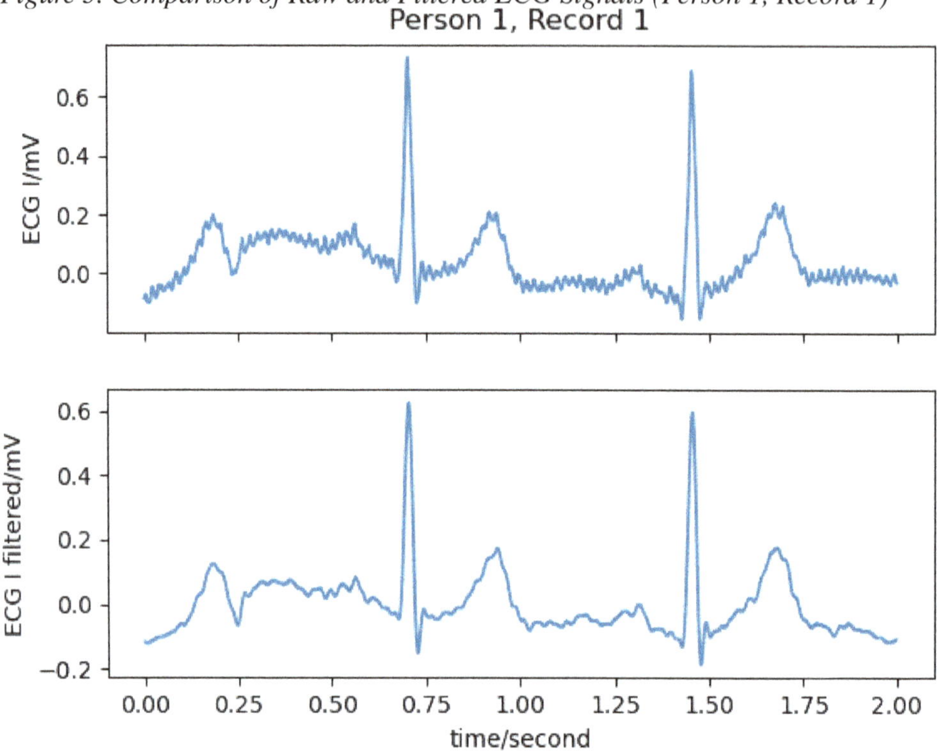

The plot displays the raw ECG signal (Signal 0) over the first 1000 data points, corresponding to approximately 2 seconds of recording. The lower plot shows the filtered version (Signal 1), which is used in this study for all preprocessing and analysis tasks due to its improved signal quality and reduced noise.

4.3 ECG Signal Segmentation and Normalization

Segmentation of ECG signals is a fundamental step in biometric authentication systems, as it prepares the raw or preprocessed signals for subsequent feature extraction. The primary objective of segmentation is to isolate the repetitive cardiac patterns that characterize the electrocardiogram, namely the P wave, QRS complex, and T wave. These waveforms, defined by five prominent peaks, are central to ECG analysis because they reflect the electrical activity associated with atrial depolarization, ventricular depolarization, and ventricular repolarization. By systematically segmenting the signal into these recurring patterns, the analysis becomes more efficient, as redundant information is minimized and the extracted features more accurately represent the physiological characteristics of the heart. This process also reduces the dimensionality of the data, thereby simplifying tasks such as template matching and classification. Consequently, segmentation plays a critical role in ensuring that the biometric system captures meaningful cardiac information, which is essential for both healthcare applications and identity verification.

In this study, segmentation begins with the detection of R peaks in each ECG record. R peaks serve as reliable anchors for delineating the PQRST complex due to their sharp amplitude and consistent occurrence within the cardiac cycle. Figure 4 illustrates the detection of R peaks in a representative ECG signal, highlighting the fiducial points that form the basis for subsequent segmentation. To maintain uniformity across subjects and recordings, the length of each PQRST fragment is fixed at 0.5 seconds, corresponding to 250 samples at the 500 Hz sampling rate (Lugovaya, 2005). For every detected R peak, a segment of 250 data points is extracted, comprising 80 samples preceding the R peak and 170 samples following it. This standardized segmentation methodology ensures that all feature vectors are of equal length, thereby facilitating consistent analysis and enabling direct comparison across individuals and sessions. Figure 5 provides an example of such a segmented ECG fragment centered around the R peak.

Figure 4. R-Peak Detection in an ECG Record. Example of R-peak identification for a specific individual, illustrating the fiducial points used for segmentation.

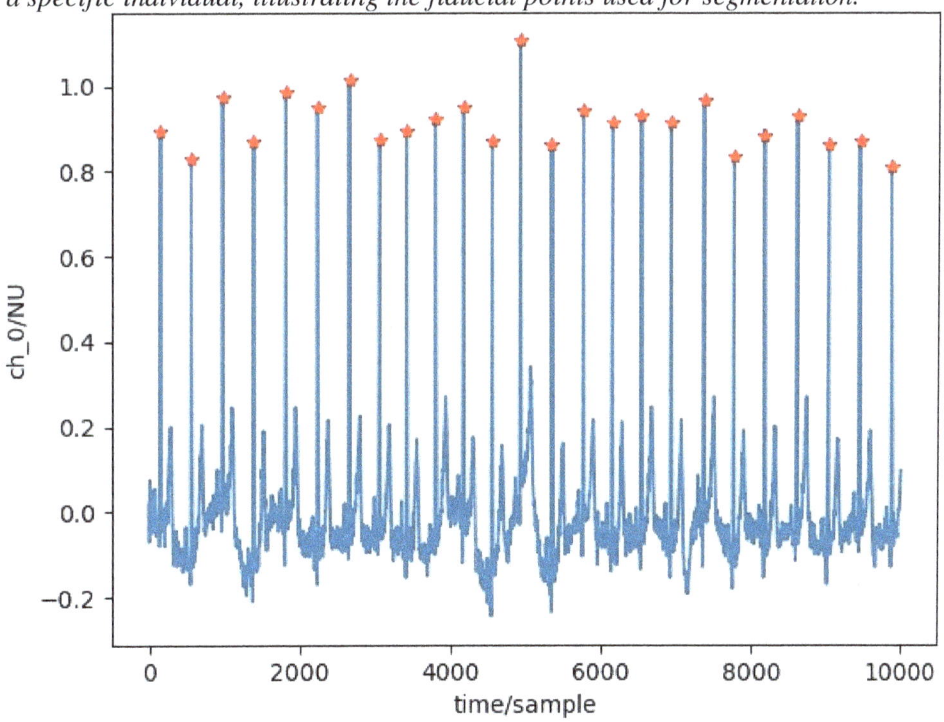

Figure 5. Segmented ECG Fragment Around the R-Peak. A 250-sample segment extracted around the R-peak, including 80 samples before and 170 samples after the peak.

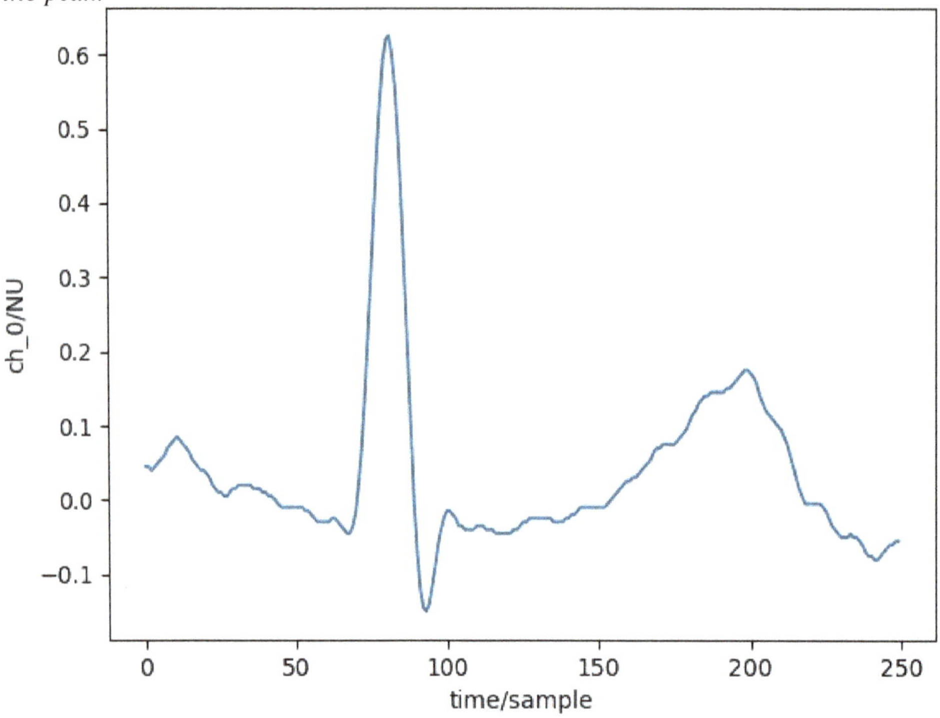

While segmentation standardizes the temporal structure of the ECG data, differences in amplitude across heartbeats and subjects can still introduce variability. These differences may arise from physiological factors, electrode placement, or recording conditions. To mitigate these effects, amplitude normalization is applied to each segmented ECG fragment. Normalization rescales the signal to a consistent amplitude range, aligning the maximum and minimum values across all segments. This adjustment eliminates inter-subject and inter-recording variability, ensuring that the extracted features are not biased by differences in signal magnitude. As a result, the normalized signals provide a more reliable basis for feature extraction and classification, enabling accurate comparison across diverse individuals. Figure 6 demonstrates the normalized ECG segments, highlighting the consistency achieved through this process.

Figure 6. Normalized ECG Segments: Illustration of amplitude normalization applied to segmented ECG signals, showing consistent scaling across all fragments.

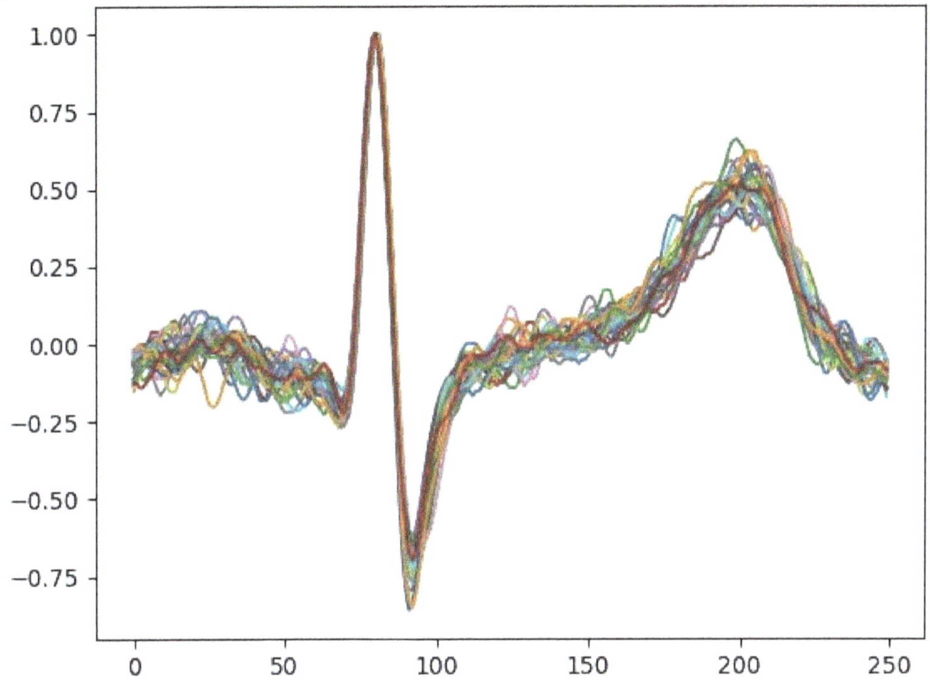

Together, segmentation and normalization establish a robust foundation for feature extraction. By isolating physiologically meaningful fragments and standardizing their amplitude, these steps ensure that the subsequent analysis captures the intrinsic cardiac dynamics rather than artifacts or recording inconsistencies. This methodological rigor enhances the reliability and discriminative power of the features used in ECG-based biometric authentication.

4.4 Feature Extraction from ECG Signals

Feature extraction constitutes a pivotal stage in ECG-based biometric authentication, as it transforms segmented signals into compact and discriminative representations suitable for classification. The primary objective of this process is to minimize intra-subject variability while simultaneously emphasizing inter-class differences, thereby improving system performance. By converting raw ECG data into a structured set of informative descriptors, feature extraction ensures that essential physiological characteristics are preserved while irrelevant or redundant details are eliminated. In the context of handcrafted approaches, feature extraction methods are broadly categorized into two groups: Fiducial point-based features and Non-fiducial features. Each category captures distinct aspects of the ECG waveform, and together they provide a comprehensive representation of cardiac activity.

Figure 7. illustrates the ECG waveform with labeled fiducial points and intervals, highlighting the regions from which features are derived.

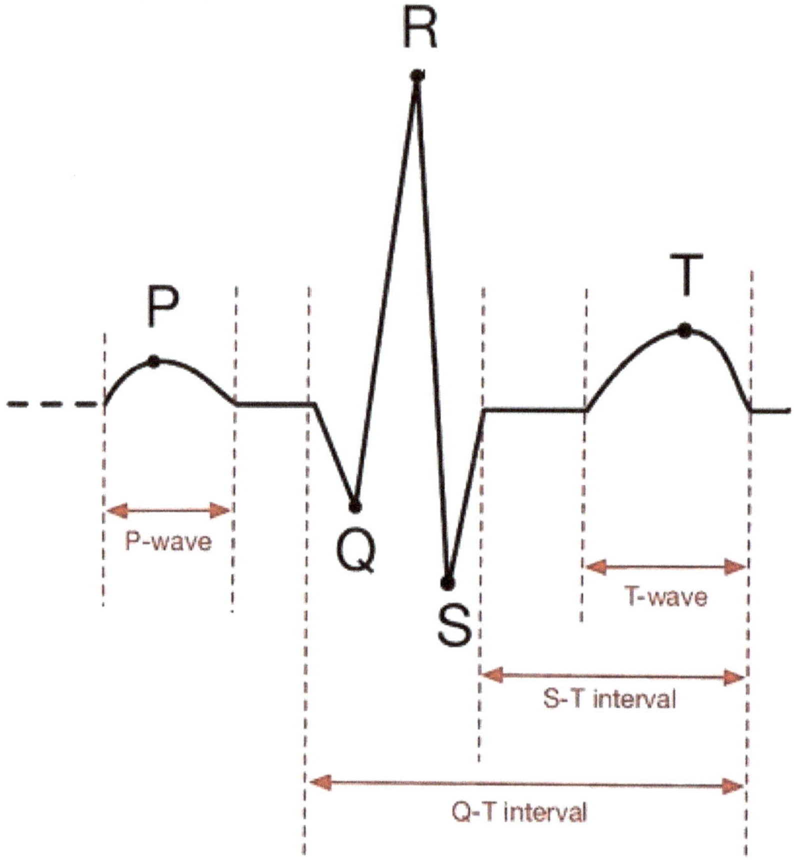

4.4.1 Fiducial Feature Extraction:

Fiducial feature extraction focuses on identifying and quantifying the unique characteristics of heartbeats by analyzing the temporal and amplitude relationships between successive fiducial points. These fiducial points serve as reference landmarks within the ECG signal, enabling precise measurement of cardiac events such as depolarization and repolarization. The extraction process involves detecting the onset, peak, and offset of the P-wave, QRS complex, and T-wave, which together define the fundamental structure of the cardiac cycle. Accurate identification of these fiducial points is critical, as they provide valuable insights into both morphological and temporal properties of the ECG waveform.

However, the process is not without challenges. ECG signals are highly sensitive to noise, and abnormal recordings may exhibit indistinct fiducial points, which can introduce errors in detection and measurement. Despite these difficulties, fiducial features remain indispensable because they directly encode physiological events. In this study, the following fiducial features were extracted:

- **P location:** First positive deflection in the ECG signal.
- **Q location:** First negative deflection following the P wave.
- **R location:** Point of maximum amplitude after the Q location.
- **S location:** Negative deflection immediately following the R peak.
- **T location:** Positive deflection occurring after the QRS complex, following the S peak.

From these fiducial points, several time intervals were computed by measuring differences between locations and dividing by the sampling frequency. These include QR, RS, PQ, QS, PS, PR, ST, QT, RT, and PT intervals. Collectively, these features capture both the morphology and timing of cardiac activity, forming a robust basis for biometric identification. The complete set of fiducial features is summarized in Table 2.

Table 2. Fiducial Features Extracted from ECG Signals

Feature	Description
P Location	First positive deflection in ECG signal
Q Location	First negative deflection after P wave
R Location	Maximum amplitude after Q location
S Location	Negative deflection following R peak
T Location	Positive deflection after QRS complex
QR Interval	(R – Q) / Sampling frequency
RS Interval	(S – R) / Sampling frequency
PQ Interval	(Q – P) / Sampling frequency
QS Interval	(S – Q) / Sampling frequency
PS Interval	(S – P) / Sampling frequency
PR Interval	(R – P) / Sampling frequency
ST Interval	(T – S) / Sampling frequency

continued on following page

Table 2. Continued

Feature	Description
QT Interval	(T – Q) / Sampling frequency
RT Interval	(T – R) / Sampling frequency
PT Interval	(T – P) / Sampling frequency

4.4.2 Non- Fiducial Feature Extraction

In contrast to fiducial features, non-fiducial features are derived without explicit reliance on landmark points. Instead, they characterize the overall statistical and structural properties of the ECG waveform, providing complementary information that enhances system robustness. Non-fiducial features are particularly valuable in cases where fiducial points are difficult to detect due to noise or abnormal morphology. By analyzing the signal in the time and frequency domains, these features capture global descriptors of shape, distribution, and variability.

For each ECG record, the following statistical measures were extracted: mean, median, variance, standard deviation, skewness, and kurtosis. Together, these descriptors provide a distinctive signature of the ECG waveform, reflecting both central tendency and higher-order distributional properties. For example, skewness indicates asymmetry in the waveform distribution, while kurtosis captures the sharpness of peaks. These statistical features are highly individual-specific and therefore play a significant role in biometric authentication, as they contribute to the uniqueness of each subject's ECG profile.

The computation of non-fiducial features was performed using Python libraries such as NumPy and SciPy, which provide efficient implementations of statistical functions. Although the explicit formulas are not detailed here, the libraries ensure accurate and reproducible calculations. The complete set of non-fiducial features is summarized in Table 3.

Table 3. Non-Fiducial Features Extracted from ECG Signals

Feature	Description
Mean	Average amplitude of ECG signal
Median	Middle value of ECG distribution
Variance	Measure of signal dispersion
Standard Deviation	Square root of variance
Skewness	Measure of distribution asymmetry
Kurtosis	Measure of peak sharpness

The proposed approach outlined in the previous section established the conceptual framework of our ECG-based biometric authentication system, detailing the sequential stages of data acquisition, preprocessing, segmentation, normalization, and feature extraction. While these stages define the logical design of the system, the true strength of the framework lies in its ability to identify the most discriminative subset of features that can maximize recognition accuracy and efficiency. To operationalize this design, the methodology section presents a comprehensive account of the implementation process, beginning with the preparation of the ECG dataset and culminating in the optimal selection of features using nature-inspired optimization algorithms (NIOAs). Since the central objective of this study is to achieve robust and efficient feature selection, the methodology emphasizes the initialization, fitness evaluation, and position update strategies employed by algorithms such as Whale Optimization, Grey Wolf Optimization, Cuckoo Search, Elephant Herding Optimization, and Bat Algorithm. This methodological focus ensures that the proposed system not only processes ECG signals effectively but also leverage advanced optimization techniques to enhance the reliability and performance of biometric authentication.

5. METHODOLOGY

The methodology of this study outlines the systematic procedures adopted to implement the proposed ECG-based biometric authentication framework. It describes the acquisition and preparation of the dataset, the preprocessing steps applied to enhance signal quality, and the extraction of both fiducial and non-fiducial features from segmented ECG signals. Furthermore, it details the construction of the feature database and the application of machine learning classifiers for preliminary evaluation. Most importantly, this section emphasizes the optimal selection of features using nature-inspired optimization algorithms (NIOAs), which represents the central contribution of this work. By presenting each stage in a structured manner, the methodology ensures reproducibility and provides a clear foundation for assessing the effectiveness of the proposed system.

5.1 Implementation of Data Processing and Feature Extraction:

For implementation, we utilized the ECG-ID database from PhysioNet, as described in the Proposed Approach. To ensure consistency, two recordings per subject were selected, and from each recording, 15 segments were retained for analysis.

This resulted in a curated dataset that balanced temporal diversity with manageable computational requirements.

Signal processing was carried out using the NeuroKit2 library in Python. R-peaks were detected with the ecg_peaks function, an extended version of the Pan–Tompkins algorithm, which provided robust peak identification across all records. Around each detected R-peak, segments of 250 samples were extracted (70 before and 180 after), ensuring that the major peaks of the ECG waveform were consistently captured. To standardize amplitude across segments, amplitude normalization was applied by dividing each data point by the maximum absolute value within the segment, thereby rescaling all signals to the range of –1 to 1.

From these normalized segments, 21 handcrafted features were extracted: 15 fiducial features derived from the principal ECG peaks and their intervals (summarized in Table 2), and 6 non-fiducial features computed as statistical descriptors (summarized in Table 3). Feature extraction was implemented using Python libraries such as NumPy and SciPy, ensuring reproducibility and efficiency.

The extracted features were compiled into a structured database comprising 2,670 rows and 22 columns. Each row represented a segment, with 21 columns corresponding to the features and the final column serving as the subject label. To establish baseline performance, preliminary classification experiments were conducted using machine learning algorithms including logistic regression, support vector machines (SVM), and random forest classifiers. These experiments provided initial accuracy estimates but were not the primary focus of this study. The central objective is the optimal selection of features using nature-inspired optimization algorithms (NIOAs), which is addressed in the following subsection.

5.2 Optimal Selection of Features using Nature-inspired Algorithms:

The optimal selection of features for the ECG-based recognition system was achieved through the application of nature-inspired optimization algorithms (NIOAs). These algorithms share a common structure consisting of three fundamental steps: initialization, fitness evaluation, and position update. While the initialization and fitness evaluation procedures are generally similar across different algorithms, the position update step varies according to the unique characteristics of each nature-inspired optimization algorithm. To provide a comprehensive understanding of the process, this section first outlines the initialization and fitness evaluation strategies, followed by a detailed discussion of the position update mechanisms employed by the NIOAs we have used in this study.

5.2.1 Initialization:

Initialization plays a critical role in NIOAs, as it establishes the foundation for the entire optimization process. A well-designed initialization ensures that the initial population covers a diverse range of the solution space, thereby promoting exploration and reducing the risk of premature convergence. In this study, initialization was performed by generating a list of 15 distinct random numbers within the range of 1 to 21, corresponding to the indices of the available features. For example, if the number 10 was generated, the feature at the 10th position was selected; similarly, the number 19 indicated the selection of the feature at the 19th position. This list was then converted into a binary vector of length 21, where a value of 1 denoted the selection of a feature and 0 indicated its exclusion. By repeating this process multiple times, a diverse feature space was constructed, forming the initial population for the optimization algorithms.

5.2.2 Fitness Evaluation:

The fitness evaluation step provides a quantitative measure of the quality of each candidate solution in the feature space. In this work, fitness values were computed using a combination of dimensionality reduction, data transformation, and classification techniques. Specifically, Principal Component Analysis (PCA) was applied for dimensionality reduction, while standard scaling was used to normalize the data. Classification was performed using multiple machine learning algorithms, including Random Forest, Logistic Regression, Support Vector Machines (SVM), Decision Trees, and k-Nearest Neighbors (kNN). The accuracy score obtained from these classifiers (denoted as *fitness_acc*) served as the primary performance metric. The fitness value was then calculated using the following formula:

$$Fitness\_value = 1 - (fitness\_acc + 1/15),$$

Here, *fitness_acc* represents the classification accuracy, and 15 corresponds to the number of randomly selected features. This formulation expresses the error rate, thereby guiding the optimization process toward feature subsets that minimize misclassification and maximize discriminative power.

5.2.3 Position Update:

The position update step in optimal feature selection using nature-inspired algorithms plays a pivotal role in directing the search process, exploring the feature space, and adapting to diverse problem landscapes. By balancing exploration and exploitation, this step enables the algorithms to escape local optima and progressively refine candidate solutions over successive iterations. Its importance lies in the ability to fine-tune feature combinations, thereby achieving more effective and efficient feature selection, which ultimately enhances recognition accuracy and model generalization.

The specific mechanism of position update varies across the algorithms employed in this study, as each is guided by its own biological inspiration. In our work, we utilized five nature-inspired algorithms: Grey Wolf Optimization (Mirjalili et al., 2014), Whale Optimization Algorithm (Mirjalili & Lewis, 2016), Cuckoo Search Algorithm (Yang & Deb, 2009), Elephant Herding Optimization (Li et al., n.d.), and Bat Algorithm (Huang & Ma, 2018). Each of these algorithms applies a distinct strategy for updating positions, reflecting its unique behavioral model.

In certain cases, the position update process produced non-binary values that were incompatible with the binary representation required for feature selection. To address this, a Sigmoid function was employed to transform continuous outputs into binary form, ensuring consistency across all algorithms and maintaining the integrity of the feature selection framework.

In summary, the methodology presented in this study establishes a complete pipeline for ECG-based biometric authentication, beginning with dataset preparation, signal processing, segmentation, normalization, and handcrafted feature extraction. These stages provided a reliable foundation for constructing a comprehensive feature database. Building upon this, the central contribution of the methodology lies in the application of nature-inspired optimization algorithms (NIOAs) for optimal feature selection. Through initialization, fitness evaluation, and algorithm-specific position update strategies, the system systematically identified the most discriminative subset of features, thereby enhancing recognition accuracy, efficiency, and generalization. Having detailed the methodological framework and optimization process, the next section presents the experimental evaluation of our approach, including performance results obtained with multiple classifiers. In addition, we provide a comparative analysis with the results reported in related studies referenced earlier, thereby situating our findings within the broader context of ECG-based biometric authentication research and highlighting the improvements achieved by the proposed system.

6 EXPERIMENTAL EVALUATION AND RESULTS

6.1 Introduction to Experimental Evaluation

In this section, we present the outcomes of our experiments, where nature-inspired optimization algorithms were employed for optimal feature selection in the ECG-based biometric recognition system. The objective of this evaluation was to examine and compare the performance of several algorithms, including the Grey Wolf Optimization Algorithm, Whale Optimization Algorithm, Cuckoo Search Algorithm, Elephant Herding Optimization Algorithm, and Bat Algorithm. Through systematic experimentation, we assessed the capability of each algorithm to identify the most relevant features from ECG data, thereby enhancing recognition accuracy and improving generalization. The outcomes of these experiments are presented in the following subsections, supported by figures and tables for clarity.

6.2 Experimental Setup for the Proposed Approach

For the optimal selection of features, we employed a set of nature-inspired optimization algorithms (NIOAs), namely the Whale Optimization Algorithm, Grey Wolf Optimization Algorithm, Cuckoo Search Algorithm, Elephant Herding Optimization Algorithm, and Bat Algorithm. Each algorithm was initialized with a population size of 250, thereby generating 250 distinct feature lists. The initialization of these lists followed the procedure described in Section 5.2.1.

To ensure rigorous exploration, 30 iterations were conducted for each algorithm. During these iterations, fitness values were calculated and feature spaces updated according to the respective position-update mechanisms of the algorithms. In the fitness evaluation stage, several machine learning classifiers were utilized as specified in fitness value of Section 5.2.2, enabling the derivation of error rates for each feature list. The error rate, representing the proportion of misclassifications, served as the primary performance metric.

For every iteration, two parameters were recorded: the error rate and the corresponding feature list. Each feature list was represented as a binary vector of length 21, where '1' indicated the selection of a feature and '0' denoted its exclusion. A lower error rate reflected superior performance, signifying higher classification accuracy.

Based on these results, graphs were generated to depict the error rate progression across iterations, with the number of selected features indicated alongside. To maintain consistency, we exclusively plotted graphs for the Random Forest classifier within each NIOA. Principal Component Analysis (PCA) was employed as a dimensionality reduction technique, while Random Forest served as the classifier in the fitness calculation function. This comprehensive setup ensured that the evaluation was both systematic and robust, ultimately facilitating the identification of compact and discriminative feature subsets to enhance the performance of the ECG-based biometric recognition system. With this setup established, the subsequent subsections present the algorithm-wise results, highlighting how each nature-inspired optimization algorithm performed in terms of error rate reduction and optimal feature selection.

6.3 Algorithm-wise Results

This section presents the outcomes of optimal feature selection using nature-inspired optimization algorithms (NIOAs) applied to the ECG-based biometric recognition system. Each algorithm was evaluated over 30 iterations to examine its ability to minimize classification error while identifying the most discriminative subset of features. For clarity, the results are reported individually for each NIOA, supported by error-rate graphs and feature selection details.

6.3.1 Whale Optimization Algorithm (WOA):

The error rate progression across 30 iterations during optimal feature selection with the Whale Optimization Algorithm is illustrated in Figure 8 below. Each iteration records both the error rate and the number of features selected, showing how the algorithm gradually converges toward an optimal subset.

- Iteration 1: Error rate of 0.3659 with 15 features selected.
- Iterations 2–3: Error rate reduced to 0.3011 with 10 features retained.
- Iterations 4–26: Stable error rate of 0.2790 with 5 features consistently selected.
- Iterations 27–30: Lowest error rate of 0.2694 achieved with only 4 features selected.

The binary feature vector corresponding to the best error rate is: (0 0 0 1 0 0 0 0 0 0 0 0 1 0 0 0 0 0 1 1), where '1' denotes a selected feature. The four optimal features identified are S-peak, RT interval, Kurtosis, and Skewness.

Figure 8. Error Rate vs. Iterations with Number of Features using Whale Optimization Algorithm

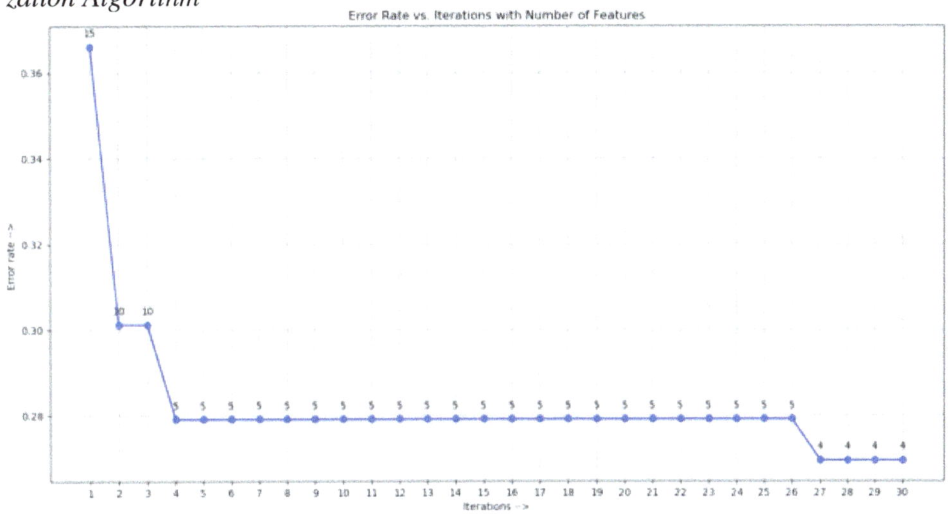

Insight: These results demonstrate that WOA effectively reduces dimensionality while preserving discriminative power. Selecting only four features from the original 21 yielded the lowest error rate, confirming the algorithm's ability to isolate key ECG-based biometric indicators. This compact subset enhances both recognition accuracy and computational efficiency.

6.3.2 Grey Wolf Optimization Algorithm (GWO):

The Grey Wolf Optimization Algorithm (GWO), as discussed in Section 3.3.2, mimics the social hierarchy and hunting behavior of grey wolves. The algorithm models a pack structure comprising alpha, beta, delta, and omega wolves, ranked from most to least dominant. In this hierarchy, the alpha wolves represent the best solution, followed by beta and delta wolves as the second and third-best candidates.

Figure 9 below illustrates the error rate progression across 30 iterations for each wolf type, along with the number of features selected in each iteration. While all wolf types are plotted, the analysis focuses on the alpha wolves, whose feature lists consistently yielded the lowest error rates.

- Iteration 1: Error rate of 0.3689 with 15 features selected.
- Iteration 2: Error rate reduced to 0.3195 with 15 features.
- Iteration 3: Error rate of 0.3193 with 12 features.
- Iteration 19 (best): Error rate minimized to 0.2627 with 8 features selected.

- Iterations 20–30: Error rate and feature count remained constant at 0.2627 and 8 features.

The binary feature vector corresponding to the best error rate is: (0 0 0 1 0 0 0 0 1 0 0 0 0 1 0 1 1 0 1 1 1), where '1' indicates a selected feature. The eight optimal features identified are S-peak, QS interval, RT interval, Mean, Median, Standard deviation, Kurtosis, and Skewness.

Figure 9. Error Rate vs. Iterations with Number of Features using Grey Wolf Optimization Algorithm

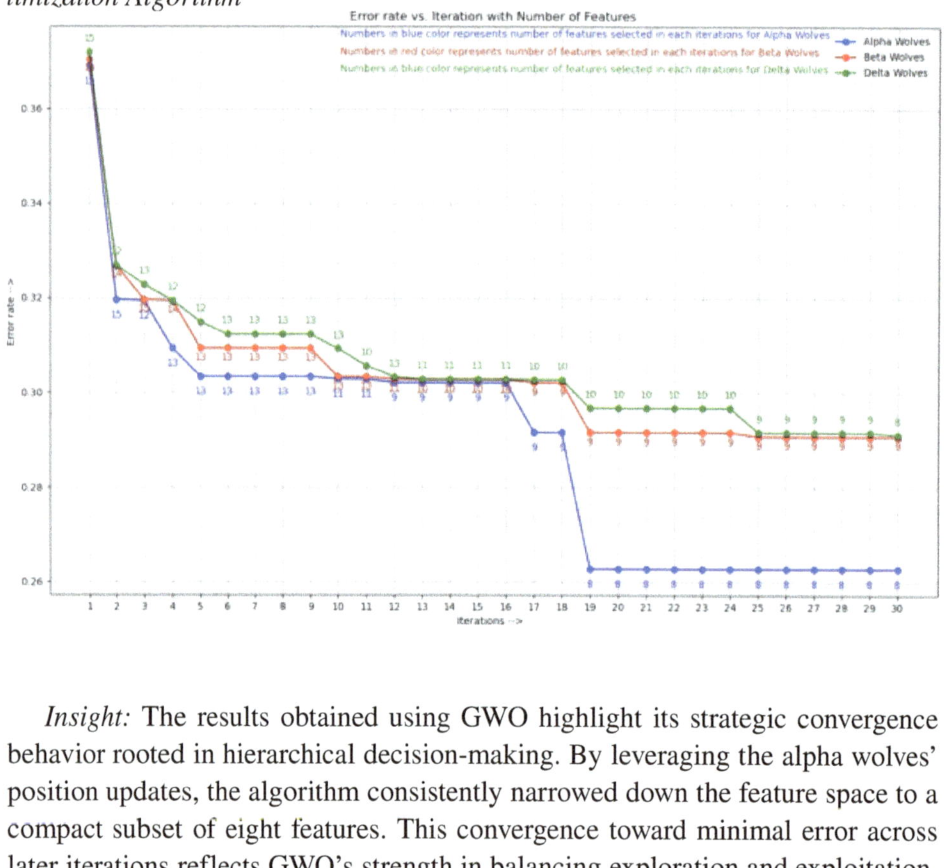

Insight: The results obtained using GWO highlight its strategic convergence behavior rooted in hierarchical decision-making. By leveraging the alpha wolves' position updates, the algorithm consistently narrowed down the feature space to a compact subset of eight features. This convergence toward minimal error across later iterations reflects GWO's strength in balancing exploration and exploitation. The selected features—spanning temporal intervals and statistical descriptors—demonstrate that GWO not only enhances recognition accuracy but also preserves interpretability by selecting physiologically meaningful ECG attributes.

6.3.3 Cuckoo Search Algorithm (CSA):

Figure 10 below illustrates the error rate progression across 30 iterations during optimal feature selection using the Cuckoo Search Algorithm. Each iteration records both the error rate and the number of features selected.

- Iteration 1: Error rate of 0.3026 with 10 features selected.
- Iterations 2–30: Error rate stabilized at 0.2745 with 10 features consistently retained.

The binary feature vector corresponding to the best error rate is: (0 1 1 0 0 0 0 1 0 1 1 0 1 0 0 1 1 0 1 0 1), where '1' denotes a selected feature. The ten optimal features identified are Q peak, R peak, PQ interval, PS interval, PR interval, QT interval, Mean, Median, Standard deviation, and Skewness.

Figure 10. Error Rate vs. Iterations with Number of Features using Cuckoo Search Algorithm

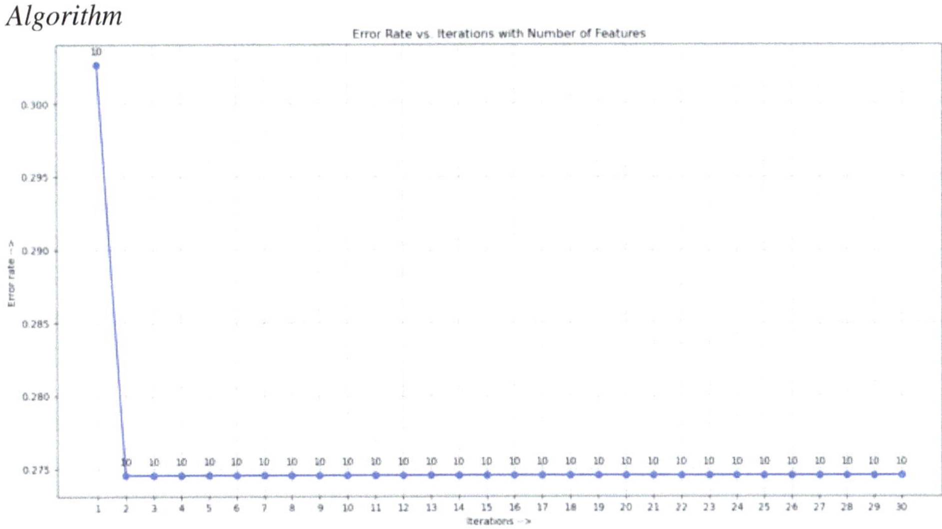

Insight: The CSA's performance reveals a rapid convergence pattern, with the error rate sharply dropping after the first iteration and remaining stable thereafter. This behavior reflects the algorithm's efficient exploration of the feature space and its ability to lock onto a high-quality solution early in the search process. The selected features span both morphological peaks and statistical descriptors, indicating CSA's capacity to balance physiological relevance with statistical robustness. Its consistent selection of the same ten features across iterations underscores its reliability in identifying a stable and discriminative subset for ECG-based biometric recognition.

6.3.4 Elephant Herding Optimization Algorithm (EHO):

Figure 11 below illustrates the error rate progression across 30 iterations during optimal feature selection using the Elephant Herding Optimization Algorithm. The best error rate is attained at the 16th iteration and remains constant for the remaining iterations at 0.2724, with 12 features selected.

The binary feature vector corresponding to this error rate is: (0 1 0 0 0 0 1 0 1 0 1 0 1 1 0 1 1 1 1 1 1), where '1' denotes a selected feature. The twelve optimal features identified are Q peak, RS interval, QS interval, PR interval, QT interval, RT interval, Mean, Median, Variance, Standard deviation, Kurtosis, and Skewness.

Figure 11. Error Rate vs. Iterations with Number of Features using Elephant Herding Optimization Algorithm

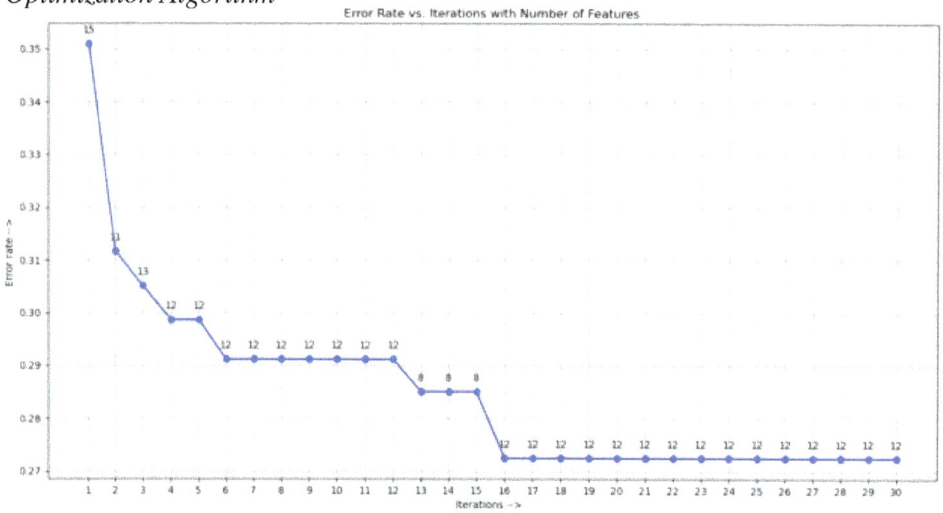

Insight: EHO demonstrates a gradual and stable convergence pattern, with the error rate minimizing mid-way through the iterations and remaining consistent thereafter. The selection of twelve features reflects the algorithm's ability to preserve a broad yet relevant subset of ECG characteristics. By combining interval-based features with statistical descriptors, EHO maintains a balance between physiological coverage and statistical depth, contributing to robust biometric recognition performance.

6.3.5 Bat Algorithm (BA):

Figure 12 below presents the error rate progression across 30 iterations for the Bat Algorithm. The best error rate is achieved at the 9th iteration and remains unchanged through iteration 30 at 0.2621, with only 6 features selected.

The binary feature vector corresponding to this error rate is: (0 0 0 0 0 1 0 0 0 0 0 1 0 1 0 0 0 0 1 1 1), where '1' denotes a selected feature. The six optimal features identified are QR interval, ST interval, RT interval, Standard deviation, Kurtosis, and Skewness.

Figure 12. Error Rate vs. Iterations with Number of Features using Bat Algorithm

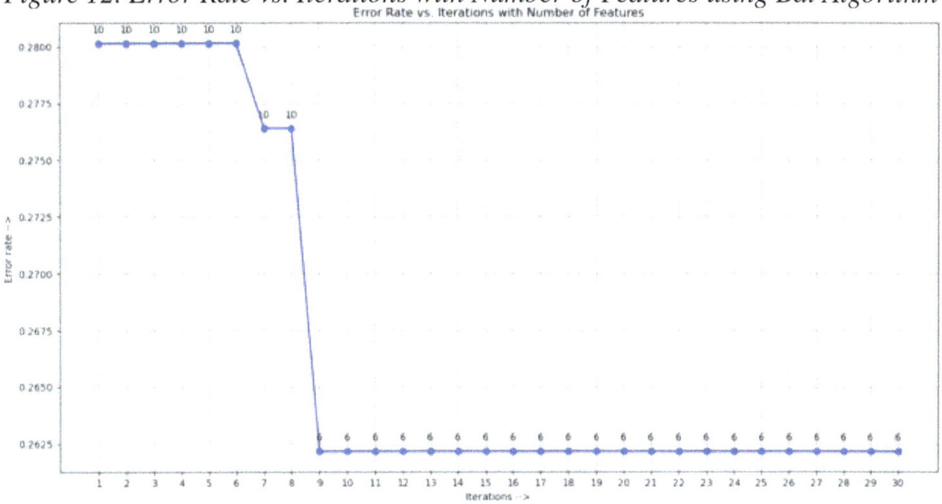

Insight: The Bat Algorithm exhibits a sharp and early convergence, achieving its lowest error rate within the first ten iterations. Its selection of only six features—focused on key temporal intervals and high-impact statistical measures—demonstrates its precision in isolating the most discriminative attributes. This minimalistic feature set not only enhances computational efficiency but also underscores BA's effectiveness in extracting core biometric signals from ECG data.

Having examined the algorithm-wise results, where each nature-inspired optimization algorithm was evaluated individually for its ability to minimize error rates and select optimal features, it is equally important to assess how these algorithms perform when paired with different classifiers. While the previous section highlighted the convergence behavior and feature selection efficiency of WOA, GWO, CSA, EHO, and BA, the next step is to compare their outcomes across multiple classification models. This classifier-wise comparison provides a broader perspective, revealing not only which algorithm achieves the lowest error rate but also which classifier consistently enhances performance when integrated into the fitness calculation function.

6.4 Classifier-Wise Comparison:

Classifier choice plays a decisive role in determining recognition accuracy, as different models vary in their ability to handle nonlinear relationships, feature dimensionality, and statistical variability. In this section, the focus shifts from evaluating algorithms in isolation to examining how their performance changes when paired with different classifiers. The objective is not only to identify which classifier achieves the lowest error rates but also to understand how classifier selection influences feature reduction and overall system robustness. By systematically applying Random Forest, Logistic Regression, Decision Trees, Support Vector Machine (SVM), and k-Nearest Neighbors (kNN) under identical preprocessing conditions (Principal Component Analysis for dimensionality reduction and Standard Scaler for normalization), we establish a comprehensive comparison framework. Presenting Tables 4–8 together allows us to observe comparative outcomes across classifiers and algorithms, while the subsequent analysis highlights consistent trends and identifies the most effective algorithm–classifier pairing for ECG-based biometric recognition.

Table 4. Performance with Random Forest Classifier

S. No.	Nature-inspired Algorithm	Error rate	Number of Selected Features
1	Whale Optimization Algorithm	0.2694	4
2	Grey Wolf Optimization Algorithm	0.2627	8
3	Cuckoo Search Algorithm	0.2745	10
4	Elephant Herding Optimization Algorithm	0.2724	12
5	Bat Algorithm	0.2621	6

Description: The Bat Algorithm achieved the lowest error rate (0.2621) among all NIOAs when paired with Random Forest, selecting only six features (QR interval, ST interval, RT interval, Standard deviation, Kurtosis, and Skewness).

Table 5. Performance with Logistic Regression Classifier

S. No.	Nature-inspired Algorithm	Error rate	Number of Selected Features
1	Whale Optimization Algorithm	0.4056	16
2	Grey Wolf Optimization Algorithm	0.4074	17
3	Cuckoo Search Algorithm	0.4042	14
4	Elephant Herding Optimization Algorithm	0.4054	12
5	Bat Algorithm	0.4089	15

Description: Logistic Regression produced relatively high error rates across all algorithms, with BA performing worst here (0.4089). This indicates limited suitability of LR for complex, nonlinear ECG data.

Table 6. Performance with Decision Tree Classifier

S. No.	Nature-inspired Algorithm	Error rate	Number of Selected Features
1	Whale Optimization Algorithm	0.3520	2
2	Grey Wolf Optimization Algorithm	0.4133	7
3	Cuckoo Search Algorithm	0.4020	7
4	Elephant Herding Optimization Algorithm	0.4207	9
5	Bat Algorithm	0.4002	7

Description: Decision Trees showed moderate performance, with WOA achieving the lowest error rate (0.3520) but at the cost of selecting only two features, which may compromise robustness.

Table 7. Performance with Support Vector Machine (SVM) Classifier

S. No.	Nature-inspired Algorithm	Error rate	Number of Selected Features
1	Whale Optimization Algorithm	0.3625	10
2	Grey Wolf Optimization Algorithm	0.3828	11
3	Cuckoo Search Algorithm	0.3697	11
4	Elephant Herding Optimization Algorithm	0.2724	12
5	Bat Algorithm	0.3868	10

Description: SVM achieved competitive results for EHO (0.2724), but overall error rates remained higher than those achieved with Random Forest. This suggests SVM can be effective in certain cases but lacks consistency.

Table 8. Performance with k-Nearest Neighbors (kNN) Classifier

S. No.	Nature-inspired Algorithm	Error rate	Number of Selected Features
1	Whale Optimization Algorithm	0.4058	7
2	Grey Wolf Optimization Algorithm	0.4241	12
3	Cuckoo Search Algorithm	0.3689	6
4	Elephant Herding Optimization Algorithm	0.3024	4
5	Bat Algorithm	0.3937	8

Description: kNN produced inconsistent results, with EHO performing best (0.3024). However, overall error rates were higher compared to Random Forest, reflecting kNN's sensitivity to feature dimensionality.

Across Tables 4–8, several consistent patterns can be observed. The Random Forest classifier emerged as the most reliable model, producing the lowest error rates across all nature-inspired algorithms. In particular, the Bat Algorithm combined with Random Forest achieved the best overall performance, yielding an error rate of 0.2621 while selecting only six features, thereby demonstrating both accuracy and efficiency. Logistic Regression consistently produced the highest error rates, indicating its limited suitability for complex, nonlinear ECG data. Decision Trees showed moderate performance, with WOA achieving a relatively low error rate but relying on only two features, which may compromise robustness. Support Vector Machines delivered competitive results in certain cases, especially with EHO, but lacked consistency across algorithms. The k-Nearest Neighbors classifier produced mixed outcomes, with EHO achieving its best performance under kNN, yet overall error rates remained higher compared to Random Forest. Taken together, these results confirm that classifier choice significantly influences recognition accuracy, and that ensemble-based methods such as Random Forest are particularly well-suited for ECG-based biometric recognition when paired with nature-inspired optimization algorithms.

The comparative evaluation across classifiers demonstrates that ensemble-based approaches, particularly Random Forest, are inherently better suited for ECG-based biometric recognition. Its ability to aggregate multiple decision paths reduces variance and captures complex nonlinear dependencies, which explains its consistent superiority across all algorithms. The Bat Algorithm's synergy with Random Forest further highlights the importance of pairing a strong feature selector with a robust classifier, as this combination achieved both minimal error and compact feature representation. In contrast, the weaker performance of Logistic Regression and k-Nearest Neighbors underscores the limitations of linear models and distance-based methods in handling high-dimensional physiological signals. Decision Trees and SVM offered partial competitiveness but lacked stability across algorithms. Overall, these findings confirm that classifier choice is not merely a technical detail but a decisive factor in system reliability, and that the BA–RF pairing establishes a benchmark configuration for optimizing accuracy and efficiency in ECG biometric systems.

Beyond classifier-wise comparisons, it is also important to examine how the system performs when no optimal feature selection using nature-inspired optimization algorithms (NIOAs) is applied. This aspect is discussed in the next section.

6.4 Error Rates of Classifiers with and Without Optimal Feature Selection:

To further evaluate the robustness of classifiers, we also examined their performance on the ECG dataset without applying optimal feature selection using nature-inspired optimization algorithms (NIOAs). This baseline comparison highlights how classifiers behave when all features are used directly, and contrasts these results with those obtained after applying optimal feature selection through NIOAs such as Whale Optimization Algorithm (WOA), Grey Wolf Optimization Algorithm (GWO), Cuckoo Search Algorithm (CSA), Elephant Herding Optimization Algorithm (EHO), and Bat Algorithm (BA). The following table presents error rates for each classifier under both conditions: without feature selection and with optimal selection using NIOAs.

Table 9. Error Rate of Classifiers

Classifier	Error rate (Without Feature selection)		Error rate feature	(During selection)	optimal	
		WHO	GWO	EHO	CSA	BA
Random Forest	0.3989	0.26946	0.26272	0.27247	0.27453	0.26217
KNN	0.5600	0.40583	0.42415	0.30243	0.36891	0.39372
Decision Trees	0.5225	0.35205	0.41332	0.42072	0.40208	0.40021
Logistic Regression	0.4776	0.40566	0.40746	0.40543	0.40422	0.40898
SVM	0.5375	0.36254	0.38287	0.36976	0.36629	0.38689

From Table 9, it is evident that when no optimal feature selection is applied, the Random Forest classifier achieves the lowest error rate (0.3989) compared to other classifiers, confirming its robustness even under full feature sets. However, when optimal feature selection using NIOAs is incorporated, performance improves significantly across all classifiers. In particular, the Bat Algorithm combined with Random Forest yields the best overall result, achieving an error rate of 0.2621 with only six selected features. This demonstrates that optimal feature selection not only reduces dimensionality but also enhances recognition accuracy, with Random Forest consistently emerging as the most effective classifier in both baseline and optimized scenarios.

The comparative evaluation and tabulated results provide a clear overview of the performance of the proposed approach. While these findings establish its quantitative effectiveness, a deeper understanding requires interpretation of outcomes, identification of strengths and limitations, and consideration of broader implications. The following Discussion section elaborates on these aspects, situating our results within the context of prior research and outlining potential directions for future work.

7 DISCUSSIONS OF RESULTS

The comparative evaluation of prior studies and our proposed approach is consolidated in Table 10. This table presents datasets, methodological choices, and reported performance, allowing a structured interpretation of how ECG biometrics have evolved and where our contribution fits.

Table 10. Comparative Summary of ECG-Based Biometric Authentication Methods

Ref–Authors	Datasets	Method	Key metrics
Ingale et al. (2020)	PTB; MIT-BIH; CEBSDB; CYBHI; ECG-ID; ECG-BG (in-house)	Filtering; segmentation; features; matching	Up to 100% accuracy; avg EER ≈ 2.11%
Noureddine et al. (2015)	MIT-BIH; ST-T; NSR; PTB; ECG-ID; in-house palm ECG	Fourier descriptors; OPF classifier; palm-based 1-lead ECG	>99% authentication accuracy
Lugovaya (2005)	ECG-ID	PCA; LDA; majority voting	Identification accuracy 96%
Biel et al. (2001)	12-lead clinical ECG (rest)	SIMCA (multivariate analysis); one-lead feasible	Near-perfect identification; 50/50 correct in reduced-feature test

continued on following page

Table 10. Continued

Ref–Authors	Datasets	Method	Key metrics
Rezgui & Lachiri (2016)	MIT-BIH Arrhythmia; MIT-BIH NSR	31 fiducial + morphological features; SVM (Gaussian/poly)	Identification ~98.8% (MIT); ~99% (NSR)
Patro et al. (2022)	ECG-ID; PTBDB	GA/PSO/LASSO/Elastic Net + SVM/RF	Best GA+RF: PTBDB 95.30%; ECG-ID 93.85%
Hammad et al. (2019)	PTB; CYBHi; MIT-BIH; MWM-HIT (new)	12-layer CNN	PTB ≈ 99.27%; CYBHi ≈ 96.48%; EER PTB ≈ 1.63%
Wang et al. (2024)	PTBDB; MITDB; ECGIDDB	CNN encoder + contrastive learning; Pearson corr.; quant/prune	PTB 99.15%; MITDB 98.67%; ECGIDDB 98.77%; edge-optimized PTB 98.67%
Mani Kumar et al. (2025)	PTB; MIT-BIH	GQRS-based feature set + 1D CNN	PTB 97.38%; MIT 100%; train Acc ~98–99%
Agrawal et al. (2023)	PTB	CNN; LSTM	CNN 98.34%; LSTM 99.69%
Prakash et al. (2023)	ECG-ID	Siamese CNN; custom activation; beat images	Acc 99.85%; Sens 99.30%; Spec 99.85%; PPV 99.76%
Proposed approach	ECG-ID (Lead I; 500 Hz; 89 subjects; 2 records each)	Handcrafted fiducial + non-fiducial features (21 total); NIOA-based optimal feature selection (WOA, GWO, CSA, EHO, BA); classifiers RF, LR, SVM, DT, kNN	Baseline RF error ≈ 0.3989; optimized BA+RF with 6 features (QR, ST, RT, std, kurtosis, skewness) error ≈ 0.2621; consistent improvement across NIOAs; RF best overall

The trajectory reflected in Table 10 shows how ECG biometrics have matured over time. Early feasibility studies such as Lugovaya (2005) and Biel et al. (2001) confirmed the potential of ECG signals but achieved only moderate accuracies, around 95–96%, due to reliance on PCA, LDA, and SIMCA. Mid-stage works like Rezgui and Lachiri (2016) and Patro et al. (2022) improved performance through feature engineering and optimization, reporting accuracies close to 99% but requiring large handcrafted feature sets. More recent contributions shifted toward deep learning: Hammad et al. (2019) employed CNNs with fiducial peaks, Wang et al. (2024) introduced self-supervised contrastive learning, and Agrawal et al. (2023) combined CNN and LSTM to capture spatial and temporal dependencies. These methods consistently reported accuracies above 98%, with Prakash et al. (2023) achieving near-perfect recognition (99.85%) using a Siamese CNN. While these pipelines deliver strong performance, they often sacrifice interpretability and demand significant computational resources.

A deeper look at these methods explains why some succeed while others face limitations. Fiducial feature-based strategies capture physiologically meaningful intervals but are sensitive to noise and segmentation errors, whereas non-fiducial statistical features scale better across populations but may lose interpretability. This trade-off highlights the value of optimization-driven feature selection, which balances physiological relevance with robustness and ensures compact yet discriminative feature sets.

In contrast, our proposed approach demonstrates that NIOA-based optimal feature selection applied to handcrafted fiducial and non-fiducial descriptors can achieve competitive performance with greater interpretability. By reducing the feature set from 21 to as few as 6 (QR, ST, RT intervals, along with statistical descriptors such as skewness and kurtosis), the system lowered the error rate from 0.3989 to 0.2621 with Random Forest. This compactness enhances physiological relevance and reduces computational overhead, situating our work as a middle ground between classical feature-based methods and modern deep learning pipelines.

These findings highlight several advantages of the proposed approach. The reduction of features without compromising accuracy demonstrates efficiency and interpretability, which are often lacking in deep learning models. The use of multi-segment ECG data further strengthens robustness, as it captures variability across different cardiac cycles and improves reliability in authentication. Moreover, the consistent improvement across multiple NIOAs confirms the adaptability of the optimization framework, with Random Forest emerging as the most effective classifier in synergy with the selected features.

These advantages translate into practical implications. In banking, ECG authentication can provide an additional layer of security against spoofing and replay attacks. In healthcare, ECG serves dual roles in patient identification and continuous monitoring, enabling secure access to medical records while simultaneously tracking cardiac health. Usability is enhanced by wearable devices that acquire signals seamlessly, though hardware variability and sensor quality remain challenges for deployment.

Nevertheless, certain limitations must be acknowledged. The evaluation was restricted to the ECG-ID dataset, which, although widely used, represents a relatively controlled population. Broader validation across diverse datasets such as PTB, MIT-BIH, and CYBHi would be necessary to confirm generalizability. Additionally, while the chosen NIOAs (WOA, GWO, CSA, EHO, BA) proved effective, the wide array of available metaheuristics suggests that further exploration could yield even more efficient or scalable solutions. Finally, the iterative nature of NIOAs introduces computational overhead, which may limit real-time deployment unless further optimized.

Despite its promise, ECG biometrics face several challenges. Signal quality is often affected by noise, motion artifacts, and sensor variability, making robust preprocessing essential. Usability can be limited by the need for continuous acquisition, while cross-population variability raises concerns about generalization beyond controlled datasets. Privacy is another critical issue, since ECG data are inherently medical and require strict safeguards against misuse. Finally, emerging trends such as deep learning, multimodal fusion, wearable devices, and federated learning highlight both opportunities and the need to balance accuracy, interpretability, and ethical deployment.

Looking ahead, future work could explore several directions. Extending validation to larger and more diverse datasets would strengthen robustness and generalization. Hybrid pipelines that combine NIOA-based feature selection with CNN or LSTM architectures could balance interpretability with the representational power of deep learning. Optimizing runtime for IoT and wearable devices would enable practical deployment in healthcare and security contexts. Furthermore, multimodal fusion of ECG with other physiological or behavioral modalities could enhance accuracy and reduce spoofing risks. Finally, incorporating privacy-preserving frameworks such as federated learning would address ethical concerns associated with biometric data.

Overall, the comparative analysis and the outcomes of our proposed approach highlight both the strengths and the areas requiring further exploration in ECG-based biometric authentication. By situating our contribution within the broader trajectory of research, it becomes evident that optimization-driven feature selection offers a meaningful balance between interpretability and performance. Having discussed the advantages, limitations, and future scope of this work, the next section provides a concise conclusion that summarizes the key contributions and implications of the study.

8 CONCLUSION

This study employed the ECG-ID database to investigate the effectiveness of electrocardiogram signals as a biometric modality, applying five nature-inspired optimization algorithms (WOA, GWO, CSA, EHO, BA) for optimal feature selection. By reducing the original 21 handcrafted fiducial and non-fiducial features to a compact subset of six, the proposed framework achieved a notable improvement in recognition performance, lowering the Random Forest error rate from 0.3989 to 0.2621. These findings confirm that ECG is a highly reliable and practical biometric tool, offering ease of acquisition, physiological interpretability, and resilience against spoofing compared to traditional modalities.

The contribution of this work lies not only in demonstrating competitive accuracy but also in showing that optimization-driven feature selection can balance efficiency with interpretability. This balance is particularly important for real-world applications, where usability, hardware variability, and computational overhead must be considered. The methodological framework presented here also has broader applicability: similar strategies could be extended to other physiological signals such as the electromyogram (EMG), or integrated into multimodal biometric systems that combine ECG with modalities like fingerprint or facial recognition to enhance robustness.

Looking forward, the integration of nature-inspired optimization with deep learning architectures, the deployment of ECG biometrics in wearable devices, and the adoption of privacy-preserving frameworks such as federated learning represent promising directions. Together, these advances position ECG biometrics as a secure, interpretable, and scalable solution for authentication in healthcare, banking, and other sensitive domains.

REFERENCES

Agrawal, V., Hazratifard, M., Elmiligi, H., & Gebali, F. (2023). Electrocardiogram (ECG)-based user authentication using deep learning algorithms. *Diagnostics (Basel), 13*(3), 439. Advance online publication. DOI: 10.3390/diagnostics13030439 PMID: 36766544

Biel, L., Pettersson, O., Philipson, L., & Wide, P. (2001). ECG analysis: A new approach in human identification. *IEEE Transactions on Instrumentation and Measurement, 50*(3), 808–812. DOI: 10.1109/19.930458

Erdogmus, N., & Marcel, S. (2014). Spoofing face recognition with 3D masks. *IEEE Transactions on Information Forensics and Security, 9*(7), 1084–1097. DOI: 10.1109/TIFS.2014.2322255

Galbally, J., Cappelli, R., Lumini, A., Gonzalez-de-Rivera, G., Maltoni, D., Fierrez, J., Ortega-Garcia, J., & Maio, D. (2010). An evaluation of direct attacks using fake fingers generated from ISO templates. *Pattern Recognition Letters, 31*(8), 725–732. DOI: 10.1016/j.patrec.2009.09.032

Hadid, A., Evans, N., Marcel, S., & Fierrez, J. (2015). Biometrics systems under spoofing attack: An evaluation methodology and lessons learned. *IEEE Signal Processing Magazine, 32*(5), 20–30. DOI: 10.1109/MSP.2015.2437652

Hammad, M., Zhang, S., & Wang, K. (2019). A novel two-dimensional ECG feature extraction and classification algorithm based on convolution neural network for human authentication. *Future Generation Computer Systems, 101*, 180–196. DOI: 10.1016/j.future.2019.06.008

Huang, J., & Ma, Y. (2018). Bat algorithm based on an integration strategy and Gaussian distribution. [Affiliation removed per APA rules.]. *Applied Soft Computing, 63*, 141–154. DOI: 10.1016/j.asoc.2017.11.040

Ingale, M., Cordeiro, R., Thentu, S., Park, Y., & Karimian, N. (2020). ECG biometric authentication: A comparative analysis. *IEEE Access : Practical Innovations, Open Solutions, 8*, 117853–117866. DOI: 10.1109/ACCESS.2020.3004464

Kose, N., & Dugelay, J.-L. (2013). On the vulnerability of face recognition systems to spoofing mask attacks. In *Proceedings of the IEEE International Conference on Acoustics, Speech and Signal Processing (ICASSP)* (pp. 2357–2361). IEEE. DOI: 10.1109/ICASSP.2013.6638076

Kumar, V. M., Phanideep, D. M., Purushotham, C., Ramesh, R., & Mohan, J. (2025). User authentication from ECG signals using 1D convolutional neural network. *Procedia Computer Science, 258*, 1262–1273. DOI: 10.1016/j.procs.2025.04.360

Li, J., Lei, H., Alavi, A. H., & Wan, G.-G. (2019). Elephant herding optimization: Variants, hybrids, and applications. [Note: standardized to journal publication; confirm if a different version is required.]. *Mathematics, 7*(12), 1–30. DOI: 10.3390/math7121140

Lugovaya, T. S. (2005). *Biometric human identification based on electrocardiogram* (Master's thesis). Electrotechnical University "LETI," Saint Petersburg, Russian Federation.

Makowski, D., Pham, T., Lau, Z. J., Brammer, J. C., Lespinasse, F., Pham, H., Schölzel, C., & Chen, S.-A. (2021). NeuroKit2: A Python toolbox for neurophysiological signal processing. *Behavior Research Methods, 53*(4), 1689–1696. DOI: 10.3758/s13428-020-01516-y PMID: 33528817

Mirjalili, S., & Lewis, A. (2016). The whale optimization algorithm. *Advances in Engineering Software, 95*, 51–67. DOI: 10.1016/j.advengsoft.2016.01.008

Mirjalili, S., Mirjalili, S. M., & Lewis, A. (2014). Grey wolf optimizer. *Advances in Engineering Software, 69*, 46–61. DOI: 10.1016/j.advengsoft.2013.12.007

Patro, K. K., Prakash, A. J., Rao, M. J., & Kumar, P. R. (2022). An efficient optimized feature selection with machine learning approach for ECG biometric recognition. *Journal of the Institution of Electronics and Telecommunication Engineers, 68*(4), 2743–2754. DOI: 10.1080/03772063.2020.1725663

Prakash, A. J., Patro, K. K., Samantray, S., Pławiak, P., & Hammad, M. (2023). A deep learning technique for biometric authentication using ECG beat template matching. *Information (Basel), 14*(2), 65. Advance online publication. DOI: 10.3390/info14020065

Rezgui, D., & Lachiri, Z. (2016). ECG biometric recognition using SVM-based approach. *IEEJ Transactions on Electrical and Electronic Engineering, 11*(S1), S94–S100. DOI: 10.1002/tee.22241

Wang, G., Shanker, S., Nag, A., Lian, Y., & John, D. (2024). ECG biometric authentication using self-supervised learning for IoT edge sensors. *IEEE Journal of Biomedical and Health Informatics, 28*(11), 6606–6618. DOI: 10.1109/JBHI.2024.3455803 PMID: 39250357

Yang, X. S., & Deb, S. (2009). Cuckoo search via Lévy flights. In *Proceedings of the World Congress on Nature & Biologically Inspired Computing (NaBIC)* (pp. 210–214). IEEE. DOI: 10.1109/NABIC.2009.5393690

APPENDIX A: MATHEMATICAL FORMULATIONS OF NIOAS

This appendix consolidates the mathematical formulations of the five nature-inspired optimization algorithms used in this study: Whale Optimization Algorithm (WOA), Grey Wolf Optimization (GWO), Cuckoo Search Algorithm (CSA), Elephant Herding Optimization (EHO), and Bat Algorithm (BA). Each equation is accompanied by a brief explanation of its role in the optimization process.

A.1 Whale Optimization Algorithm (WOA)

Encircling prey: The distance between a whale and the prey (best solution) is calculated as:

$$\vec{D} = \left|\vec{c}\,.\,\vec{Y}'(i) - \vec{Y}(I)\right| \tag{1}$$

This measures how far the current whale is from the best solution.

The whale then updates its position toward the prey:

$$\vec{Y}(i+1) = \vec{Y}'(i) - \vec{A} \bullet \vec{d} \tag{2}$$

This ensures convergence toward the best candidate solution.

Coefficient vectors:

$$\vec{A} = 2 \bullet \vec{a} \bullet \vec{r} - \vec{a} \tag{3}$$

$$\vec{c} = 2 \bullet \vec{a} \bullet \vec{r} \tag{4}$$

Here, $\vec{a}$ decreases linearly from 2 to 0, controlling the balance between exploration and exploitation. and $\vec{r}$ is a random vector in [0, 1].

Spiral updating (bubble-net attack):

$$\vec{Y}(i+1) = \vec{d'} \bullet e^{b\Theta} \bullet \cos(2\pi\,\Theta) + \vec{Y'}(i) \tag{5}$$

This models the helix-shaped movement of whales around prey, where $\vec{d'}$ is the distance to prey, b defines spiral shape, and Θ is a random value.

Probabilistic exploitation:

$$\vec{Y}(i+1) = \begin{cases} \vec{Y}'(i) - \vec{A} \bullet \vec{d}, \ if\, p < 0.5 \\ \vec{d'} \bullet e^{b\Theta} \bullet \cos(2\pi\,\Theta) + \vec{Y}'(i), \ if\, p \geq 0.5 \end{cases} \tag{6}$$

This ensures whales alternate between shrinking encircling and spiral updating with probability p.

Exploration:

$$\vec{d} = \left|\vec{c} \bullet \overrightarrow{Y_{rand}} - \vec{Y}\right| \tag{7}$$

$$\vec{Y}(\mathrm{i}+1) = \overrightarrow{Y_{rand}} - \vec{\mathrm{A}} \bullet \vec{\mathrm{d}} \tag{8}$$

Here, whales move relative to randomly chosen agents when $|\vec{A}|>1$, diversifying the search space.

A.2 Grey Wolf Optimization Algorithm (GWO)

Encircling prey: Grey wolves estimate the prey's position and surround it:

$$\vec{d} = \left|\vec{c} \bullet \vec{x}_p(i) - \vec{x}(i)\right| \tag{9}$$

This measures the distance between a wolf and the prey (best solution).

Position update toward prey:

$$\vec{x}\left(i+1\right) = \vec{x}_p\left(t\right) - \vec{A} \bullet \vec{d} \tag{10}$$

This ensures wolves move closer to the prey.

Coefficient vectors:

$$\vec{\mathrm{A}} = 2 \bullet \vec{\mathrm{a}} \bullet \vec{r}_1 - \vec{\mathrm{a}} \tag{11}$$

$$\vec{c} = 2 \bullet \vec{\mathrm{a}} \bullet \vec{r}_2 \tag{12}$$

Here, $\vec{a}$ decreases linearly from 2 to 0, controlling exploration/exploitation balance. r1, r2 $\in [0,1]$ are random vectors.

Hunting (leader guidance): The three best wolves (α, β,δ) guide the population:

$$\vec{d}_\alpha = \left|\vec{c}_1 \bullet \vec{x}_\alpha - \vec{x}\right|, \vec{d}_\beta = \left|\vec{c}_2 \bullet \vec{x}_\beta - \vec{x}\right|, \vec{d}_\delta = \left|\vec{c}_3 \bullet \vec{x}_\delta - \vec{x}\right| \tag{13}$$

Position updates relative to leaders:

$$\vec{x}_1 = \vec{x}_\alpha - \vec{a}_1 \bullet \left(\vec{d}_\alpha\right), \vec{x}_2 = \vec{x}_\beta - \vec{a}_2 \bullet \left(\vec{d}_\beta\right), \vec{x}_3 = \vec{x}_\delta - \vec{a}_3 \bullet \left(\vec{d}_\delta\right) \tag{14}$$

Combined update rule:

$$\vec{x}\left(i+1\right) = \frac{\vec{x}_1 + \vec{x}_2 + \vec{x}_3}{3} \tag{15}$$

This ensures wolves converge by following the collective guidance of the top three leaders.

Attacking prey (Exploitation): As a→0, the coefficient vector $\vec{A}$ shrinks within [−1,1]. Wolves then update their positions between their current location and the prey, simulating the attack phase and intensifying exploitation.

Searching for prey (Exploration): When $|\vec{A}|>1$, wolves move away from the prey, diversifying the search. The coefficient vector $\vec{c}$ introduces stochasticity by randomly weighting the prey's influence, amplifying, or reducing its effect. Unlike $\vec{A}$, parameter $\vec{c}$ is not decreased linearly, encouraging randomness and helping GWO avoid local optima.

A.3 Cuckoo Search Algorithm (CSA)

New solution generation (Lévy flights):

$$Y_c^{(i+1)} = Y_c^{(i)} + \alpha \quad \text{Lévy}(\lambda) \tag{16}$$

Here, $\alpha>0$ is the step size (commonly set to 1), and ⊗ denotes entry-wise multiplication. This stochastic equation represents a random walk, enabling both local and global search.

Lévy distribution:

$$\text{Lévy} \sim u = i^{-\lambda},\ (1 < \lambda \leq 3) \tag{17}$$

The Lévy distribution is characterized by infinite mean and variance, producing occasional long jumps that enhance global exploration.

Algorithm rules:

1. Each cuckoo lays one egg in a randomly chosen nest.
2. Nests with high-quality eggs (better solutions) are retained for the next generation.
3. A fraction p_a of nests are discovered by host birds and replaced with new random solutions.

Explanation:

- Equation (16) defines how new candidate solutions are generated using Lévy flights.
- Equation (17) specifies the distribution governing step lengths, ensuring a mix of short local moves and long global jumps.

- The three rules model brood parasitism and nest discovery, balancing exploitation (retaining good solutions) and exploration (random replacement).

A.4 Elephant Herding Optimization Algorithm (EHO)

Clan updating operator: The matriarch influences the position of each clan member:

$$Y_{new,d_k,l} = Y_{d_k,l} + \alpha * \left(Y_{best,d_k} - Y_{d_k,l}\right) * r \tag{18}$$

Here, $Y_{d_k,l}$ is the current position of elephant l in clan d_k, Y_{best,d_k} is the fittest elephant in the clan, $\alpha \in [0,1]$ is a scaling factor, and $r \in [0,1]$ is a random number. This equation moves elephants closer to the clan leader.

The fittest elephant is further updated toward the clan center:

$$Y_{new,d_k,l} = \beta * Y_{center,d_k} \tag{19}$$

where $\beta \in [0,1]$ controls the influence of the clan center.

The clan center for dimension h^{th} is defined as:

$$Y_{center,d_k,h} = \frac{1}{N_{d_k}} * \left(\sum_{l=1}^{d_k} Y_{d_k,l,h}\right) \tag{20}$$

Here, $Y_{d_k,l,h}$ is the hth component of elephant l, and N_{d_k} is the number of elephants in the clan. This ensures the leader aligns with the average clan position.

Separating operator: The least fit elephant in each clan is reinitialized to simulate mature males leaving the herd:

$$Y_{worst,d_k} = Y_{min} + \left(Y_{max} - Y_{min} + 1\right) * rand \tag{21}$$

Here, Y_{worst,d_k} is the position of the least fit elephant, rand $\in [0,1]$ is a stochastic variable, and Y_{max}, Y_{min} are the bounds of the search space. This introduces diversity and enhances global search.

Explanation:

- Equations (18–20) define how elephants update positions within clans, guided by leaders and clan centers.
- Equation (21) models' separation, ensuring exploration by reinitializing poor solutions.
- Together, these operators balance exploitation (clan guidance) and exploration (separation), preventing premature convergence.

Chapter 5
Knuckle Crease Patterns in Forensic Biometrics:
A Novel Trait for Identity Verification and Investigative Applications

Prachi Sharma Biswas
https://orcid.org/0009-0006-2262-4190
Shri Vaishnav Vidyapeeth Vishwavidyalaya, Indore, India

Swati Dubey Mishra
https://orcid.org/0000-0001-7501-7665
Shri Vaishnav Vidyapeeth Vishwavidyalaya, Indore, India

ABSTRACT

Knuckle crease patterns, formed by natural skin folds over finger joints, are emerging as a valuable biometric trait in forensic science and digital identity. Often visible in photographs and surveillance footage, they aid identification when fingerprints or DNA are unavailable. This chapter reviews their anatomical foundations, persistence, and distinctiveness, and examines automated recognition methods from handcrafted descriptors to deep learning, alongside manual classification approaches emphasizing reproducibility. Forensic applications include criminal investigations, child sexual abuse material analysis, disaster victim identification, and multimodal biometrics. Ethical, legal, and evidentiary issues are discussed, and future challenges for validation and responsible implementation are highlighted.

DOI: 10.4018/979-8-3373-4972-5.ch005

INTRODUCTION

Human identification has always been at the heart of forensic science. From the earliest adoption of anthropometry in the nineteenth century to the revolutionary impact of fingerprinting and the later development of DNA profiling in the twentieth century. Forensic practice has consistently advanced through the search for markers that are both scientifically reliable and legally admissible. Each innovation has addressed the limitations of its predecessors, expanding the evidentiary base and improving accuracy and judicial trust.

In parallel, biometrics has developed as a scientific discipline dedicated to recognizing individuals based on measurable biological and behavioural traits. Initially rooted in applications for security, border control, and administration, biometrics has increasingly intersected with forensic science, particularly as the demands of investigation and security have grown more complex (Jain et al., 2016).

Within this converging landscape, traditional biometric modalities such as fingerprints, facial recognition, and iris patterns continue to dominate; yet, each presents its own limitations. Fingerprints may degrade due to occupational wear, scarring, or deliberate alteration; facial recognition is vulnerable to occlusion, poor lighting, or disguise; and iris recognition requires specialized capture devices. These constraints have motivated the exploration of alternative or "soft" biometric traits, features that may not rival fingerprints in uniqueness but can provide supplementary evidentiary value, especially when conventional modalities are unavailable or compromised.

Knuckle crease patterns (KCPs), the natural folds on the dorsal surface of the fingers formed over interphalangeal joints, have recently attracted attention as an emerging biometric modality. Anchored to skeletal articulation and established during intrauterine development (Kimura et al., 1990), these creases are assumed to persist throughout life with remarkable resilience against environmental and occupational influences. Unlike fingerprints, which require cooperative collection or specialized scanners, KCPs are often incidentally captured in real-world imagery, even under non-cooperative conditions. This makes them particularly relevant in forensic contexts such as child sexual abuse material (CSAM) investigations, covert surveillance analysis, and disaster victim identification.

The research trajectory of KCPs differs from traditional identifiers. While fingerprints and DNA initially gained forensic acceptance before the advent of computational adaptation, KCPs were initially explored as a biometric trait. Kumar and Ravikanth (2009) demonstrated their feasibility for automated recognition, and later studies refined these methods through transform-based techniques and multimodal fusion. Deep learning has since improved accuracy, while explainable AI approaches address the transparency needed for forensic use. Only afterwards did forensic studies examine KCPs manually, focusing on classification, reproducibility, and stability (Baines, 2011; Chattopadhyay & Sukul, 2012; Dan et al., 2023; Biswas & Mishra, 2024). This reverse trajectory highlights both opportunities and challenges. Computational distinctiveness is evident, but forensic application requires reproducibility, validated error rates, and methodological clarity.

The forensic value of KCPs lies not only in their unique nature but also in their ability to complement other identifiers. Crease configurations can be integrated with tattoos, scars, vein patterns, or dorsal skin texture to strengthen evidentiary conclusions. This is particularly relevant when partial, degraded, or low-quality imagery is recovered, where conventional identifiers may be absent. Documented forensic casework has already shown that dorsal hand features, including crease patterns, can withstand judicial scrutiny under certain conditions, though challenges in standardization, dataset availability, and courtroom admissibility remain.

Despite their potential, several hurdles must be addressed before KCPs achieve routine forensic use. Large-scale population databases are lacking, limiting statistical validation. Manual classification suffers from subjectivity and limited reproducibility. Automated systems show promise but must prove explainability and reliability on real-world forensic data. Courts also require transparent methods, documented error rates, and compliance with admissibility standards. Ethical and privacy concerns arise due to the incidental visibility of dorsal hands in everyday imagery, requiring safeguards against misuse.

In light of these considerations, this chapter positions knuckle crease biometrics as an emerging modality situated between forensic science and biometric technology. It offers an integrated perspective that draws on anatomical foundations, biometric recognition methods, forensic case studies, and, legal and ethical frameworks. The chapter outlines the anatomical and developmental basis of knuckle creases and reviews advances in computational recognition systems, from handcrafted descriptors to deep learning and explainable AI. It also examines manual classification approaches along with reproducibility studies and observer reliability. In addition, the chapter assesses forensic applications in criminal investigation, CSAM analysis, disaster victim identification, and surveillance. Ethical, legal, and evidentiary considerations, such as data protection, privacy, and admissibility, are also evaluated. The chapter further highlights current limitations and identifies future directions. By bringing these elements together, it contributes to the wider discourse on forensic biometrics and shows how overlooked anatomical features can develop into scientifically robust and ethically governed identifiers. Ultimately, knuckle crease patterns demonstrate how the convergence of anatomy, technology, and forensic practice can expand the evidentiary toolkit and support human identification and justice in the digital era.

ANATOMICAL AND DEVELOPMENTAL BASIS OF KNUCKLE CREASES

Knuckle crease patterns (KCPs) represent distinct skin folds that form at sites of maximal flexion over the finger joints, reflecting the underlying anatomical and developmental structure of the hand. Each primary crease aligns with a specific articulation: the distal interphalangeal (DIP), proximal interphalangeal (PIP), and metacarpophalangeal (MCP) joints (Kimura et al., 1990). These creases originate during prenatal development and are shaped by joint anatomy, biomechanical forces, and repetitive flexion. Over the lifespan, they tend to deepen with age and mechanical use, yet their overall configuration remains relatively stable, which supports their utility as reliable long-term markers in forensic and biometric analyses. While generally robust, KCPs morphology can be altered by congenital anomalies, scarring, or pathological conditions, necessitating careful interpretation when applied in investigative contexts.

Anatomy of the Finger Joints

The human hand is a complex anatomical structure adapted for fine motor skills and functional versatility. Each finger, excluding the thumb, consists of three bones: the proximal, middle, and distal phalanges, which are connected by hinge-like synovial joints (Moore et al., 2018). Figure 1 depicts the anatomical arrangement of the phalanges in human digit. The articulations of the human finger include the metacarpophalangeal (MCP) joints, connecting the metacarpal bones of the palm to the proximal phalanges; the proximal interphalangeal (PIP) joints, located between the proximal and middle phalanges; and the distal interphalangeal (DIP) joints, found between the middle and distal phalanges. Collateral ligaments, a fibrous capsule, and an overlying extensor mechanism stabilize each joint, ensuring both mobility and structural integrity.

Figure 1. Anatomical Arrangements of Phalanges in the Human Digit. (Created by the author using original research data).

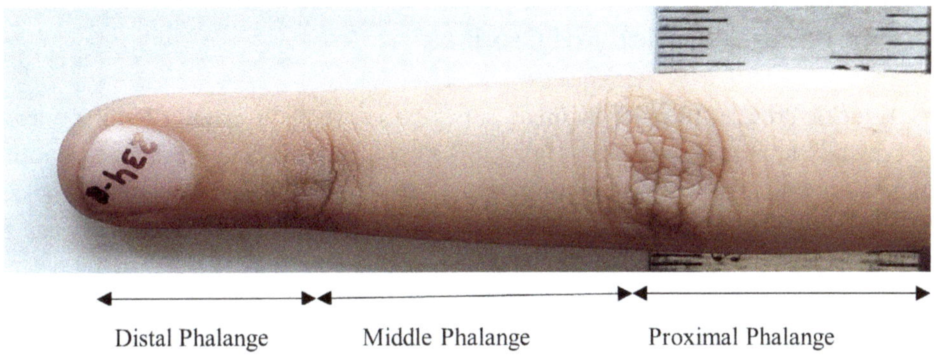

When the finger flexes, the skin anchored to the underlying structures folds along predictable lines, producing visible transverse creases on the dorsal surface (Standring, 2016). The PIP joint typically displays the most prominent crease, often referred to as the "major knuckle", while creases over the MCP and DIP joints are shallower and are sometimes described as "minor knuckles" (Kulkarni and Rout, 2012; Diviya et al., 2015). Figure 2. depicts the anatomical positions of the major and minor knuckles in the human finger.

Figure 2. Anatomical Positions of Minor and Major Knuckles in the Human Digit. (Created by the author using original research data)

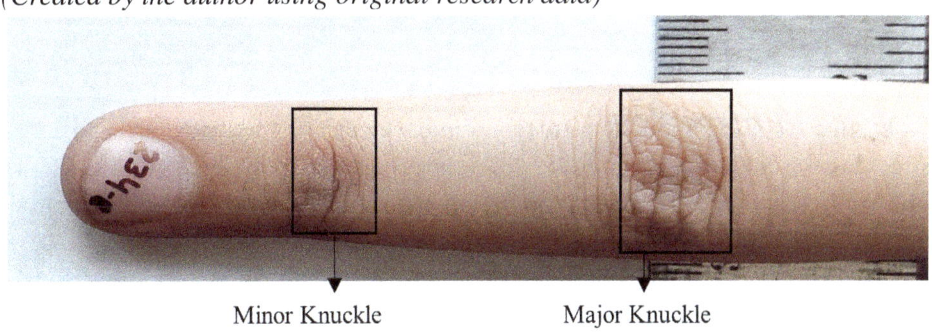

The formation of these creases is not merely a superficial skin feature but is closely aligned with joint articulation. Their location corresponds to zones of maximum skin folding, dictated by repetitive motion and attachment of the dermis to deeper connective tissues. Over time, repetitive flexion during growth and daily activities deepens these creases, making them permanent anatomical landmarks that can survive even after soft tissue damage. Understanding this anatomical alignment is essential before considering the prenatal development and stability of these creases.

Developmental Biology of Creases

Knuckle creases, like palmar creases, begin to develop prenatally. Palmar creases typically appear between the 10th and 13th weeks of gestation, shaped by joint movement and differential growth patterns (Stevens et al., 1988). Dorsal knuckle creases follow similar biomechanical principles, generally developing between 8 and 13 weeks of gestation and well established by the end of the first trimester (Kimura et al., 1990).

Once formed, dorsal creases remain largely stable throughout life. Although crease depth may vary slightly due to changes in skin elasticity, hydration, or occupational wear, the overall configuration is preserved. Because these creases are closely linked to joint anatomy and underlying connective tissues, they often reappear along similar anatomical planes even after superficial injury or dermatological changes. This resilience supports their utility as documentable and comparable forensic markers.

Longitudinal studies and biometric research indicate strong stability over several years (Kumar, 2014). Over time, creases deepen with age and repeated use, but their fundamental morphology remains intact. Exceptional cases, such as congenital malformations, scarring, or disease, can alter crease patterns, highlighting the need for cautious interpretation. The combination of developmental origin, anatomical alignment, and long-term stability makes KCPs especially attractive for forensic and biometric applications.

Individual Variation

Distinctiveness is a fundamental requirement for any biometric feature, and dorsal knuckle creases exhibit pronounced differences between individuals (Kumar,2014). These differences result from a combination of genetic, developmental, and environmental factors that shape the morphology of each person's creases (Kadane, 2022). A notable aspect is the number of creases on each finger, which can vary significantly. In addition, the length and prominence of the creases differ, with some extending clearly across the joint while others remain shorter or less defined. The orientation of the patterns also varies, ranging from horizontal, vertical, and oblique to more intricate arrangements, reflecting subtle anatomical characteristics. Symmetry, defined as the degree to which crease patterns mirror each other on corresponding fingers, adds further individuality, capturing both inter and intra-personal distinctions. While some individuals display relatively simple crease arrangements, others present complex intersecting or branching patterns. Together, these characteristics highlight the potential of knuckle creases as reliable biometric markers, underscoring the importance of careful observation, accurate documentation, and systematic classification in forensic analysis.

These differences are influenced not only by joint morphology and biomechanical forces during growth but also by genetic factors. Moreover, pathological or acquired conditions such as scarring or connective tissue disorders can modify crease appearance, although such cases are uncommon and may even contribute to their forensic distinctiveness (Dan, 2023). Despite the evident diversity, comprehensive population-level studies evaluating demographic, environmental, or occupational influences on crease patterns are still limited. Therefore, further research is necessary to establish reliable correlations and strengthen the scientific foundation for applying knuckle crease analysis in both forensic and biometric contexts.

Imaging and Visualization

Accurate documentation of knuckle crease patterns requires the use of appropriate imaging techniques. In most forensic and biometric applications, conventional digital photography is sufficient, provided that resolution and lighting conditions are carefully controlled. However, to improve consistency and enhance recognition accuracy, researchers have explored several advanced modalities. Early texture-based feature extraction approaches, for instance, relied on high-resolution imaging to capture fine ridge and crease details with greater precision (Kumar & Ravikanth, 2009). In addition to this, near-infrared imaging has been utilised to simultaneously record both creases and dorsal vein patterns, allowing for their integration into multimodal recognition systems and thereby strengthening identification reliability (Kumar & Zhou, 2012). More recently, 3D scanning technologies have been introduced to overcome limitations caused by pose variation, illumination inconsistencies, and perspective distortion, offering more geometrically accurate representations of dorsal hand anatomy (Cheng et al., 2023).

Despite these advancements, real-world forensic practice often relies on opportunistic imagery obtained from CCTV footage, social media content, or seized digital devices, where image quality is frequently suboptimal. This challenge has driven ongoing research toward developing recognition methods that remain effective under varying resolutions, lighting conditions, and hand orientations, thereby bridging the gap between controlled acquisition and practical investigative scenarios.

Forensic Significance of Anatomical Features

The anatomical and developmental foundations of knuckle creases highlight three key properties that underscore their forensic value: universality, persistence, and distinctiveness. Universality refers to the presence of knuckle creases in all individuals, regardless of age, sex, or ethnicity. Persistence indicates that, once formed, the patterns remain stable over time, anchored to joint anatomy and repetitive movement. Distinctiveness arises from inter-individual variability in crease number, orientation, and branching, creating unique patterns that can be used for identification.

In forensic scenarios, these features hold particular value because knuckle creases are often incidentally captured in digital imagery. Unlike fingerprints, which typically require cooperative collection, dorsal hand views are frequently visible in photographs and videos, including covert or non-cooperative contexts. For example, in cases of online child exploitation, perpetrators often obscure their faces but inadvertently reveal parts of their hands. In such situations, crease analysis provides investigators with valuable biometric evidence (Dan et al., 2023).

Thus, the biological basis of knuckle creases directly supports their forensic application. Their universality, stability, and variation align with the essential principles of biometric science, while their visibility in real-world digital media offers unique advantages for modern forensic practice.

AUTOMATED BIOMETRIC DEVELOPMENTS

Research on knuckle crease patterns began primarily with automated approaches rather than manual observation. Over the years, techniques have evolved from handcrafted descriptors to advanced deep learning models, with increasing emphasis on robustness, accuracy, and forensic applicability. The following section reviews these developments chronologically under key methodological phases.

Early Automated Approaches

The first documented attempt to use finger knuckle patterns as a biometric trait was by Woodard and Flynn (2005), who captured data with a 3D sensor (Minolta 900/910) and applied linear regression with normalization coefficients for template generation. Their system achieved an Equal Error Rate (EER) of 5.5%, providing initial evidence of the feasibility of KCPs for recognition.

Kumar and Ravikanth (2009) further advanced the field by employing subspace methods such as PCA, LDA, and ICA on textured patterns formed during finger bending. Their experiments achieved an EER of 1.39% and a Decidability Index of 2.35, demonstrating significant discriminative potential.

Transform-Based Methods

Subsequent work focused on transform-based techniques that could capture orientation and phase information of crease textures.

- Kumar and Zhou (2009) introduced the concept of *Knuckle Codes* using the Localized Radon Transform. Matching was performed by minimum distance measures, achieving a rank-one recognition rate of 98.6% with an EER of 1.14%.
- Nanni and Lumini (2009) combined Radon transform and Haar wavelet features, processed through a non-linear Fisher transform, and classified using a Parzen window classifier. Their system achieved a very low EER of 0.3%.
- Zhang et al. (2009) proposed a dedicated capturing device and applied the Band-Limited Phase-Only Correlation (BLPOC) technique for simultaneous image registration and matching. Their method reported a genuine acceptance rate of 97%, a false acceptance rate of 0.02%, and an EER of 1.09%.

These studies established transform-based methods as highly effective for knuckle pattern analysis.

Feature-Based and Fusion Approaches

Badrinath et al. (2011) employed Scale Invariant Feature Transform (SIFT) and Speeded Up Robust Features (SURF) descriptors, combining their matching scores with a weighted sum rule. Evaluation on the PolyU FKP database yielded an EER of 0.215%, confirming the effectiveness of local feature-based fusion strategies.

Kumar (2014) expanded the scope by conducting a longitudinal study with images collected over 4–7 years. Both major and minor knuckle creases were analysed, and fusion of features improved recognition accuracy. The results confirmed the permanence of crease patterns, with verification accuracy and EER values demonstrating long-term stability of the trait.

Automated Localization of Knuckle Regions

Vyas et al. (2022) proposed an ensemble-based approach utilizing multiple object detector frameworks to localize the knuckle regions effectively. Their method achieved high Intersection over Union (IoU) scores, enhancing the reliability of subsequent feature extraction, particularly for forensic images captured under uncontrolled conditions.

Deep Learning Approaches

The advent of deep learning brought a shift from handcrafted descriptors to data-driven feature learning.

- Vidhyapriya and Rose (2019) utilized Gabor-based texture extraction combined with the Expectation-Maximization (EM) algorithm for finger knuckle print recognition, achieving 98% recognition performance.
- Trabelsi et al. (2020) developed a multimodal CNN framework combining multiple finger modalities, which outperformed single-modality systems and achieved an EER of 0.000% for open-set identification.
- Cheng et al. (2023) introduced 3D knuckle recognition techniques, integrating surface topology with crease information to enhance robustness against pose and illumination variation.
- Gao et al. (2024) proposed the Dual-Order Texture Competition Network (DOTCNet), which leverages dual-order texture modeling and multi-scale feature learning, achieving state-of-the-art accuracy in finger-knuckle biometric recognition.

Explainable and Forensic-Friendly AI

Zhou et al. (2025) developed an explainable AI framework trained on a large-scale dataset (805768 images). Their system generated interpretable key point templates, allowing visualization of crease regions contributing to recognition. Importantly, it also estimated the uniqueness of patterns, aligning the technology with forensic requirements for transparency and individuality.

Benchmarks and Performance Trends

The development of automated knuckle biometrics has progressed through distinct methodological phases, each addressing the limitations of earlier approaches and enhancing recognition performance. Early handcrafted descriptors demonstrated feasibility but were limited by small datasets and higher error rates. Transform-based methods, including Rado and Fourier-based techniques, significantly improved accuracy and robustness, achieving sub-1% error rates on moderate datasets.

Feature-based fusion strategies, such as the combination of SIFT and SURF, further enhanced performance and validated the long-term stability of crease patterns. The shift to deep learning introduced Convolutional Neural Networks (CNNs), multimodal fusion, and 3D imaging, resulting in near-perfect recognition accuracy under controlled conditions. More recently, explainable AI frameworks have emphasized transparency by generating interpretable key point templates and estimating the uniqueness of crease patterns, thereby aligning recognition systems with forensic requirements.

Overall, benchmarks show a steady trend toward lower error rates, greater robustness to variability, and improved forensic readiness. Table 1 summarizes this trajectory, highlighting the evolution from early automated descriptors to advanced explainable deep learning systems.

Table 1. Evolution of Automated Knuckle Biometric Methods.

Phase / Period	Key Methods	Performance Metrics	Dataset / Context	Forensic Relevance
Early Automated Approaches (2005–2009)	3D imaging, PCA, LDA, ICA	EER ≈ 5.5% (Woodard & Flynn, 2005); EER ≈ 1.39%, DI = 2.35 (Kumar & Ravikanth, 2009)	Small datasets, initial proof-of-concept	Demonstrated feasibility of KCPs as biometric identifiers
Transform-Based Methods (2009–2012)	Knuckle Codes (Radon transform), Haar wavelets, BLPOC	Rank-1 accuracy 98.6%, EER ≈ 1.14% (Kumar & Zhou, 2009); EER = 0.3% (Nanni & Lumini, 2009); GAR = 97%, FAR = 0.02% (Zhang et al., 2009)	Moderate dataset; controlled environments	Improved robustness and accuracy, established discriminative power
Feature Fusion & Longitudinal Studies (2011–2014)	SIFT + SURF fusion; multi-joint crease fusion	EER = 0.215% (Badrinath et al., 2011); Long-term stability confirmed, EER ≈ 3.62% (Kumar, 2014)	PolyU FKP DB; longitudinal dataset (503 subjects, 4–7 years)	Validated permanence of crease patterns and benefits of feature fusion
Deep Learning Approaches (2018–2024)	CNNs (Siamese, multimodal, 3D, DOTCNet)	EER < 1% (Liu et al., 2018); Recognition ≈ 98% (Vidhyapriya & Rose, 2019); EER = 0.000% (Trabelsi et al., 2020); Enhanced robustness in 3D/pose-invariant recognition (Cheng et al., 2023); State-of-the-art accuracy (Gao et al., 2024)	Larger experimental datasets, multimodal systems	Achieved near-perfect recognition under experimental conditions

Limitations of Automated Systems

Despite technical progress, early automated systems faced important drawbacks from a forensic standpoint. Most relied on small, laboratory-controlled datasets that did not reflect the variability of real-world evidence, such as blurred CCTV stills or partial hand images. Their primary goal was commercial deployment, not forensic admissibility, meaning they rarely addressed issues like observer verification, error rate documentation, or courtroom transparency.

Deep learning methods, although accurate, introduced "black-box" concerns that courts may find unacceptable without interpretability. These limitations highlighted the need for forensic experts to validate the biological distinctiveness and legal defensibility of crease patterns. This gap motivated the emergence of manual classification studies.

While automated methods are advancing, manual studies remain crucial, particularly in forensic casework where subtle patterns may not be captured by algorithms. The following section explores these manual approaches and their contribution to knuckle crease research.

MANUAL CLASSIFICATION AND RELIABILITY STUDIES

Manual classification studies emerged as the forensic community's approach to validating knuckle crease patterns in ways that automated recognition could not. While automated systems demonstrated technical feasibility, forensic science required human-observable categories, reproducible methods, and transparent criteria that could withstand legal scrutiny. These studies confirmed that crease patterns are distinctive and persistent, while also exposing challenges of observer subjectivity and classification reliability. By showing that the trait is biologically distinctive and systematically categorizable, manual research established the foundation for recognizing knuckle creases as a forensic identifier, bridging the gap between biological uniqueness and practical admissibility in investigative and judicial contexts.

Early Descriptive Classification – Baines (2011)

The earliest systematic attempt to classify dorsal knuckle crease patterns was conducted by Baines (2011), who studied the left index fingers of 158 individuals. He proposed an 11-type classification system that included Star, Yin/Yang, Vertical, Offset/S-shaped, Doughnut, C-type, three ridges, two ridges, one ridge, E-type, and Circle-with-lines enclosed. Figure 3 depicts all 11 patterns. In this classification, in most of the patterns, instead of crease, fleshy folds were considered; for example, in 3 ridges, it is not the ridges, but actually the three fleshy folds along with four ridges. This framework was highly descriptive and captured a wide range of morphological variations. However, subsequent testing revealed that it was difficult to apply consistently. Independent observers achieved reliable agreement only for a limited subset of categories, namely, Star, Yin/Yang, and Vertical, while the remaining types suffered from significant overlap and ambiguity. As a result, although Baines' system highlighted the potential distinctiveness of knuckle creases, it was regarded as too subjective and inconsistent for forensic application.

Figure 3. The eleven knuckle crease pattern types described by Baines (2011) illustrating major morphological variations. (Created by the author).

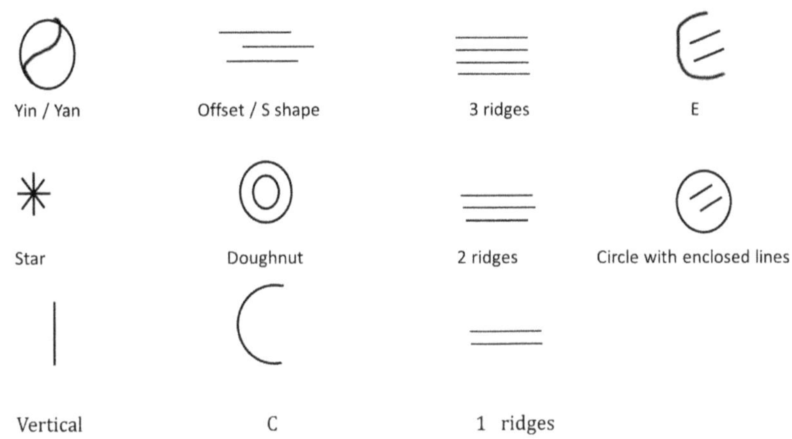

Five-Type Framework – Chattopadhyay and Sukul (2012)

To improve reproducibility, Chattopadhyay and Sukul (2012) proposed a simplified five-type classification system based on digital photographs of 257 Bengali individuals. The framework categorized knuckle creases into five types: horizontal, vertical, oblique, semilunar, and mixed. This approach made classification more practical and easier to apply across different observers. Horizontal creases were the most frequently observed, whereas semilunar creases were the least common. However, the system relied entirely on visual inspection without incorporating quantitative measures such as crease angle or spacing, leaving room for subjectivity and variability in observer interpretation.

Expansion into Six Types – Biswas and Mishra (2024)

Building on the five-type model, Biswas and Mishra (2024) expanded the classification system by introducing a sixth category, the Cross type, characterized by intersecting creases forming an "X"-shaped pattern. Their study utilized a larger and more diverse dataset beyond a single regional population, enhancing the generalizability of the classification. Figure 4 illustrates all six knuckle crease pattern types.

The key findings of the study indicated that Horizontal creases were the most common pattern, while Cross creases, though relatively rare, were consistently identifiable in both sexes. Additionally, the analysis revealed clear sexual dimorphism in knuckle crease patterns, with males exhibiting a greater average distance between adjacent creases than females. This difference was statistically significant (t-test, 95% confidence level), underscoring the potential of knuckle crease patterns not only for personal identification but also for sex estimation in forensic anthropology and criminal profiling.

Figure 4. The six-type classification system includes (a) Horizontal, (b) Vertical, (C) Oblique, (d) Semi-lunar, (e) Mixed, (f) Cross. (Created by the author using original research data).

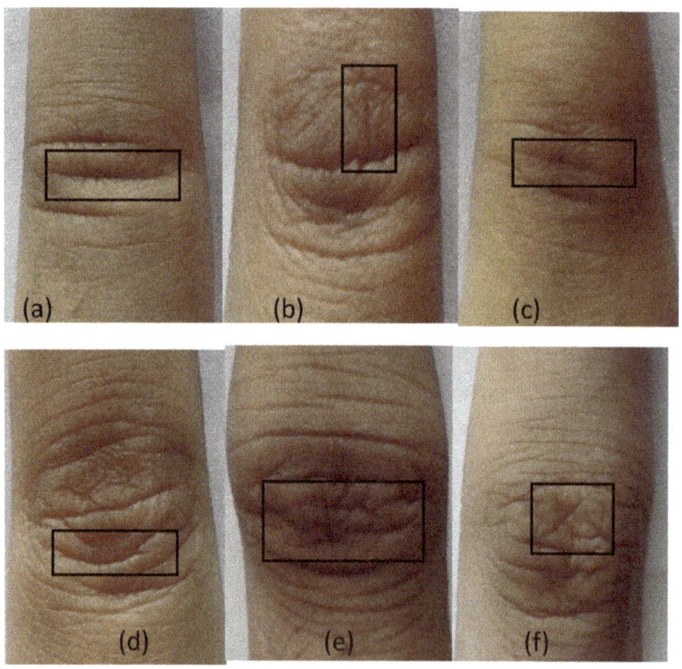

Reliability and Observer Variation

A classification scheme can only be forensically valuable if it achieves consistent results across observers. Therefore, inter and intra-observer reliability studies have been critical in evaluating knuckle crease classifications.

Dakshinamurthy (2018) critically evaluated Baine's 11-type classification system for classifying knuckle crease patterns to assess its reliability and repeatability. Inter-observer tests with 49 untrained participants showed extremely low agreement (Fleiss' kappa = 0.0045), while intra-observer tests after a 10-day interval also demonstrated negligible consistency (kappa = 0.00538). A subsequent test with a participant having anatomical knowledge yielded a kappa of -0.00756, confirming low repeatability. Contributing factors included difficulty in accurately identifying creases, lack of standardized procedures for marking crease lines, limited anatomical understanding, and variability in image quality. Dakshinamurthy concluded that the method shows promise but recommended improving reliability through participant training or automating key steps such as crease highlighting and central axis drawing to reduce subjectivity and human error.

Biswas and Dubey further investigated observer agreement using 6 type classification schemes, reporting that while simplified systems enhanced reproducibility, certain patterns, such as horizontal and semi-lunar, and mixed and cross types, remained prone to misclassification. These findings collectively demonstrate that observer variation is an inherent consideration in knuckle crease analysis and underscore the importance of standardized methodology and training.

Persistence and Imaging Protocols – Dan et al. (2023)

Dan et al. (2023) investigated the persistence of dorsal knuckle creases during finger flexion to assess their reliability for forensic identification from digital hand images. The study utilized the *Knuckle Down ID* app, a citizen science initiative, to collect high-quality images of knuckle creases at various flexion angles (0°, 20°, 45°, and 90°). A total of 972 images from 81 participants were analysed.

The researchers adapted the modified Baines classification method to evaluate the impact of finger flexion on the appearance of knuckle creases. Results indicated that while knuckle creases are generally persistent, their visibility and the frequency of specific features can vary with finger position. Significant differences were observed when comparing 0° to 45° flexion ($p = 0.04$) and 90° flexion to other positions ($p < 0.001$). However, comparisons between 0° and 20° hyperflexion or 20° and 45° were not statistically significant, suggesting these positions are comparable.

The study underscores the importance of standardized imaging protocols in forensic investigations, highlighting that variations in finger flexion can influence the analysis of knuckle creases. These findings are crucial for forensic practitioners when conducting 1:1 comparison of knuckle creases in casework involving unconstrained hand images.

Comparative Evaluation of Existing Methods

Taking these major contributions together, the trajectory of manual classification highlights a shift from complex to simplified, operationally viable systems:

- Baines (2011): Highly descriptive but impractical (11 types, low reproducibility).
- Chattopadhyay & Sukul (2012): Simplified and operational (five types).
- Biswas & Mishra (2024): Expanded, balanced, and biologically validated (six types with sexual dimorphism).

Overall, manual classification studies demonstrate that knuckle creases can be systematically categorized, but reproducibility varies depending on pattern type. Simpler patterns tend to be more consistently recognized, while complex patterns remain challenging. This inconsistency underscores the need for examiner training, clearer decision rules, and eventually, integration with automated support systems to ensure forensic admissibility.

FORENSIC APPLICATIONS OF KNUCKLE CREASE PATTERNS

The forensic applicability of a biometric trait depends not only on its individuality but also on its practicality in real-world scenarios where identity needs to be established. Knuckle crease patterns (KCPs) offer a unique combination of advantages that distinguish them from traditional biometric traits such as fingerprints and facial recognition. They are permanent anatomical features, formed early in life and remaining stable across time. They are also universal, present in every individual, and resistant to circumvention, since the dorsal surface of the hand is rarely in contact with external objects and cannot easily be replicated or stolen. These properties make KCPs suitable for use in a wide range of forensic and biometric applications, from criminal investigations to disaster victim identification, surveillance, and secure personal authentication systems.

Table 2. Comparative Features of Fingerprints and Knuckle Crease Patterns (KCPs) as Forensic Biometrics

Characteristic	Fingerprints	Knuckle Crease Patterns (KCPs)
Biological origin	Formed from friction ridges on volar pads during foetal development	Formed by flexion of the dorsal skin over the finger joints. Its formation also takes place during foetal development.
Research maturity	Well-established, with extensive global studies	Emerging area with limited but growing studies
Classification systems	Clear and universally adopted (arches, loops, whorls)	Preliminary systems proposed; no global consensus
Database availability	Large international databases for forensic and civil use	No standardized large-scale databases yet
Permanence	Proven stability across the lifespan	Early evidence supports stability, but long-term data are limited
Distinctiveness	Legally recognized as unique	Believed to be distinctive; large-scale validation is still required
Ease of capture	Requires fingerprint scanners or ink impressions	Can be collected without contact using digital cameras
Resistance to circumvention	Vulnerable to theft via latent prints, as the palmar surface of the hand more often comes in contact with surfaces and objects	More resistant, as the dorsal surface of the hand rarely comes into contact with any surface or object.
Forensic expertise	Established examiner community and legal precedents	Few specialists; Methods are still under development

Knuckle Crease Patterns versus Fingerprints

Fingerprints have long been regarded as the "gold standard" of forensic identification, but they also present limitations. Fingerprint ridges can become worn or damaged in manual labourers, older adults, or individuals with dermatological conditions (Lee et al., 2013), (Bharadwaj et al., 2014). They are also vulnerable to theft: because people constantly touch surfaces, fingerprints are frequently deposited and can be lifted from crime scenes or stolen through latent print recovery. This creates both forensic opportunities and security vulnerabilities, since latent prints can be misused in spoofing attacks.

In contrast, KCPs occupy the dorsal aspect of the hand, a region that rarely comes into direct contact with external surfaces (Biswas, 2024). As a result, they are not routinely left behind in the environment, and thus cannot easily be stolen or replicated by malicious actors. Unlike fingerprints, which can be deliberately altered or degraded, KCPs are not typically affected by abrasion, burns, or heavy manual work (Li et al., 2022). This makes them a more robust identifier in populations where fingerprint quality is compromised. The comparative resilience of KCPs to external wear and theft strengthens their candidacy as an alternative or supplementary biometric trait in forensic and security domains. Table 2 summarizes the comparative features of fingerprints and knuckle crease patterns (KCPs) as forensic biometrics.

Criminal Investigations and CSAM Cases

One of the most direct forensic applications of KCPs is in criminal investigations where suspects' hands, rather than their faces, are captured in evidentiary material. In cases of child sexual abuse material (CSAM), perpetrators often conceal their faces but inadvertently reveal their hands in photographs or videos. The dorsal surface, including knuckle creases, becomes a critical point of comparison.

Beyond CSAM, KCPs are useful in other criminal contexts. In robbery, assault, and homicide investigations, CCTV footage frequently captures partial hand views while suspects interact with objects. Even when fine crease details are obscured, the overall crease configuration can provide additional lines of evidence. Such observations do not replace traditional identifiers but serve as supporting evidence, especially when corroborated with other anatomical markers such as scars, veins, or tattoos.

Disaster Victim Identification (DVI)

In the aftermath of mass disasters, such as earthquakes, plane crashes, or fires, the challenge of victim identification is immense. Standard methods such as fingerprints, dental records, and DNA remain the foundation of disaster victim identification (DVI). While these techniques are highly reliable, they have certain limitations: fingerprints may be compromised due to tissue damage, and DNA analysis is often time-consuming. In such scenarios, knuckle crease patterns (KCPs) can serve as a supplementary or alternative means of identification.

Knuckle creases, being deeply embedded anatomical folds, are often preserved even when superficial skin ridges are lost. Moreover, with the widespread availability of personal and social media images, many victims have ante-mortem photographs where the dorsal surface of the hand is visible. By comparing these with post-mortem photographs, forensic experts can identify distinctive crease configurations.

KCPs are not a replacement for DNA or fingerprints, but they function as a valuable secondary identifier, especially in resource-limited or urgent situations where rapid triage is required. Their universality ensures that they can be applied to all victims, and their permanence provides confidence in comparisons even across long time spans.

Surveillance, Intelligence, and OSINT Investigations

Knuckle crease patterns (KCPs) are gaining prominence as a biometric tool in surveillance and intelligence operations. With the proliferation of CCTV systems in public spaces such as ATMs, transportation hubs, and retail environments, the dorsal surface of the hand is frequently captured during routine interactions. Unlike facial recognition, which can be obstructed by masks or poor angles, and fingerprints, which require physical contact, KCPs are naturally visible. They can be recorded at a distance without the subject's awareness.

In intelligence and open-source investigations (OSINT), individuals engaged in criminal or extremist activities often share images or videos online in which their faces are concealed but their hands are visible. These instances, such as holding weapons, making hand signs, or displaying contraband, offer investigators an opportunity to extract and compare knuckle crease configurations. By analysing these features across different appearances, analysts can establish links between individuals even in the absence of facial or fingerprint data.

The ability to collect KCPs contactlessly through ordinary video or photographic footage enhances their value in covert and remote operations. However, the increasing accessibility of this biometric identifier also raises important ethical and legal considerations. The use of hand imagery from publicly shared content must be guided by strict forensic protocols to avoid misuse and protect privacy rights. As such, integrating KCP analysis into legitimate investigative frameworks requires both scientific validation and regulatory oversight to ensure responsible application.

Personal Identification and Alternative Biometric Role

KCPs have emerged as a reliable biometric trait for personal identification, either as a standalone modality or as an effective supplement to existing systems. Like any dependable biometric, they fulfil the core criteria of universality, permanence, and distinctiveness, as every individual possesses knuckle creases, the patterns remain largely unchanged throughout life, and their variations are sufficient to distinguish one person from another. Their utility is further evident in populations where fingerprints are difficult to capture due to damage, ageing, or occupational wear. For instance, manual laborer's often experience erosion of fingerprint ridges through repeated physical work, yet their knuckle creases remain clearly visible. Likewise, in elderly individuals, fingerprint recognition may fail as skin becomes loose and additional creases form, leading to false rejections by biometric devices, while dorsal folds remain prominent. In such circumstances, KCPs provide a practical alternative, offering greater inclusivity for applications such as national identity programs, border control, and law enforcement databases.

Banking, Security Check-Ins, and Civil Authentication

In the civilian domain, KCPs can play a role in secure authentication systems for banking, airport check-ins, and access control in sensitive facilities. Their primary advantage lies in their contactless capture through standard cameras. Unlike fingerprint scanners, which require physical contact and can be compromised by dirt, moisture, or occupational wear, dorsal hand images can be taken quickly and hygienically.

In banking, where identity theft is a growing concern, KCPs offer resilience because they cannot be easily stolen. Fingerprints are frequently deposited on surfaces and can be reproduced using latent print techniques, but dorsal creases are rarely left behind in daily life. This makes them less vulnerable to spoofing or replication. As a result, KCP-based authentication provides an additional layer of security, particularly when used in conjunction with a multimodal system that incorporates fingerprints or facial recognition.

Contactless Capture and Digital Forensics

The demand for contactless biometrics has increased significantly in recent years, especially during the COVID-19 pandemic. KCPs lend themselves naturally to contactless capture using existing digital infrastructure. High-resolution cameras, even those embedded in mobile devices, are sufficient to record crease configurations.

In digital forensic investigations, KCPs can be extracted from images and videos stored on seized devices or shared online. Unlike fingerprints or DNA, which require physical samples, KCPs can be retrospectively analysed from digital media. This adaptability is particularly valuable in transnational investigations, where physical access to suspects may be limited but large amounts of imagery are available to investigators.

Permanence, Universality, and Resistance To Theft

From a forensic science perspective, KCPs satisfy the key requirements for a robust biometric. Their permanence ensures that they remain stable across a lifetime. Their universality ensures that they can be applied across populations without exclusion. Their distinctiveness provides discriminatory potential between individuals.

Crucially, they are resistant to theft. Fingerprints are frequently stolen because individuals touch countless surfaces every day, leaving behind latent impressions that can be collected and misused. By contrast, the dorsal surface of the hand is rarely in contact with external objects, and thus crease patterns are not routinely deposited in the environment. This makes them significantly harder to steal or replicate.

Combined with their immunity to occupational wear, this property reinforces their value as a secure, alternative biometric for both forensic and civil use.

CASE STUDIES

The forensic value of knuckle crease patterns (KCPs) has been most convincingly illustrated in real-world cases where conventional identifiers such as fingerprints, DNA, or facial recognition were unavailable. In these situations, investigators turned to the dorsal surface of the hand, examining creases, scars, freckles, and other anatomical markers to establish identity. The following examples demonstrate how KCPs have been used in practice, the degree to which they have withstood judicial scrutiny, and the limitations that remain.

Dean Hardy Case (United Kingdom, 2007–2009)

In 2007, Dean Hardy was apprehended after authorities discovered a large number of indecent images of children on his computer. Five of these images captured the dorsal aspect of his hand, notably the thumb. Professor Sue Black, a forensic anthropologist, examined these photographs, focusing on features that are difficult to disguise, such as knuckle crease patterns, skin pigmentation, and freckles. A detailed comparison with Hardy's hand confirmed a match, leading to a positive identification. This case was notable as it was the first in the United Kingdom where dorsal hand features, including knuckle creases and freckles, were successfully used as forensic evidence in court. The methodology was upheld during legal proceedings, the defense challenges were unsuccessful, and Hardy ultimately pleaded guilty, setting an important precedent for the admissibility of hand feature analysis in forensic investigations (Black et al., 2009).

Operation Algebra – Neil Strachan Case (Scotland, 2009)

Shortly after the Hardy case, another landmark investigation reinforced the evidentiary potential of dorsal hand features. Operation Algebra was a large-scale inquiry into an organized child abuse network in Scotland. Authorities recovered more than 7,000 indecent images, including one containing a partial image of an adult hand.

Forensic experts compared this image to known photographs of Neil Strachan, one of the suspects. The analysis focused on the morphological characteristics of the thumb and knuckle crease orientation. The comparison revealed strong consistencies, which were presented as corroborative evidence at trial.

Strachan was convicted at the High Court in Edinburgh and received a life sentence (Carrell, 2009). The case underscored two key points. First, that dorsal hand features, including knuckle creases, could serve as reliable supporting identifiers in serious criminal cases. Second, that courts were willing to consider such evidence when it was presented by qualified experts and placed within the broader evidentiary framework

Devin Whitfield Case (United States, 2020)

A decade later, the forensic use of knuckle crease patterns resurfaced in the United States. In 2020, Devin Whitfield was charged with possession of child pornography after investigators recovered incriminating material from a mobile phone. One image prominently displayed the dorsal aspect of a hand, showing visible knuckle creases.

A fingerprint expert initially testified that the crease pattern matched Whitfield's hand. However, during cross-examination, the expert was unable to clearly articulate the methodology or provide statistical support for the conclusion. The defence highlighted the lack of validation studies and population-level data, leading the prosecution to withdraw this specific piece of evidence.

Although Whitfield was ultimately convicted on other charges, the case highlighted the current limitations of dorsal hand evidence. It demonstrated that while knuckle creases may provide valuable investigative leads, their courtroom admissibility depends on the existence of standardized protocols, robust statistical backing, and peer-reviewed validation studies (Kadane, 2020).

Table 3. Landmark Forensic Cases Involving Knuckle Crease Pattern (KCP) Evidence

Case / Year	Jurisdiction	Key Details	Role of Knuckle Creases	Outcome
Dean Hardy (2007–2009)	United Kingdom	Thousands of indecent images recovered; included offender's dorsal hand features	Forensic anthropologists compared knuckle creases, freckles, and pigmentation with reference photographs.	Evidence admitted; Hardy pleaded guilty and received a custodial sentence
Operation Algebra – Neil Strachan (2009)	Scotland	Child abuse network with >7,000 indecent images, one showing a partial hand	Experts analysed thumb morphology and crease orientation to link to the suspect	Evidence accepted; Strachan convicted and sentenced to life imprisonment
Devin Whitfield (2020)	United States	Child pornography case; image displayed a dorsal hand with visible creases	Crease comparison attempted, but expert testimony was challenged due to a lack of validation studies	Though not accepted as evidence but Whitefield was later convicted based on other evidence.

Reflections on Casework

Taken together, these cases highlight both the promise and the challenges of using knuckle creases as forensic evidence. The Hardy and Strachan cases showed that dorsal features, including crease patterns, can be compelling when supported by expert testimony and corroborative evidence. The Whitfield case, however, revealed the vulnerabilities of presenting such evidence without standardized methodologies and statistical validation.

Importantly, these cases shaped the trajectory of research in this field. They spurred efforts to develop systematic classification systems, improve inter-observer reliability, and design automated recognition tools capable of handling real-world imagery. They also emphasized the need for population-level databases and formal validation studies before KCPs can achieve the level of legal acceptance enjoyed by fingerprints or DNA.

The broader lessons are clear. Knuckle crease patterns can play a valuable role in investigations, particularly when conventional identifiers are missing. They have already proven useful in leading to convictions in serious criminal cases. At the same time, their forensic future depends on addressing issues of standardization, reliability, and admissibility. These challenges are explored in the following section. A concise summary of these landmark forensic cases is provided in Table 3, outlining their investigative context, the role of knuckle crease evidence, and judicial outcomes.

CHALLENGES IN THE FORENSIC AND BIOMETRIC USE OF KNUCKLE CREASE PATTERNS

Although knuckle crease patterns (KCPs) have demonstrated their utility in investigations and even in courtroom settings, their widespread acceptance as a reliable biometric still faces significant challenges. These challenges arise from scientific, technical, legal, and ethical domains, and they must be addressed before KCPs can be routinely used in forensic practice or large-scale biometric systems.

Absence of Large-Scale Databases

Unlike fingerprints or DNA, which are supported by global databases encompassing millions of profiles, KCPs lack comprehensive repositories. Most published studies rely on relatively small sample sizes, often a few hundred individuals. This limitation makes it difficult to understand the full spectrum of crease variability across populations and to establish robust statistical models of uniqueness. Without large-scale databases, it is challenging to quantify error rates or to provide probability-based evidence in court.

Variability in Image Sources and Quality

Forensic cases rarely provide high-quality images. Investigators often rely on CCTV footage, social media posts, or seized digital media, where dorsal hand images are captured unintentionally. These images may suffer from low resolution, poor lighting, compression artefacts, motion blur, or partial visibility. While crease patterns are permanent features, their appearance is sensitive to pose, camera angle, and skin tension. Such variability complicates both manual and automated analysis, reducing the reliability of comparisons in real-world scenarios.

Observer Subjectivity in Manual Classification

Manual classification systems remain central to KCP analysis. While certain categories achieve relatively high agreement among observers, others, particularly Horizontal and sei lunar, and Mixed and Cross, are prone to misinterpretation. Disagreements arise from subtle variations in crease branching, orientation, or overlap. Without standardized decision rules and examiner proficiency testing, subjective interpretation can undermine reproducibility. This subjectivity also poses a risk in courtroom contexts, where defence attorneys may challenge the scientific reliability of the examiner's conclusions.

Limited Validation and Standardization

The forensic sciences are under increasing pressure to demonstrate validation, reproducibility, and known error rates. Fingerprints and DNA enjoy global acceptance largely because of decades of standardization and empirical validation. In contrast, KCPs lack universally accepted Standard Operating Procedures (SOPs) for image capture, annotation, classification, and reporting. Laboratory methods vary between research groups, and no international body has yet issued standardized guidelines. This lack of consistency prevents uniform practice across jurisdictions and weakens the scientific credibility of KCP analysis.

Challenges in Courtroom Admissibility

Legal frameworks, such as the Daubert standard in the United States and Section 45 of the Indian Evidence Act, require that expert testimony be founded on reliable principles, validated methods, and known error rates. Early cases, including Hardy

(2007), demonstrated that dorsal hand features and knuckle crease patterns could be admitted successfully in court. However, the admissibility of such evidence is contingent upon the robustness of the methodology. In the absence of large-scale validation studies, population statistics, and proficiency testing, knuckle crease pattern evidence remains vulnerable to legal challenge. Consequently, courts are more likely to treat such evidence as corroborative rather than as primary identification evidence.

Ethical and Privacy Concerns

The accessibility of KCPs introduces ethical dilemmas. Unlike fingerprints, which must be collected through deliberate procedures, KCPs can be captured passively from ordinary photographs or videos. This raises concerns about consent and privacy, particularly if KCP analysis is used in mass surveillance or social media monitoring. Public acceptance of new biometric systems depends heavily on addressing such ethical questions, establishing clear boundaries, and ensuring accountability in the use of dorsal hand imagery.

Integration with Existing Biometric Systems

Another challenge is institutional. Forensic laboratories and biometric systems are heavily invested in established modalities such as fingerprints, iris, and face. Introducing KCPs requires new infrastructure, examiner training, and possibly software development. Unless KCPs can demonstrate a clear advantage, such as superior reliability in populations where fingerprints fail, agencies may be reluctant to invest resources in their adoption. The inertia of existing systems poses a practical barrier, even if scientific progress continues.

Technical Limitations in Automation

Automated recognition of KCPs is still in its infancy. Algorithms must deal with pose variation, inconsistent illumination, occlusion by accessories, and differences in skin tone. While machine learning methods show promise, they require large annotated datasets for training, which are not yet available. Additionally, forensic contexts demand explainability; algorithms must not only match crease patterns but also provide interpretable outputs that can be explained in court. Meeting this demand remains an open challenge.

Resistance to Over-Reliance

Finally, there is the risk of over-reliance. Just as fingerprints were once considered infallible before error cases emerged, there is a danger in treating KCPs as standalone evidence without sufficient safeguards. Overstating their accuracy or uniqueness without supporting data could damage the credibility of the discipline. The forensic community must balance enthusiasm for innovation with caution, ensuring that KCPs are used responsibly and in conjunction with other identifiers.

FUTURE DIRECTIONS AND RECOMMENDATIONS

The challenges identified in the forensic and biometric applications of knuckle crease patterns (KCPs) highlight the need for sustained research, standardization, and policy development. To progress from promising casework demonstrations to routine forensic practice, a structured roadmap must address scientific validation, technical development, ethical safeguards, and institutional adoption. The following sections outline the key directions for advancing the field.

Building Large-Sale Databases

The lack of large and diverse databases remains a major obstacle to the wider use of KCPs. Future work should prioritize creating collections that represent different ages, sexes, ethnic groups, and occupational backgrounds so that natural variation can be properly studied. These repositories should also record additional dorsal features such as veins, scars, and pigmentation to support multimodal analysis. Including both high-quality laboratory images and low-quality opportunistic images is important to reflect real forensic conditions. Although open public access may not be feasible, controlled-access databases for researchers and authorized agencies would support reproducibility and algorithm development. Building such resources is essential for generating data on distinctiveness, error rates, and population trends, which are crucial for legal admissibility.

Standardizing Capture and Classification Protocols

Developing standardized operating procedures for image capture and crease classification is another critical priority. These SOPs should outline consistent protocols for camera distance, lighting conditions, and the use of scaling markers in both laboratory and field settings. Clear annotation guidelines are needed to define crease types, branching points, and overlapping creases, ensuring uniform classification across examiners.

Explicit decision rules should distinguish similar categories, such as Horizontal and Semilunar creases, to reduce subjective interpretation. Proficiency testing, structured training, and examiner certification programs will further enhance consistency. Achieving international consensus through recognized forensic and biometric standard setting bodies would significantly strengthen credibility and reproducibility across jurisdictions.

Advancing Automated Recognition Systems

Automated recognition can reduce subjectivity and make KCP analysis feasible for large datasets. Future research should prioritize developing algorithms that remain reliable despite oblique angles, uneven lighting, or partial occlusion. Improving explainability is equally important, with models producing interpretable outputs such as saliency maps or marked key points that examiners can justify in court. Work on cross domain learning will help systems trained on both high and low quality images, perform better in real forensic settings. Integrating KCPs with other biometrics, including fingerprints, veins, or facial features, will further strengthen multimodal systems. Ultimately, advances in explainable AI are essential, as opaque models are unlikely to meet forensic admissibility standards.

Enhancing Courtroom Admissibility

For KCPs to achieve legal recognition, the field will need a strong body of evidence demonstrating reliability and clearly defined error rates. This requires large, multi-center validation studies capable of evaluating accuracy, reproducibility, and measurable error margins across laboratories. Alongside this, the discipline must establish standardized ways of reporting error rates so that these figures can be incorporated into expert testimony in a manner comparable to established fingerprint evidence. As cases begin to incorporate KCP findings, it will be important to document and publish judicial decisions to gradually build a recognizable body of case law. Equally essential is the training of examiners in legal principles so that they are prepared to explain their methodology confidently, particularly under cross-examination. By meeting these admissibility expectations in a systematic way, KCP research can strengthen its credibility and acceptance within both national and international legal systems.

Addressing Ethical and Privacy Concerns

Because dorsal hand images can be captured in routine environments, the use of KCPs introduces unique ethical considerations. Future frameworks should prioritize explicit and informed consent for civil or research datasets, ensuring transparency in how crease images are collected and used. Regulations must also address the application of KCP analysis in surveillance contexts, establishing limits that prevent overreach and ensure accountability.

Strong data protection measures, including secure storage and controlled access, are essential to reduce misuse. Public awareness and clear communication about data handling will be central to maintaining trust. Ethical use must remain a core principle as the technology expands into forensic and security applications

Integrating Knuckle Crease Patterns into Multimodal Systems

Rather than functioning as a standalone biometric, KCPs are more effective when integrated into multimodal systems. In security settings, crease based recognition can complement facial or fingerprint verification for banking, airport screening, and access control. Within forensic investigations, KCPs can serve as corroborative evidence alongside fingerprintsand other established methods. They may also support disaster victim identification by contributing an additional comparison layer when used with dental records, genetic profiles, and ante-mortem photographs. Presenting KCPs as a complementary modality emphasizes their practical value without overstating their independence.

Expanding Forensic Training and Awareness

For the effective adoption of KCPs in forensic practice, examiners must be trained in both manual and automated approaches. This includes structured workshops and certification programs covering classification, reliability assessment, and reporting standards, as well as interdisciplinary training that brings together forensic scientists, biometric engineers, and legal experts. Regular blind proficiency testing is also essential to monitor examiner accuracy and strengthen confidence in the technique. Equally important is broader awareness among law enforcement and the judiciary, ensuring that investigators, prosecutors, and judges understand the capabilities and limitations of KCPs so the evidence is applied appropriately.

Encouraging International Collaboration

The global nature of forensic challenges, particularly online crime and disaster victim identification, demands strong international collaboration. Shared transnational databases of crease images, joint multicenter validation studies to assess inter-laboratory reliability, and harmonized policies on admissibility and ethical safeguards are essential steps. Such coordinated efforts will accelerate scientific progress, improve methodological consistency, and enhance the global credibility and acceptance of KCP evidence.

Roadmap for the Next Decade

In summary, future progress in KCP research requires a coordinated agenda. Key priorities include building demographically diverse databases, establishing standardized SOPs, and developing transparent, explainable automated systems supported by strong validation and clear error rates. Ethical and privacy safeguards must guide data use, while multimodal integration will enhance practical applicability. Expanding training for forensic and legal professionals and strengthening international collaboration for resource sharing and methodological alignment will further support the acceptance of KCPs as a reliable biometric identifier.

CONCLUSION

Knuckle crease patterns have emerged as a promising yet underexplored biometric trait with distinct value for both forensic investigation and personal identification. This chapter traced their development from anatomical foundations and early manual classification systems to recent automated approaches and real-world applications. The evidence demonstrates that KCPs are permanent, universal, and resilient identifiers that can provide investigative leads where conventional biometrics such as fingerprints, DNA, or facial recognition are unavailable or compromised.

The review of forensic applications illustrated that KCPs can play a vital role in diverse contexts. In criminal investigations, particularly cases involving child sexual abuse material, they have already contributed to convictions by linking suspects to incriminating imagery. In surveillance and intelligence operations, their visibility in everyday imagery provides investigators with an additional tool. In civil and security domains, their resistance to occupational wear, permanence across age, and contactless capture make them attractive as an alternative biometric, particularly for populations where fingerprint quality is poor.

At the same time, the chapter highlighted the current limitations of KCP research and practice. The absence of large-scale databases, variability in image quality, observer subjectivity, lack of standardized protocols, and insufficient validation studies restrict their reliability and admissibility in court. Ethical and privacy concerns also arise from the fact that dorsal hand features are frequently visible in ordinary photographs and may be captured without consent. These challenges explain why KCPs, despite their potential, remain largely supplementary rather than primary identifiers in forensic and biometric systems.

The case studies discussed, ranging from the Hardy and Strachan convictions to the Whitfield trial, underscore both the promise and the caution required in this domain. While early successes demonstrated that dorsal hand features can withstand judicial scrutiny, subsequent cases revealed the risks of presenting such evidence without robust scientific grounding. These experiences have shaped the trajectory of research, encouraging systematic classification, improved reliability testing, and the pursuit of automated solutions capable of handling real-world variability.

Looking ahead, the future of KCPs depends on coordinated global efforts to overcome current barriers. Large-scale, diverse databases must be developed to establish statistical distinctiveness and error rates. Standardized capture, annotation, and reporting protocols are needed to reduce subjectivity and enable reproducibility. Automated recognition systems should prioritize explainability to ensure forensic defensibility. Ethical frameworks and privacy safeguards must be integral to any implementation, ensuring public trust and preventing misuse. Finally, integration with multimodal systems will likely be the most effective path, positioning KCPs as a complementary biometric that strengthens the overall reliability of forensic and security applications.

In conclusion, knuckle crease patterns represent a valuable but still maturing field within forensic science and biometrics. They are unlikely to replace fingerprints or DNA as primary identifiers in the near future, but their role as a supplementary, corroborative, and contactless biometric is already evident. With sustained research, international collaboration, and clear ethical oversight, KCPs have the potential to evolve into a widely recognized and reliable biometric identifier. Their uniqueness lies in bridging the gap between anatomical permanence and practical visibility, making them especially suited to the digital age, where identity must often be established from images rather than physical traces.

REFERENCES

Badrinath, G. S., Nigam, A., & Gupta, P. (2011). An efficient finger-knuckle-print based recognition system fusing SIFT and SURF matching scores. In Vol. 7043, pp. 374–387). Lecture Notes in Computer Science. Springer., DOI: 10.1007/978-3-642-25243-3_30

Baines, K. N. (2011). *Dorsal knuckle crease pattern: Classification and inter-observer consistency* (Master's thesis, University of Dundee). University of Dundee Research Portal. https://discovery.dundee.ac.uk/files/90057557/Nicoll_Baines_msc_2011.pdf

Bharadwaj, S., Kirchbacher, M., Ras, Z. W., & Harrison, M. (2014). *Biometric quality: A review of fingerprint, iris, and face. EURASIP Journal on Image and Video Processing, 2014*(34). DOI: 10.1186/1687-5281-2014-34

Biswas, P. S., & Mishra, S. D. (2024). Novel approach for personal identification using dorsal knuckle crease patterns: A pilot study. *Mapana Journal of Sciences, 23*(4), 83–97. DOI: 10.12723/mjs.71.5

Biswas, S., & Mishra, A. (2024). Emerging biometric modalities for disaster victim identification. *Journal of Forensic Sciences*, *69*(2), 311–320.

Black, S.. (2009). Forensic analysis of the hand in child abuse investigations. *Journal of Forensic Sciences*, *54*(1), 20–25. PMID: 19018931

Carrell, S. (2009, May 7). Scottish paedophile ring found guilty of more than 50 charges. *The Guardian*. https://www.theguardian.com/uk/2009/may/07/scottish-paedophile-ring-guilty-child-abuse

Chattopadhyay, S., & Sukul, B. (2012). Identification from dorsal finger pattern: A new approach. *Medicine, Science, and the Law*, *52*(1), 17–21. DOI: 10.1258/msl.2011.010145 PMID: 22399024

Cheng, K. H. M., Xu, C., & Zhao, G. (2023). Advancing 3D finger knuckle recognition via deep feature learning. *arXiv*. https://arxiv.org/abs/2301.02934

Dakshinamurthy, S. (2018). *The classification of knuckle creases and their prevalence within a population: Inter-observer and intra-observer validation* (Unpublished master's thesis). School of Science and Engineering, University of Dundee.

Dan, L., Hackman, L., Rinaldi, V., & Trucco, E. (2023). The persistence of knuckle creases during finger flexion for the identification of perpetrators from digital images of their hands. *Forensic Science International*, *348*, 111603. DOI: 10.1016/j.forsciint.2023.111603 PMID: 36842847

Diviya, P., Logapriya, K., Febiyana, G. N., Sivashankari, M., & Kumar, R. D. (2015). Identification of suspects using finger knuckle patterns in biometric fusions. *International Journal of Research in Computer Science*, *2*(2), 5–9. https://researchscript.in/wp-content/uploads/2015/03/IJRCS020202.pdf

Gao, C., Yang, Z., Teoh, A. B. J., & Zhu, M. (2024). Beyond first-order: A multi-scale approach to finger knuckle print biometrics. *arXiv preprint arXiv:2406.19672*. https://doi.org//arXiv.2406.19672DOI: 10.48550

Jain, A. K., Nandakumar, K., & Ross, A. (2016). 50 years of biometric research: Accomplishments, challenges, and opportunities. *Pattern Recognition Letters*, *79*, 80–105. DOI: 10.1016/j.patrec.2015.12.013

Kadane, J. B. (2020). Knuckles and fingerprints: A comparison and case study. *Journal of Forensic Identification*, *70*(3), 265–272. https://dr.lib.iastate.edu/bitstreams/5c66d011-8161-4b8b-bc68-b9b27cc4bb55/download

Kimura, S., Schaumann, B. A., Plato, C. C., & Kitagawa, T. (1990). Embryological development and prevalence of digital flexion creases. *The Anatomical Record*, *226*(2), 249–257. DOI: 10.1002/ar.1092260214 PMID: 2301741

Kulkarni, S. S., & Rout, R. D. (2012). Secure biometrics: Finger knuckle print. *International Journal of Advanced Research in Computer and Communication Engineering*, *1*(10), 852–854. https://www.ijarcce.com/upload/december/21-Secure%20Biometrics.pdf

Kumar, A. (2014). Importance of being unique from finger dorsal patterns: Exploring minor finger knuckle patterns in verifying human identities. *IEEE Transactions on Information Forensics and Security*, *9*(8), 1288–1298. DOI: 10.1109/TIFS.2014.2328869

Kumar, A., & Ravikanth, C. (2009). Personal authentication using finger knuckle surface. *IEEE Transactions on Information Forensics and Security*, *4*(1), 98–110. DOI: 10.1109/TIFS.2008.2011089

Kumar, A., & Zhou, Y. (2009). Human identification using knuckle codes. In *Proceedings of the IEEE Biometrics Symposium* (pp. 1–6). IEEE. DOI: 10.1109/BTAS.2009.5339021

Lee, C. K., Chang, C. C., Johar, A., & Wong, P. F. (2013). Fingerprint changes and verification failure among patients with hand dermatitis. *JAMA Dermatology*, *149*(3), 294–299. DOI: 10.1001/jamadermatol.2013.1425 PMID: 23682364

Moore, K. L., Dalley, A. F., & Agur, A. M. R. (2018). *Clinically oriented anatomy* (8th ed.). Lippincott Williams & Wilkins.

Nanni, L., & Lumini, A. (2009). A multi-matcher system based on knuckle-based features. *Neural Computing & Applications*, *18*(1), 87–91. DOI: 10.1007/s00521-007-0160-4

Standring, S. (Ed.). (2016). *Gray's anatomy: The anatomical basis of clinical practice* (41st ed.). Elsevier Limited.

Stevens, C. A., Carey, J. C., Shah, M., & Bagley, G. P. (1988). Development of human palmar and digital flexion creases. *The Journal of Pediatrics*, *113*(1 Pt 1), 128–132. DOI: 10.1016/S0022-3476(88)80546-8 PMID: 3385521

Trabelsi, S., Samai, D., Meraoumia, A., & Bensid, K. (2020). Finger-knuckle-print recognition using deep convolutional neural network. In *Proceedings of the 1st International Conference on Communications, Control Systems and Signal Processing (CCSSP)* (pp. 1–6). IEEE. DOI: 10.1109/CCSSP49278.2020.9151531

Vidhyapriya, R., & Rose, L. (2019). Personal authentication mechanism based on finger knuckle print using Gabor filter and expectation-maximization algorithm. *Journal of Medical Systems*, *43*(8), 232. DOI: 10.1007/s10916-019-1332-3 PMID: 31203468

Vyas, R., Williams, B. M., Rahmani, H., Boswell-Challand, R., Jiang, Z., Angelov, P., & Black, S. (2022). Ensemble-based bounding box regression for enhanced knuckle localization. *Sensors (Basel)*, *22*(4), 1569. DOI: 10.3390/s22041569 PMID: 35214467

Woodard, D. L., & Flynn, P. J. (2005). Finger surface as a biometric identifier. *Computer Vision and Image Understanding*, *100*(3), 357–384. DOI: 10.1016/j.cviu.2005.06.003

Zhou, Z., Dong, C., & Kumar, A. (2025). Towards explainable and unprecedented accuracy in matching challenging finger crease patterns. In *Proceedings of the IEEE/CVF Conference on Computer Vision and Pattern Recognition (CVPR)* (pp. 6212–6221). IEEE. DOI: 10.1109/CVPR52734.2025.00582

Chapter 6
Voice/Speaker Recognition in Criminal Justice

Tamara Phillips Fudge
https://orcid.org/0000-0002-8682-9711
Purdue University Global, USA

ABSTRACT

Recordings of the human voice are found in many contexts today, such as musical recordings, social media videos, television shows, and movies. Voice recordings used for proving identity are most often stored in databases as voiceprints, and can be the cause of some concern, however. They are often employed for accessing personal accounts but can also potentially be used in illegal or unethical ways. It is helpful to understand some of the basics of the technology and the human voice, and to explore laws from different parts of the world in regard to this biometric. Some historical and recent cases assist in seeing the progression of this technology. Cloning (deepfakes), masking, financial fraud, loss of jobs or reputation, and a concern for public figures' safety, along with the fact that the purpose of collecting some voice data is unknown is included, with surveillance and human rights issues surveyed. Recommendations include thoughtful application in the courtroom, focused legislation, continued technical research, and education for best protection.

INTRODUCTION

Biometrics are considered personal identifying information (PII) and so there are some limits as to how and what is allowed in forensic science. Fingerprints left at crime scenes can prove a particular person was present and touched weapons

DOI: 10.4018/979-8-3373-4972-5.ch006

and other items, and facial and gait recognition on captured video can be used in public pleas for potential perpetrator identification. These physical and visual components may be favored by some prosecutors, but there is also the possibility of auditory evidence: identifying a subject's voice via recording. The technology itself, its applications, relevant laws, noted issues, and the benefits and concerns are worthy of exploration.

LITERATURE REVIEW: THE TECHNOLOGY

Biometrics are human characteristics that might be used to identify individuals. Physical biometrics such as facial recognition and fingerprints are constant throughout a person's life, while behavioral biometrics like gait analysis and a person's speech might change over time; this fact can affect trust in their use within the court systems (Krishna Prakasha & Sumalatha, 2025). However, technologies have improved in recent years, and voice recognition can often be used in the criminal justice system to attempt to identify perpetrators.

Jansen et al. (2021) claims the 1950s saw the beginning of voice technology, and research tended to focus on how it could be used to bolster forensic investigations. Other sources attribute the beginnings to Swedish professor Gunnar Fant, who in 1960 analyzed x-rays of people uttering different sounds (Federal Bureau of Investigation, 2013). Voiceprints were met with much skepticism, citing slight differences in the speaker's linguistic abilities and dialect; a university report stated that spectography, like polygraph testing, could not be trusted since "stable guidelines for interpreting its graphic records appear permanently beyond reach" (Vanderslice, 1969, p. 386). Indeed, technology as it is employed today was still in its infancy. In the 1970s, speech technologies worked to match sounds via analog filtering (Federal Bureau of Investigation, 2013).

The field can be divided into two basic technologies: assistive and identification-based. Speech Recognition is what can be used to *dictate* to computer programs instead of relying on keyboarding skills; this can make input faster and assist those who have disabilities (Bodepudi et al., 2019). It can be text-dependent, where the technology learns from the person's input, or text-independent, which does not rely on any individual (Federal Bureau of Investigation, 2013). On the other hand, Voice or Speaker Recognition – sometimes called Biometric identity systems – is different, as its purpose is to *identify* the speaker instead (Jansen et al., 2021). There are several subsets of these technologies, including Far-Field Speech Recognition wherein voices can be captured from a distance and background noise is reduced ("12 types," 2024).

Voice/Speaker Recognition cultivates what is called a voiceprint, which can be categorized in terms of language, accent, channel, and text-independence ("What is voice biometrics," n.d.). Tests at Michigan State University back in 1971 were underway to ensure accuracy. By that time, there had already been nearly a dozen cases using voiceprints in court, covering military offenses, bomb threats, perjury, and other instances in the United States, England, and Switzerland ("Voiceprint," 1971).

The three main features of voiceprints are frequency, amplitude, and time ("Voiceprint," 1971). Frequency is the measurement of a sound wave's vibrations per second; lower frequencies result in lower pitches (Gully et al., 2022). People with shorter vocal folds will have a higher pitched range; some facets of the voice are purely biological, however, the way in which the voice is used is also a factor. This can include dialectical differences, pacing, and even a person's emotional state. Combined with physical actions for articulation, these make each voice unique (Gully et al., 2022).

A few of the technicalities are appropriate to describe. First, a microphone collects sounds waves as analog signals, and then with the assistance of a converter, transforms it into digital data (Bäckström et al., 2022). The differences between these sound waves is shown below; natural speech is analog, with smooth, continuous transitions, and digital sound is made up of binary values called bits that Butler (2025) describes as "slices of the sound ... like the individual frames of a movie" (para. 6).

Figure 1. Simple representations of analog and digital sound waves

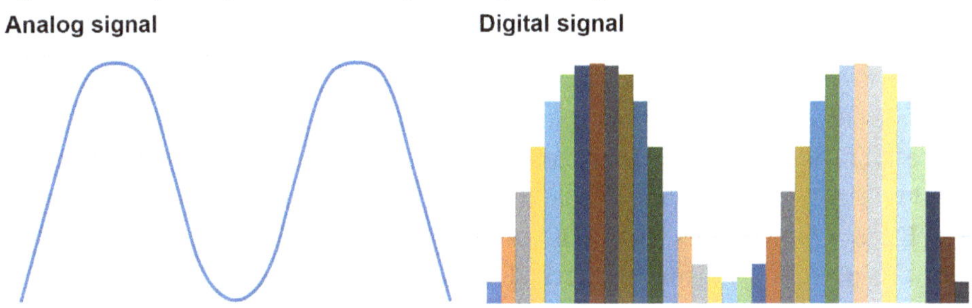

This in turn is divided into pieces called "windows" and consecutively-arranged windows produce a timeline called a "movie." Spectral bands indicate energy and are measured in Hertz, which can be translated into decibel levels (Bäckström et al., 2022).

The audio is divided into fifteen frequency bands, each within 0-5000 MHz, and each band's amplitude is then "sampled and digitized at 2-millisecond intervals, then averaged to produce a digital value representing the speech pattern at 100-millisecond intervals" ("Voiceprint," 1971).

The audio can be shown formally as a waveform or a spectrogram. In spectrograms, the brighter/lighter hues indicate wave amplitude and the vertical aspect represents frequency. Duration (time) is shown horizontally in both waveforms and spectrograms ("What is voice biometrics," n.d.). Figure 1 shows these two formats for the same audio recording:

Figure 2. Wave form (on top) and spectrogram (on bottom) of a woman's voice asking "Can voiceprints be reliable forensic evidence?"

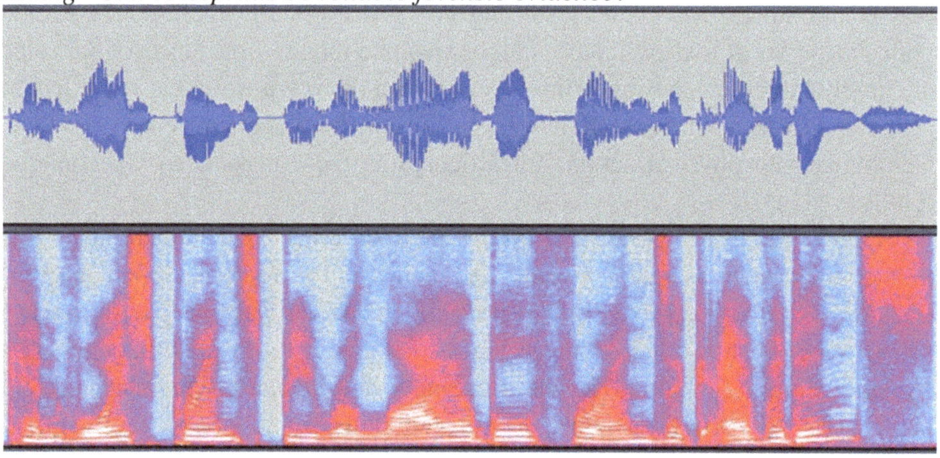

(Created in Audacity)

The technology is complicated, but a few of the main concepts are worthy of mention to demonstrate the intricacies. Bäckström et al. (2022) state that short-time signals are analyzed with Fourier transform, which van Drongelen (2007) explains is where "transforms data from the time (or spatial) domain into the frequency domain" (p. 91). Windowing allows the separation of the data to avoid poor transitions and establishes a "compromise between frequency selectivity and amplitude accuracy" ("Fast Fourier, Transformation," n.d., Time window section).

Mel-Frequency Cepstral Coefficients (MFCCs) are the most commonly used audio analysis method (Bäckström et al., 2022; Kyi & Lin, 2019). MFCC data is captured from the signal's spectrum version (Abdul & Al-Talabani, 2022). It is calculated using a complicated mathematical and trigonometry process based on a range of pitches, harmonic frequencies, vocal tone, time intervals, and other factors, including extraneous noise reduction (Hansen et al., 2005). It is an accepted technique for dealing with audio, being able to map vocal tract features, understand isolated words common in spontaneous speech, pick up on the speaker's emotional state, and identify subtle dialectical differences (Abdul & Al-Talabani, 2022). Once extracted, these features are saved as templates for using in comparisons (Abdulhameed et al., 2025); this in turn can "reconstruct the voice accurately" (Qian et al., 2018, p. 93). MFCC can also be used to remove extraneous sounds including other human voices in the background and white noise (Kyi & Lin, 2019). The depth of MFCC can be seen in the experiments conducted by Khare & Srivastava (2023), where 58 such traits were employed; it must be said the amount of data is thus extensive. Other mathematical methods can also be applied such as Short Time Energy (STE) to ensure clarity, Perceptual Linear Prediction (PLP), and others (Kyi & Lin, 2019).

Background noises in recordings can complicate analysis, creating the possibility for both false positives and false negatives in finding matches (Joseph, 2017). Qian et al. (2018) performed a study that tested speaker recognition in various locations and movement, and found the accuracy degraded but not significantly so except when the person was running. The amount of words spoken are significant, as text-dependent systems may require more input for accuracy (Alver, 2007). Wang et al. (2024) agree that recordings made inside and outdoors, as well as the number of speakers, make a difference with automatic speech recognition (ASR), and frameworks are being proposed and developed to assist with analysis.

Voice recordings can be used to detect "soft biometrics" such as the speaker's approximate age, gender, language, and regional accent in the quest to identify suspected criminals (Jansen et al, 2021). Where a person grew up, their height, and even race and nationality may also be identifiable through the voice, according to Phonexia ("What is forensic voice comparison," n.d.).

Some uses outside of law enforcement can include voice authentication over the phone instead of passwords for banking, security in ordering goods and services, and remote employee verification ("What is voice biometrics," n.d.). It has been long understood people may not always speak words the same way, so the idea has been to seek the closest possible match rather than a perfect match ("Voiceprint," 1971). Jansen et al. (2021) note the person's emotional state, the time of day, and external issues such as smoking can also make a difference. Overall, however, voiceprints are considered scalable and cost-effective (Abdulhameed et al., 2025).

When a voiceprint request is made, the user/customer "enrolls" by speaking something the company has specified, such as an account number or the person's name. This is converted into a computer algorithm and stored in a database, from where it will be pulled for matching whenever the user returns to the authentication system (Alver, 2007). The audio stream itself is made of dilated convolutional layers (Altyar et al., 2022), which essentially means the input is filtered into layers; this aids in scalability and analysis, but can increase computational costs. Extraction can include steps such as transforming audio files into float values and then putting them back together again, and the program must then be trained to understand the algorithms (Altyar et al., 2022).

CONSIDERATIONS REGARDING THE HUMAN VOICE

The sounds in human language are divided into "phonemes," the smallest unit of speech ("How can speech recognition," 2023). International Phonetic Alphabet uses phonemic symbols to dissect pronunciations, with one symbol for each sound: [lus] is "loose," [slip] is "sleep," and [ju] is "you." There are additional symbols for sounds not clearly identified in Latin or Roman alphabets, such as for the Umlaut vowels and the fricative "ch" consonants in German and the nasal vowels in French. There are language and dialectical differences in abundance; as just one example, the German language uses the long "o" without the diphthong heard in English. It must also be noted that singers, actors, linguists, and speech pathologists may be more in tune with these subtleties than the general populace (Fudge, n.d.).

Not unsurprisingly, much of a person's sound output depends on the physical shape of each person's vocal mechanisms, with use of the breath through the larynx and then shaped further by the tongue, teeth, jaw, soft palate, gums, lips, and resonating centers in the head – with the length of the vocal cords as another mitigating factor (Federal Bureau of Investigation, 2013; "What is voice biometrics," n.d.).

Of some concern since the beginning of voice printing is the vocal tract might change with age and/or not be as unique as fingerprints, which are consistent throughout a person's lifetime (Vanderslice, 1969). Females in their late teens tend to use a breathiness and may have unstable pitch (Baker et al., 2024); this "voice change" occurs later than boys' changes, which are far less subtle, and the human voice – depending on the source – may not fully mature until the person is in their late 20s. As the voice ages, production and articulation can also be expected to change due to "physiological changes in the larynx, vocal tract, and the respiratory system" (Tremblay et al., 2023, p. 4332).

Permanent voice changes might also occur due to accident (Accurso, 2023) or medical issues such as chronic sinus infections and acid reflux (Tafiadis et al., 2017). Learning to sing formally may also alter a voice's trajectory over the years. Tremblay et al. (2023) claim "singers can operate on a wider range of vocal intensities and frequencies" and may have better frequency stability as the voice ages (p. 4333). It is interesting to note experienced singers and those who have no training both tend not to change jaw-position patterns when speaking but trained male singers might speak with a wider intonation range (Tremblay et al., 2023).

Lifestyle choices may be more clearly connected to changes in the voice. Smoking, for example, typically lowers the person's range (pitch) and creates a raspiness due to vocal fold damage. Smoke-induced inflammation in the larynx can affect the voice's jitter (frequencies) and shimmer (amplitude), according to Tafiadis et al. (2017). Overuse or misuse of the voice causes strain; Jander et al. (2025) rationalize that professions in which speaking is a major factor, such as teaching, may result in more vocal distress, absences, and health care costs. Alcohol use and stress may also generate mostly temporary changes ("What is forensic voice comparison," n.d.).

Lungova and Thibeault (2020) indicate vocal problems can be due to physical abnormalities with the vocal folds present from birth. Congenital imperfections are lifelong challenges, however, and would not affect matches between voiceprints made by the same person within a reasonable amount of time. Most of the changes mentioned above happen over time, which is a major point: when comparing two voiceprints, the longer timespans between recordings may render matches less reliable.

PRE-TECHNOLOGY VOICE IDENTIFICATION CASES

Before today's technology, the courts were already using "earwitness" accounts to support other evidence and testimony. Two famous and troubling examples illustrate problems of relying on what someone may have heard: the Lindbergh baby kidnapping and a Canadian murder.

In the U.S., the kidnapping of aviator Charles Lindbergh's baby in 1932 hinged greatly on eight handwriting experts and examination of the wooden ladder, but also included the voice of the supposed kidnapper when meeting a go-between named Dr. John Condon. Lindbergh's recollection of a heavy-accented German voice heard in a cemetery was also weighed; while there have been many conspiracy theories since, Bruno Richard Hauptmann was executed for the crime (Durina, 2021).

In Canada, the 1984 rape and murder of 9-year-old Christine Jessop led to the arrest of a neighbor, Guy Paul Morin, suspected partially because of a "weird" personality (Harland-Logan, 2023). He was acquitted but authorities declared the trial needed to be redone due to some mistakes. As in the first trial, similar hairs and fibers were key forensic elements. Morin's voice became another piece of the prosecution's puzzle, as at the child's funeral, her mother claimed to hear Morin's voice screaming. He was declared guilty in the second trial and sent to prison. In 1995, it was determined the hair and fiber evidence may have been contaminated, both hearsay and earwitness accounts were unreliable, and no DNA evidence linked him to the situation, and so he was set free. DNA research finally identified a family acquaintance named Calvin Hoover as the murderer in 2020, five years after Hoover had committed suicide (Harland-Logan, 2023).

TODAY'S LAWS REGARDING BIOMETRICS

Fingerprints, facial recognition, and DNA provide much of the data sought for criminal justice databases (Jansen et al., 2021). Voice matching can be useful as an additional biometric identifier, especially when there are poor visual environments or even none at all, such as when a crime is committed over the phone or internet (Kramer et al., 2021). In an experiment conducted by Kramer et al. (2021), it was surmised that longer time intervals may mean larger differences, but the finding was that there was no correlation between time and consistency.

European Union

The 27-country membership of the European Union has been working to protect citizens' data since 1995 when the Data Protection Directive was passed (Rubin, 2025). Today, the General Data Protection Regulation (GDPR) provides strict control being emulated worldwide. In effect since May 25, 2018, the law includes in Article 4(1) that data with "one or more factors specific to the physical, physiological, genetic, mental, economic, cultural or social identity of that natural person" is protected, whether it is direct or indirect identification, while only photos of a person's face and fingerprints are specifically mentioned in Article 4(14) as examples. The Principles declared in Article 5 explain personal data must be transparently and lawfully processed, the purpose must be specified, collection must be restricted to what is truly needed, accuracy is crucial, retention is limited, and the data must be secured. The law requires consent and the individual may demand removal of their biometrics from databases. Violations may result in sanctions of "tens of millions of Euros" (Rubin, 2025, p. 265).

The rules set forth in the GDPR are far-reaching, as it is effective for residents of the European Union as well as companies outside the EU who process biometric data of EU citizens (Claypoole, 2021). It also is invoked by the European Economic Area consisting of Iceland, Liechtenstein, and Norway (Rubin, 2025).

The United States of America

There are few federal laws in the U.S. affecting biometrics, but the Children's Online Privacy Protection Act (COPPA) can be applied to voiceprint circumstances. The Federal Trade Commission (2013) declared that "greater protections" for children's biometrics needed to be in place; it relates to the *potential* for individual identification, whether or not the child is identified; parental consent is required even if the child is the person uploading a voice recording, image, or video. This law is intended to protect children under the age of 13 (Federal Trade Commission, 2013). While not really necessary, Oregon's Consumer Privacy Act (OCPA) includes reminders of the need to follow COPPA mandates (Oregon Department of Justice, 2024).

Deskus and Fattal (2019) suggest the Fourth Amendment may apply to the use of voiceprints. Others such as the Purdue Global Law School say biometric data used as "searches" would be governed by the Fourth Amendment, whereas something like a fingerprint needed to unlock a device might be covered under the Fifth Amendment instead, as it could lead to self-incrimination; courts are still grappling with where biometrics truly fit ("Does the Fifth Amendment," 2022). While federal governance may be uncertain in some respects and lacking in others, there is a growing number of U.S. states with laws directly addressing biometric collection, use, and storage.

Illinois

Biometric Information Privacy Act (740 ILCS 14): BIPA is strongly focused on collection and use of biometric data such as facial scans, fingerprints, and voiceprints. Requirements include written notification about what is collected and/or stored and for what purpose, plus written consent is required (ACLU Illinois, 2025). The Act was set in place on October 3, 2008 and declares biometric information is unique; unlike Social Security Numbers, it is intrinsic to the individual and cannot be changed; Section 5(f) also states, "The full ramifications of biometric technology are not fully known." Voiceprints are specifically included in Section 10's definition of "biometric identifiers," although the definitions curiously exclude handwriting, photographs, and data collected in healthcare situations (Illinois General Assembly, 2008). Claypoole (2021) explains this law is triggered when a voice recording is analyzed with the purpose of identifying the individual; simply recording a voice is not necessarily a violation but there may be questions as to what constitutes "analysis."

The law originally allowed for each separate violation to incur additional monetary penalties, but in mid-2024, it was amended to limit a company's liability to one penalty for multiple instances of the same biometric and same person (Doran et al., 2024).

California

California Consumer Privacy Act (CCPA) is meant for California residents and includes rights to know, delete, opt-out, correct, and limit the use of personal data, which includes biometrics. Businesses cannot sell data about children under the age of 13 without parental opt-in; the child can do the opt-in if age 13-15 (State of California, 2024). The state's Department of Justice proclaims, "Businesses cannot deny goods or services, charge you a different price, or provide a different level or quality of goods or services just because you exercised your rights under the CCPA" (State of California, 2024, Right to Non-Discrimination section).

Colorado

Privacy of Biometric Identifiers & Data amendment to the Colorado Privacy Act was effective as of July 1, 2025, and identifies the need to establish timelines for retention of biometric data as well as incident response, and obtain consent. It puts limitations on companies in obtaining biometric data (Colorado General Assembly, 2024). Martinez (2025) specifies that this law affects many stakeholders, including "employees, job applicants, contractors, interns, and fellows" (para. 6), and reiterates the law applies to instances wherein individuals are intended to be identified, rather than the collection of metadata.

Oregon

Consumer Privacy Act (OCPA) was signed in 2023 and enacted mid-2024, provides control for consumers such as the fundamental rights to know, correct, and/or delete data collected about them, "opt-in" consent requirements, and a reminder that minors are to be protected under COPPA (Oregon Department of Justice, 2024).

The state recommends people protect themselves by following the acrostic LOCKED: List where they allowed data to be collected, Opt out when possible, obtain Copies of what has been collected, Know what kind of data is being collected, Edit and Delete incorrect or unwanted data (Oregon Department of Justice, n.d.).

Texas

Capture Or Use Of Biometric Identifier Act (Tex. Bus. & Com. Code Ann. §503.001): CUBI specifically includes voiceprints as one of the affected biometrics governed. As can be expected, it restricts collection, sale, and disclosure of personal data, and asks for a practical timeframe for deletion; it also calls for civil penalties up to $25,000 per incident (Texas Attorney General, n.d., para. 1). Law enforcement purposes are mentioned, as is the fact it might be individuals as well as companies who collect biometric data. Purposes include the need to "identify an individual in the event of the individual's disappearance or death if the individual consents" but clearly, that consent must be given before disappearance or death (Texas Attorney General, n.d.).

Washington State

Title 19.375 (House Bill 1493): Business Regulations: Biometric Identifiers includes rights to disclosure and control as with other states' laws, and insists the collection of biometric data must be for a specific purpose. The law focuses on commercial purposes, which excludes law enforcement needs for searches and seizures executed within other laws. Penalties are not defined (Washington State Legislature, 2017). It specifically omits images and voiceprints collected for health care payments, citing HIPAA as the main governing factor for that purpose (Washington State Legislature, 2017).

My Health My Data Act (MHMDA): A new privacy law enacted March 31, 2024 strengthens the state's stance on biometrics, including voice recordings, gait, keystroke rhythms, and other such data. The right to receive information about any third parties involved is included. This act protects state residents and anyone whose data is collected in the state, but excludes "employment context" (Quinn-Barabanov et al., 2024, para. 4). Like Illinois' biometric law, MHMDA allows for consumer action in the courts.

A Sampling of Other Countries

The data laws of a few countries outside of the European Union and United States are provided below. They demonstrate the similarities to the GDPR that new legislation is trying to accomplish as well as the need for some countries to begin work on formulating protective laws.

Argentina

The Personal Data Protection Law of Argentina was first passed in 2000 and has been amended as technology advances. It includes rules for dealing with processing and risk management and mitigation, requiring parental consent for children's online data, and other facets of data collection, storage, and sharing. Importantly, the law regulates data traversing international borders and is intended to align with Europe's GDPR – and there are rigorous rules for biometrics ("Behind the PDPL," 2024). Biometric data is considered sensitive when it is used to identify ethic origin, health, or other facets that can be used to discriminate ("Comparing privacy laws," 2023).

Cambodia

There is an E-Commerce Law in Cambodia covering some data regulations, and their Constitution, Civil Code, and Criminal Code may be useful in court

("Cambodia," 2025), but in mid-2025, Aw and O'Leary (2025) indicate the country's lawmakers are working to perfect and pass a Law on Personal Data Protection (LPDP). Seven other nations of the Association of Southeast Asian Nations have already enacted such laws. Cambodia's proposed legislation is modeled largely on the GDPR, will affect both Cambodians and outside entities that work with citizens of the country, and outlines both fines and potential imprisonments for violations (Aw & O'Leary, 2025).

Mexico

Several data laws were recently aligned in Mexico, including the General Law on Transparency and Access to Public Information, the General Law on the Protection of Personal Data Held by Public Sector Entities, the General Law on the Protection of Personal Data Held by Public Sector Entities, and the Federal Law on the Protection of Personal Data Held by Private Parties (Espejel et al., 2025). Documentation requirements, disclosure, and procedures for complaints and appeals were made clearer. Principles such as accountability, minimization, limitations, timelines, and citizens' rights to access, correct, and delete are similar to other countries' laws (Espejel et al., 2025).

Mexico's government has begun a rather controversial mandatory biometric ID card program. The card will include a photo and a QR code leading to iris and fingerprint data (Borak, 2025). There is no mention of voiceprints, although QR codes could lead to any kind of data in the future. Some fear surveillance of ordinary activities, but the stated premise is for law enforcement to be assisted in searching for missing citizens and to find controls for organized crime activity (Baptista, 2025).

Egypt

This country's Personal Data Protection Law No.151 of 2020, like other such laws, covers biometric data; rules for collecting, processing, correcting, and retention of data; and requires data protection officers. Its Article (1) includes voice as a potential identifier and therefore protected under this law ("Egypt," 2024). There are other laws which include voice recordings as well: the Egyptian Criminal Procedures Law,, Article 95, indicates courts may authorize surveillance related to certain criminal offenses, including the analysis of recorded conversations (Basheer Aljbour et al., 2025).

Other Legal Matters

Budnik and Evpak (2024) identified copyright rather than trademark as a preferred method to protect voice-related technologies, and use the phrase "vocal identity" within a legal context.

Comparison of Three Major Laws

The GDPR is the strongest of all the laws, demonstrated by countries outside the EU working to create similar laws. In the U.S.A., BIPA is the most focused law regarding biometric data, and CCPA is included in the following chart due to its emphasis on data collection in general.

Table 1. Overview comparison chart of three relevant laws

	GDPR (European Union)	**BIPA (Illinois, U.S.A.)**	**CCPA (California, U.S.A.)**
Purpose	All data collected must be for a specific purpose. All EU organizations must abide, no matter where the individual resides.	Focuses on biometric data that can be used as "identifiers" and applies to individuals, companies, and any other kind of group in Illinois.	Data collection is for a specific purpose unless prior notification is provided. The law focuses on commercial data in California businesses with annual gross income of $25m or those that meet certain thresholds of income derived from sale of data.
Consent	Consent is required and must be specific. Service cannot be denied if consent is not provided.	Consent is required for collection or disclosure of biometric identifiers. This data is not allowed to be sold.	Companies must tell consumers if their data is to be "monetized" at the time of data collection and offer an opt-out.
Accountability	A data protection officer must be appointed to oversee compliance. Notification of breaches is required.	There is no requirement for a data protection officer through this law. Breaches must be reported to residents of Illinois and the Attorney General.	There is no requirement for a data protection officer. Notification of breaches is required.

(ACLU Illinois, 2025; Buresh, 2023; Nasser, 2024; Polsinelli, 2024; State of California, 2024; "Staying in compliance," 2020)

TYPES OF ABUSES

Pustovoitov et al. (2024) clearly identify two purposes for biometrics: authentication (to allow a person access to something), and identification. Personal data is collected by electronic communication such as social media, smartphones, smart home assistants, medical instruments, and elsewhere, and as Buresh (2023) says, the individual's “comfort level” with this depends on how the data is used. Notably, Command and Control Speech Recognition systems understand select voice input and perform an action as a result of that input (“12 types,” 2024), enabling fraud and other problems for the individual.

Krishna Prakasha and Sumalatha (2025) remind that biometric data is not changeable, unlike passwords, passcodes, and other authentication types. The fact biometrics – including fingerprints, facial recognition, and voice recognition – are often used for multiple factor authentication on smart phones and other internet-connected logins means the average technology user has already freely given out some personal data (Rebovich, 2017). Artificial Intelligence spoofing attacks, replay attacks, and other kinds of cybersecurity failures can leak information to parties who may be profiling or tracking individuals for suspicious purposes (Krishna Prakasha & Sumalatha, 2025). As Rebovich (2017) says, “there is always the possibility someone may be able to record another person’s voice and then play it back to gain entry to sensitive data systems” (pp. 239-240).

Voiceprint collection has been going on for quite some time. Over a dozen years ago, the New Zealand Inland Revenue Service already had voiceprints for 10% of the country's population, claiming it was “extremely secure” (Mayhew, 2021, para. 4). Australia followed suit, asserting it could save as much as 45 seconds with every incoming phone call to the Taxation Office, which in turn could save many thousands of hours each year (Grubb, 2014). Experts said this was “fairly secure” but suggested the phrase that needed to be spoken for the recording should change (Grubb, 2014, para. 11). Since voice data has been collected now for many years and current technologies such as AI were not understood by the public at the time, there are strong concerns about ethical use of the data today, whether in the court system or any other part of a person's life.

Cloning

Sedláček (2018) argued several years ago that because there are different methods for creating voiceprints, the potential for fraud using this biometric is nil. However, technology has made strides since that time, including voice cloning or deepfakes, with various nefarious purposes.

Deepfake audio uses Machine Learning to imitate a real person's voice, and can be convincing enough to spread false information, steal identities, and attempt blackmail or extortion (Rabbi et al., 2024).

Voice cloning, also called voice synthesis or retrieval-based voice conversion, is when technology attempts to recreate a person's voice. This uses Machine Learning, pulls from a few minutes of the target human speaking, and because it is considered "sequential input," may include the use of the recurrent neural network (RNN) model (Genelza, 2024). Parmar (2025) explains this can be completed using open-source AI algorithms and even a very short original recording could be cloned with up to 95% accuracy.

A University of South Wales (Australia) scientist, Toby Walsh, experimented and found it only took about five minutes to create his voice clone; the recommendation is to use multifactor authentication so that voice cloning is not the only identification process (Evershed & Taylor, 2023). Budnik and Evpak (2024) concur, saying 15 seconds of a human's voice is enough for an adequate AI impersonation using OpenAI. Within this context, "impersonation is understood as a situation of taking someone's name or personality without permission with the purpose of harming, cheating, intimidating or threatening" (Budnik & Evpak, 2024, p. 35).

Gu et al. (2024) describe four kinds of faked audio, including text-to-speech models, a spoofing technique termed voice conversion that attempts to emulate the person's rhythmic patterns and tone color (also called "timbre"), replay which inserts the person's voice into an automatic system, and partially fake wherein select words are replaced. They also list impersonation as a real human mimicking another's speech rhythm, pitch, dialect, etc. (Gu et al., 2024). Any of these factors could be used against a defendant in a court of law or an incarcerated individual.

Masking

Hautamäki et al. (2018) point out people can attempt to disguise their voice through the use of masks, whispering, using an accent, talking while chewing gum, manipulating the mouth or nose, or trying to sound like a child or an elderly person. Some of these, particularly child-like speaking by an adult, may cause a mismatch with previously recorded voiceprints of the same person (Hautamäki et al., 2018). There is also an AI technology that modifies the speaker's accent and pronunciations, purportedly to make the speaker feel less self-conscious; however, this may be constructed to perpetuate the idea of white supremacy, as Artificial Intelligence systems are known for such bias (Payne et al., 2024).

Financial Fraud

Alver (2007) noted almost two decades ago that identity theft has a "significant emotional impact on its victims, comparable to that of violent crime" (p. 75). While voices are supposed to be unique, close family members may be able to impersonate one another. In a planned experiment in 2017, the fraternal twin of BBC reporter Dan Simmons was able to get into his brother's HSBC bank account and was given the capability to transfer money from one account to another with the same login (Simmons, 2017).

The "grandparent" scam can also be perpetrated through the use of cloned voices. Scammers impersonate a child or grandchild's voice in a phone call, saying they have been arrested or have some other emergency; they need cash immediately. The clones are created by harvesting voices from videos posted in social media (Vellani & Common, 2025). In mid-2025, thirteen people in the Dominican Republic used this ruse to scam several hundred senior citizens of more than $5 million, according to the charge levied in Boston (United States Attorney's Office, 2025). While this operation may have relied more on the elderly not being able to readily recognize a grandchild's voice and it is not clear if voiceprints were used or just phone-distorted human voices, it showcases the potential for audio-generated fraud.

Job or Reputation Losses

A voice clone's impersonation can perpetrate fraud in many ways, including blackmail and job loss, which can lead to lawsuits. AI-generated voices, whether clones of real people or created using the typical Artificial Intelligence deep learning models that pull tiny pieces from large recording datasets (which means they are still based on human input), have the potential to take jobs away from real humans. The voice-over industry has in the past used human actors to "dub" movies into different languages; Reuters claims a rather large percentage of viewers prefer dubbed soundtracks compared to subtitles (Pineau et al., 2025). Not only could an AI voice be employed instead, a specific actor's voice clone could be used. Boris Rehlinger, for example, has a voice well known in his country as the French voice of Joaquin Phoenix, Ben Affleck, and others, and fears unregulated theft of his livelihood (Pineau, 2025).

A Russian bank was sued by Alyona Andronova, a voice actress who recorded audio for the bank to train an AI voice assistant but did not give permission for it to be used in advertisements; this incident resulted in an online petition demanding "legal acknowledgment of the human voice as an intangible commodity" (Pustovoitov et al., 2024).

Bollywood playback singer Arijit Singh was the victim of voice cloning and won a Bombay High Court lawsuit in 2024 against the company who impersonated his voice. With the audio and stolen visual likeness, Codible Ventures fraudulently claimed Singh's endorsement of performances, took over 450 of his recordings to upload into AI tools, and used their own websites to promote the falsified items (Parmar, 2025). As noted in a lawsuit won by Bollywood actor Anil Kapoor in the previous year, endorsements may be major sources of income for an artist (Parmar, 2025).

Public Personality Identity Theft and Safety

Celebrities, public servants, and other famous "personalities" may be the object of identity theft for any number of reasons, including reputational harm. Examples include a deepfake video showing Meta CEO Mark Zuckerberg appearing to endorse data regulation, an audio file pretending to be Vladimir Putin of Russia admitting to U.S. election interference, and a video of former U.S. President Barack Obama saying offensive words he never actually said (Pustovoitov et al., 2024).

Pustovoitov et al. (2024) add that individual surveillance using biometrics such as voice cloning might encourage attacks on individuals. Buresh (2023) fears too much cyber surveillance challenges the Constitutional right to privacy in the United States. Unauthored tracking and cyber surveillance also threaten personal autonomy and physical security (Krishna Prakasha & Sumalatha, 2025). Some may think deepfakes are harmless entertainment, but there is a victim involved, and again, this may end up in court or with violence.

Unknown Data Collection and Surreptitious Use

Information collection must be clearly defined to avoid "function creep," where data is used for a purpose other than the intended use (Office of the Victorian Information Commissioner, n.d.). In Guy-Powell et al. v. Applebee's Restaurants LLC et al. class-action complaint (2022), several plaintiffs sued a large number of restaurant chains including Applebee's, Blaze Pizza, Chipotle, Noodles & Company, Portillo's, Red Lobster, and others. The dispute centers on an automated voice ordering system created by a company named SYNNQ3 and implemented by these restaurants in Illinois and other states starting in 2017. Phone orders, customer questions, and other information was commandeered by Artificial Intelligence and voiceprints collected, all without the express consent or knowledge of customers. The court case states,

> *Critically, the Defendants' AVO systems collect, store, and use customers' voiceprints and biometric information to understand, interpret, analyze, and/or make use of speech signals produced by customers; engage in speech recognition and other functions; and fulfill machine learning processes, which, among other things, help the Defendants' automated voice ordering systems to become more accurate over time; compile guest-specific data and/or identify customers; provide a tailored experience; and allow for expeditious reordering. (Guy-Powell et al. v. Applebee's Restaurants LLC et al. 2022, Section A.16)*

This Cook County, Illinois complaint included that the main offices for the restaurants and the technology company were in California, Colorado, Florida, and Massachusetts, and some have legal registration in Delaware and/or Nevada. Portillo's alone had Illinois as its main place of business. The court decided the case was indeed relevant to Illinois law (BIPA in particular) because the plaintiffs reside in Illinois and their data was collected at restaurants in Illinois (Guy-Powell et al. v. Applebee's, 2022). As of mid-2025, there has not been any publicly-shared update for this case.

Voiceprint collection concerns are not limited to customers. Whole Foods was sued over employee voiceprints collected via headset use without their consent in a Chicago-area company warehouse; a settlement of $297k was subsequently earmarked for both current and previous workers since the action was in violation of the Illinois BIPA (Bitter, 2023).

Class action lawsuits and settlements due to BIPA violations are numerous. McConvey (2025) describes that in one settlement, a parent settled with Amazon regarding their minor child's face biometrics collected when playing certain features of a video game. Cameras in cabs and "selfies" for in-app identification were also subjects of other suits (McConvey, 2025); while these are focused on visual biometrics, audio surely could be included in cam recordings.

Illinois' BIPA may reach outside the state's boundaries thanks to the internet. In a case involving Meta – which is headquartered in California – a federal judge allowed a case to continue as a potential Illinois violation that a user's verbal input into Facebook or its related Messenger program was used to train the system to recognize the speaker (Gennaro, 2024). Despite this outcome, not all cases will side with the home state of the complainant. Texas-based owner of the dating website Tinder, Match Group, was sued by a resident of Illinois, who argued the website did not require consent to gather biometrics. Burt (2024) explained the federal judge's decision to side with Match Group was based on Texas having a larger stake than the complainant's home state.

MORE ABOUT LAW ENFORCEMENT AND THE COURTS

Major concerns include how voiceprint data is collected, stored, and used; permissions and other privacy rights must be considered. In U.S. Courts, the Frye Test or Frye Standard might be applied. Based on Frye v. United States, 293 F. 1013 (D.C. Cir. 1923), the intent of this standard has been to ensure that new methods are accepted by experts in the specific field as acceptable practices before being allowed as evidence ("Frye Standard," 2022). Frye may be usurped in many states today by the Daubert Standard instead. Daubert v. Merrell Dow Pharmaceuticals Inc., 509 U.S. 579 (1993) allows for judges to make the determination about the reliability of scientific evidence presented during trial (Robinson, 2023).

According to Reynolds and Webber (1979), voice spectrographs were first considered in court as evidence in 1967 during United States v, Wright, 17 C.M.A.

183, 37 C.M.R. 447 (a military court) and in State v. Cary, 49 N.J. 343, 230 A.2nd 384 (a state appellate court). Notably, these failed the Frye Test.

Edward Lee King was convicted in 1965 of arson and looting during the infamous Watts riots in Los Angeles, at least partially due to a television documentary, *Watts, Riot or Revolt* (Accurso, 2023; People vs. King, 1968). While CBS did not show the face, a man admitted to wrongdoing; police suspected King and secretly recorded him speaking. The recording was compared to that of the man in the broadcast interview, and the Bell Labs engineer who analyzed the results said King was one and the same. Ultimately, other engineers pushed for and received a reversal of the conviction due to the analysis methodology used (Accurso, 2023). The engineer's qualifications were also questioned his background was electrical engineering and physics, but not physiology or medical research (People vs. King, 1968).

Miller (1973) reported that in 1972, bomb threats were traced to a phone booth; while the calls were anonymously made, the alleged perpetrator was in proximity to the booth and arrested. The ensuring trial included recordings of the calls. On appeal, the voiceprints were scrutinized but it was decided the conviction would stand, as the prints were not the only evidence presented (Miller, 1973).

In more recent years, the U.S. Coast Guard received an SOS call in 2020 about an alleged boat in distress with three occupants (Accurso, 2023). Private citizens in boats, the Marine Patrol and Coast Guard, and helicopters spent 4-5 hours searching in vain, as the call ended up being a hoax (Betts, 2021). A dock worker named Nathan Libby became a person of interest and to try to prove his involvement, investigators secretly recorded his voice and that of a coworker. A Carnegie Mellon University associate professor used Artificial Intelligence to assert Libby was the person who made the fraudulent distress call, despite a noticeable lack of comparison with several other voices which may have been more comparable to Libby's than the one coworker. Libby eventually took a plea deal (Accurso, 2023; Betts, 2021).

Surveillance and Ethics

While other biometrics such as fingerprinting and facial recognition clearly require the person's presence at the time and location of data capture, voiceprints might be made far from the person; phone or video calls enable recording anywhere (Claypoole, 2021). The question arises then about the ethics of "passive" extraction, where the speaker is not actively involved in the recording process ("What is voice biometrics," n.d.).

Audio used in law enforcement could include wiretapped conversations, emergency call center recordings, and in-person or video suspect interviews ("How can speech recognition," 2023). Forensic voice comparisons can help law enforcement track down perpetrators of threatening phone calls, and similarly, voiceprints could rule out innocent speakers (Gully et al., 2022). Audio recognition might also be implemented via eavesdropping surveillance (Altyar et al., 2022), but conversation snippets might be misused since the context of full conversations need to be known (Basheer Aljbour et al., 2025).

In November of 2019, a Boston resident was captured on video selling narcotics to an undercover police officer. The judge allowed the video but suppressed the audio portion of the recording. The police argued that recordings should be allowed, as there was reasonable suspicion of organized criminal activity, providing an exception to the state's wiretapping laws. Since the phone was out in the open during the exchange, defendant Thanh Du may have known that recording was taking place (Smith, 2024). During appeal, the defense asked for the video also be suppressed, but the appellate court ruled that both audio and video were admissible (Massachusetts Appellate Courts, 2024). It is not clear if voiceprint technology was employed, but this case demonstrates some courts' resistance to admitting any audio in court.

Human Rights

The question of rights makes a difference when considering whether or not voiceprints are going to be admissible in court. They may be considered “a unique type of electronic surveillance” that may need prior authorization from a judge (Basheer Aljbour et al., 2025, p. 348). Recordings may be mined from social media, videos, smart phone apps, intercepted phone calls, and other electronic surveillance (Jansen et al., 2021; Basheer Aljbour et al., 2025).

In prisons, use of biometrics in general for surveillance poses threats, however, particularly to the relationships between keepers and inmates, as well as psychological effects of this kind of privacy invasion in a captive circumstance (Joseph, 2017). While some might say the purpose of recorded phone calls and conversations is to surveil for gang connections and other illicit activities, and prisoners may expect less privacy while incarcerated, family members' voices are also captured and security of those records are unknown (Deskus & Fattal, 2019). According to Quinton and Lipton (2022), audio captured in prisons unleashed a flurry of patent wars, but more importantly, the systems might be trained to detect certain words or topics, including a database of known slang vocabulary; detect illicitly possessed devices, and even profile everyone on phone calls including children of inmates. Where the data is stored, with whom such data is shared, and what happens to this biometric data once an inmate has served their time are good questions (Quinton & Lipton, 2022). Deskus and Fattal (2019) additionally express unease that one voiceprint alone could be used by law enforcement to “automatically identify limitless future conversations” (para. 18).

A European venture named Speaking Identification Integrated Project (SiiP) holds voice biometrics in an internationally-available database accessed by Interpol and other law enforcement agencies (Jansen et al., 2021). Several fundamental human rights might be compromised by voice recognition technologies – in particular, "the right to privacy, the right to freedom of association and assembly, and the right not to be discriminated against" (Jansen et al., 2021, p. 2). Basheer Aljbour et al. (2025) explain there are several opinions about using voiceprints: they might be considered simply tools to be used by investigators, as legal documentation or evidence, or perhaps not usable at all since they are typically made without prior consent of the individual whose voice is being recorded.

The ACLU warned over a decade ago that false positives and false negatives were possible, and there is a concern that the liberty for anonymous speech, "a core American speech tradition at least since the publication of *The Federalist Papers* and other anonymous colonial writings," may be compromised (Stanley, 2014, para. 6).

RECOMMENDATIONS AND CONCLUSION

Basu et al. (2022) conducted a study in which judges and other "lay" persons attempted to identify speakers by listening. The results were compared to technological assessments, and it was found the speaker-recognition technology was far more accurate than human listening, which could include issues of unfamiliar language and accents. Their recommendation is to obtain expert testimony instead of relying on human listening skills (Basu et al., 2022). This means the technology is superior to perception of the human ear, especially in court. Abdulhameed et al. (2025) strongly recommend voiceprints be used in conjunction with other authentication such as fingerprinting; similarly, Altyar et al. (2022) recommend combining voice and facial recognition for accuracy.

Legislation may need to be more specific with voiceprint technology, but must also consider the ramifications of any new laws or amendments. In May of 2025, the U.S. House of Representatives inserted verbiage into a larger bill that would "impose a ten-year federal moratorium on state and local regulation of AI" (Kimery, 2025, para. 1). Considering the bill was more than 1100 pages and that it is known legislators do not always read bills thoroughly, this is more than problematic. It would mean deepfakes, bias, and other "innovations" would be allowed to make the country's citizenry vulnerable to fraud and identity theft (Kimery, 2025). Fortunately, this was removed from the bill before it was signed into law. The final version includes some restrictions regarding licensing, royalties, and other facets of AI that affect foreign entities, and it adds compliance, reporting, and training requirements for technologies (Penti & Adams, 2025).

Research into improving the technology itself and methods of analysis of voiceprints must continue; for example, Recurrent Neural Networks (RNNs), Convolutional Neural Networks (CNNs), and other technologies are being explored for use in deepfake audio detection (Rabbi et al., 2024). Siddika (2025) provides one possible solution to the templated, database-stored biometric authentication issues that can lead to instances of hacking and misuse: privacy-preserving biometrics, wherein encryption is immediately applied, leaving hashes and encrypted data only.

For individuals to protect themselves and their families, Vellani and Common (2025) suggest having a family "code" word for being able to uncover suspicious phone call requests for money. Hally (2021) recommends not speaking with unknown callers until they speak first so the purpose of the call can be known first.

Companies and policy makers need to be aware of the harm that can happen from requiring voice authentication, and know of laws in the company's locality as well as those in other domains. The development of a biometric policy would be prudent for the protection of employees, customers, and partnerships. Components of such a document might include:

- Statement of the policy's purpose, including which laws are applicable.
- Definitions of biometrics to be collected.
- Data collection:
    - Uses of the intended data collection, including whether or not this data is legally allowed to be sold.
    - Details about what exactly is to be collected, how the data will be stored, and the timeline showing when such data will be destroyed.
- Parameters for consent, with consideration of stakeholders' roles, plus any minor children.
- Identification of training for employees.
- Business plans:
    - Incident response plans in the case of a biometric data breach.
    - Communication plans to reach various stakeholders.
- Responsibilities:
    - Outline of the technical requirements and who is responsible for maintenance of the database.
    - Designation of an employee or a team of employees as gate-keepers of this data, responsible for communication and maintenance of the policy.

Voiceprint technology is still an emergent field in terms of use in the court systems, but it also a growing threat to civil liberties and personal identity. Unethical uses include but are not limited to impersonation for defamation or job theft, financial fraud, unwelcome surveillance, and uses that remain unknown to the person whose biometric data is stolen. Personal safety issues and human rights for prison populations must be mitigated, as is the fear that the technology will either fail and produce a false positive or the technology will not be allowed in the courtroom, making some cases more difficult.

Thoughtful application in the courtroom, development and enforcement of legislation that clearly aligns with ethical principles and international laws, continued efforts in both analysis techniques and cloning detection, and educating individuals will all be important in preventing misuse and regulating complications. In addition, further exploration into countries' prison system rules for voice recording use and storage should be done to assist in creating stronger protections.

REFERENCES

12 Types of Speech Recognition. (2024, January 17). Transkriptor. https://transkriptor.com/speech-recognition-types

Abdul, Z. K., & Al-Talabani, A. K. (2022, November). Mel Frequency Cepstral Coefficient and its applications: A review. *IEEE Access : Practical Innovations, Open Solutions*, *10*, 122136–122158. DOI: 10.1109/ACCESS.2022.3223444

Abdulhameed, A. A., Ssemakula, J. B., & Abbas, O. O., AbdulAzeez, A. A., & Ahmed, A. A. (2025, January). Design and implementation of a voiceprint recognition system for hybrid security and authentication in a university organizational infrastructure. *International Journal of Engineering Technology Research & Management*, *9*(1), 140–149.

Accurso, A. W. (2023, October). Digital voiceprinting is not ready for court. *Criminal Legal news*, *6*(10), 47. https://www.criminallegalnews.org/news/2023/oct/1/digital-voiceprinting-not-ready-court

ACLU Illinois. (2025, May 15). *Biometric Information Privacy Act (BIPA)*. https://www.aclu-il.org/sites/default/files/field_documents/protect_bipa_5.8.24.pdf

Altyar, S. S., Hussein, S. S., & Mohammed, M. J. (2022, March). Human recognition by utilizing voice recognition and visual recognition. *International Journal of Nonlinear Analysis and Applications*, *13*(1), 343–351. DOI: 10.22075/ijnaa.2022.5501

Alver, C. (2007). Voice biometrics in financial services. *The Journal of Financial Services Technology*, *1*(1), 75–81.

Aw, C., & O'Leary, C. (2025, July 28). *Cambodia moves to enact comprehensive data privacy law*. Hogan Lovells. https://www.hoganlovells.com/en/publications/cambodia-moves-to-enact-comprehensive-data-privacy-law

Bäckström, T., Räsänen, O., Zewoudie, A., Pérez Zarazaga, P., Koivusalo, L., Das, S., Gómez Mellado, E., Mansali, N. B., Ramos, D., & Vali, M. H. (2022). *Introduction to speech processing* (2nd ed.). Aalto University., https://speechprocessingbook.aalto.fi

Baker, C. P., Sundberg, J., Purdy, S. C., & Rakena, T. O. (2024, July). Female adolescent singing voice characteristics: An exploratory study using LTAS and inverse filtering. *Logopedics, Phoniatrics, Vocology*, *49*(2), 83–95. DOI: 10.1080/14015439.2022.2140455 PMID: 36322641

Baptista, D. (2025, July 14). *Will Mexico's new biometric ID card harm digital privacy?* Context News. https://www.context.news/digital-rights/will-mexicos-new-biometric-id-card-harm-digital-privacy

Basheer Aljbour, K. S., Aleshoush, M., El-Manaseer, S., Aloun, D. M., Alkayid, J. H. H., & Namous, A. A. A. Y. A. (2025). In A. Hannoon and A. Mahmood (Eds.), *Intelligence-Driven Circular Economy: Regeneration Towards Sustainability and Social Responsibility, Volume 2*. From Studies in Computational Intelligence 1174 (pp. 343-353), Springer. DOI: 10.1007/978-3-031-74220-0_26

Basu, N., Bali, A. S., Weber, R., Rosa-Aguilar, C., Edmond, G., Martire, K. A., & Morrison, G. S. (2022, October). Speaker identification in courtroom contexts – Part I: Individual listeners compared to forensic voice comparison based on automatic-speaker-recognition technology. *Forensic Science International, 341*, 111499. DOI: 10.1016/j.forsciint.2022.111499 PMID: 36283276

Behind the PDPL: Navigating Argentina's Personal Data Protection Law. (2024, February 2). Pandectes. https://pandectes.io/blog/behind-the-pdpl-navigating-argentinas-personal-data-protection-law

Betts, S. (2021, January 30). Rockland man charged with making false marine distress call. *Portland Press Herald*. https://www.pressherald.com/2021/01/29/rockland-man-charged-with-making-false-marine-distress-call

Bitter, A. (2023, January 6). Whole Foods must pay $300,000 over accusations that a system meant to track productivity captured warehouse workers' voices without their consent. *Business Insider*. https://www.businessinsider.com/whole-foods-pay-300k-recording-warehouse-employees-voices-2023-1

Bodepudi, A., Reddy, M., Gutlapalli, S. S., & Mandapuram, M. (2019). Voice recognition systems in the cloud networks: Has it reached its full potential? *Asian Journal of Applied Science and Engineering, 8*(1), 51–60. DOI: 10.18034/ajase.v8i1.12

Borak, M. (2025, July 18). *Mexico makes biometric identifier mandatory for all citizens*. Biometric Update. https://www.biometricupdate.com/202507/mexico-makes-biometric-identifier-mandatory-for-all-citizens

Budnik, R. A., & Evpak, E. G. (2024). Human voice: Legal protection challenges. *Legal Issues in the Digital Age, 5*(4), 28–45. DOI: 10.17323/2713-2749.2024.4.28.45

Buresh, D. L. (2023, March/April). BIPA: The Illinois' Biometric Information Privacy Act what every Illinois attorney should know. *DuPage County Bar Association Brief, 35*(5), 6-13. https://www.dcba.org/mpage/v35-Donald-L-Buresh

Burt, C. (2024, November 8). *Biometric privacy law in Texas close enough to BIPA to protect Match.* Biometric Update. https://www.biometricupdate.com/202411/biometric-privacy-law-in-texas-close-enough-to-bipa-to-protect-match

Butler, S. (2025, January 15). *Digital vs. analog audio: The key differences and do they matter?* Yahoo Tech. https://tech.yahoo.com/audio/articles/digital-vs-analog-audio-key-150015487.html

Cambodia: Data protection laws of the world. (2025, January 20). DLA Piper. https://www.dlapiperdataprotection.com/?c=KH

Claypoole, T. F. (2021, September 28). *Voice analysis complicates personal privacy.* Womble Bond Dickinson. https://www.womblebonddickinson.com/us/insights/blogs/voice-analysis-complicates-personal-privacy

Colorado General Assembly. (2024). *HB24-1130 Privacy of biometric identifiers & data.* https://leg.colorado.gov/bills/hb24-1130

Comparing privacy laws: GDPR vs. Personal Data Protection Act. (2023). OneTrust DataGuidance. https://www.dataguidance.com/sites/default/files/gdpr_v._personal_data_protection_act_argentina.pdf

Deskus, C., & Fattal, J. R. (2019, October 17). *A Fourth Amendment framework for voiceprint database searches.* Just Security. https://www.justsecurity.org/66571/a-fourth-amendment-framework-for-voiceprint-database-searches

Does the Fifth Amendment protect biometrics? (2022, March 2). Purdue Global Law School. https://www.purduegloballawschool.edu/blog/constitutional-law/fifth-amendment-biometrics

Doran, B. M., Fordyce, T. S., Valdetero, J. M., & Pestine, Z. (2024, August 14). *BIPA update: Illinois limits liability and clarifies electronic consent for biometric data collection.* Greenberg Traurig. https://www.gtlaw.com/en/insights/2024/8/bipa-update-illinois-limits-liability-and-clarifies-electronic-consent-for-biometric-data-collection

Durina, M. (2021, December). 90 years later: Revisiting the Lindbergh kidnapping case. *Journal of the American Society of Questioned Document Examiners, 24*(2), 3–14. DOI: 10.69525/jasqde.276

Egypt: Data protection laws of the world. (2024, January 19). DLA Piper. https://www.dlapiperdataprotection.com/index.html?t=law&c=EG

Espejel, E., & Garcia Gómez, F. Mosquera, A., Oacheco Mendoza, J. J., Luna, W., Arellano, C. A., & Torres Montalvo, R. (2025, March 7). *Mexico enacts new data protection regime.* White & Case. https://www.whitecase.com/insight-alert/mexico-enacts-new-data-protection-regime

Evershed, N., & Taylor, J. (2023, March 16). AI can fool voice recognition used to verify identity by Centrelink and Australian tax office. *The Guardian.* https://www.theguardian.com/technology/2023/mar/16/voice-system-used-to-verify-identity-by-centrelink-can-be-fooled-by-ai

Fast Fourier Transformation FFT: Basics. (n.d.). NTI Audio. https://www.nti-audio.com/en/support/know-how/fast-fourier-transform-fft

Federal Bureau of Investigation. (2013, June 3). *Speaker recognition.* Biometric Center of Excellence. https://ucr.fbi.gov/fingerprints_biometrics/biometric-center-of-excellence/files/speaker-recognition.pdf

Federal Trade Commission. (2013, January 17). 16 CFR Part 312, Children's Online Privacy Protection Rule ("COPPA"). *Federal Register, 78*(12), 3972–4014. https://www.ftc.gov/system/files/2012-31341.pdf

Frye Standard. (2022, December). Cornell School of Law Legal Information Institute. https://www.law.cornell.edu/wex/frye_standard

Fudge, T. P. (n.d.). *The International Phonetic Alphabet.* Fudgelink: Voice lessons. https://fudgelink.com/IPA

Genelza, G. G. (2024, October 29). A systematic literature review on AI voice cloning generator: A game-changer or a threat? *Journal of Emerging Technologies, 4*(2). Advance online publication. DOI: 10.57040/ag587791

Gennaro, M. (2024, February 28). Privacy class action over unauthorized voiceprint collection to proceed against Meta. *Courthouse News Service.* https://www.courthousenews.com/privacy-class-action-over-unauthorized-voiceprint-collection-to-proceed-against-meta

Grubb, B. (2014). Australian Taxation Office uses 'voiceprint' technology to speed up calls. *The Sydney Morning Herald.* https://www.smh.com.au/technology/australian-taxation-office-uses-voiceprint-technology-to-speed-up-calls-20140909-10edli.html

Gu, H., Yi, J., Wang, C., Ren, Y., Tao, J., Yan, X., Chen, Y., & Zhang, X. (2024, October 28). Utilizing speaker profiles for impersonation audio detection. [Conference proceedings]. *32nd ACM International Conference on Multimedia.* pp. 1961-1970. DOI: 10.1145/3664647.3681602

Gully, A., Harrison, P., Hughes, V., Rhodes, R., & Wormald, J. (2022, February 14). How voice analysis can help solve crimes. *Frontiers for Young Minds*, *10*, 702664. DOI: 10.3389/frym.2022.702664

Guy-Powell v Applebee's Restaurants LLC et al., case number 2022-CH-08365. Class Action. https://www.classaction.org/media/guy-powell-et-al-v-applebees-restaurants-llc-et-al.pdf

Hally, L. (2021, September 21). *Voiceprint hack*. Cyber Security: Luke Hally. https://www.lukehally.au/data-security-privacy/voiceprint-hack

Hansen, J. H. L., Sreenivasa Rao, K. S., D'Urso, P., & Kim, W. (2005). *Cepstral*. Science Direct Topics. https://www.sciencedirect.com/topics/computer-science/cepstral

Harland-Logan, S. (2023, November 14). *Exonerations: Guy Paul Morin*. Innocence Canada. https://www.innocencecanada.com/the-latest/exoneration/guy-paul-morin

Hautamäki, R. G., Kanervisto, A., Hautamäki, V., & Kinnunen, T. (2018, June 26-29). *Perceptual evaluation of the effectiveness of voice disguise by age modification*. Speaker Odyssey 2018: The Speaker and Language Recognition, Les Sables d'Olonne, France (pp. 320-326). https://www.isca-archive.org/odyssey_2018/hautamaki18_odyssey.pdf

How can speech recognition be used in law enforcement? (2023, May 23). Amberscript. https://www.amberscript.com/en/blog/how-can-speech-recognition-be-used-in-law-enforcement

Illinois General Assembly. (2008). *Civil Liabilities (740 ILCS 14/) Biometric Information Privacy Act*. https://www.ilga.gov/Legislation/ILCS/Articles?ActID=3004&ChapterID=57

Jander, N., Hutter, N., Mueller, T., Immerz, A., Stritt, F., Traser, L., Spahn, C., & Richter, B. (2025, June 27). Protective factors for vocal health in teachers: The role of singing, voice training, and self-efficacy. *International Journal of Environmental Research and Public Health*, *22*(7), 1018. Advance online publication. DOI: 10.3390/ijerph22071018 PMID: 40724085

Jansen, F., Sánchez-Monedero, J., & Dencik, L. (2021, July). Biometric identity systems in law enforcement and the politics of (voice) recognition: The case of SiiP. *Big Data & Society*, *8*(2), 20539517211063604. Advance online publication. DOI: 10.1177/20539517211063604

Joseph, J. (2017). Technoprison: Technology and prisons. In Moriarty, L. J. (Ed.), *Criminal Justice Technology in the 21st Century* (3rd ed., pp. 172–205). Charles C. Thomas.

Khare, P., & Srivastava, S. (2023, June). Enhancing security with voice: A comprehensive review of AI-based biometric authentication systems. *International Journal of Research and Analytical Reviews, 10*(2), 398–403. https://ijrar.org/papers/IJRAR23B4732.pdf

Kimery, A. (2025, May 23). *Move in House to block state AI laws draws bipartisan fire; Senate support questionable.* Biometric Update. https://www.biometricupdate.com/202505/move-in-house-to-block-state-ai-laws-draws-bipartisan-fire-senate-support-questionable

Kramer, R. S. S., Jones, A. L., & Gous, G. (2021). Individual differences in face and voice matching abilities: The relationship between accuracy and consistency. *Applied Cognitive Psychology, 35*(1), 192–202. DOI: 10.1002/acp.3754

Krishna Prakasha, K., & Sumalatha, U. (2025, January). Privacy-preserving techniques in biometric systems: Approaches and challenges. *IEEE Access : Practical Innovations, Open Solutions, 13*, 32584–32616. DOI: 10.1109/ACCESS.2025.3541649

Kyi, A. S. S., & Lin, K. Z. (2019, November). Detecting voice features for criminal case. *Proceedings of the 2019 International Conference on Advanced Information Technologies (ICAIT)*, Yangon, Myanmar, 212-216. DOI: 10.1109/AITC.2019.8921212

Lungova, V., & Thibeault, S. L. (2020, October). Mechanisms of larynx and vocal fold development and pathogenesis. *Cellular and Molecular Life Sciences : CMLS, 77*(19), 3781–3795. DOI: 10.1007/s00018-020-03506-x PMID: 32253462

Martinez, A. (2025, June 27). Colorado's biometric privacy law takes effect July 1: Are you ready? *Forbes.* https://www.forbes.com/sites/alonzomartinez/2025/06/27/colorados-biometric-privacy-law-takes-effect-july-1-are-you-ready

Massachusetts Appellate Courts. (2024). *Commonwealth v Thanh Du SJC-13557.* https://www.ma-appellatecourts.org/docket/SJC-13557

Mayhew, S. (2021, October 5). *One in 10 New Zealanders have a voiceprint.* Biometric Update. https://www.biometricupdate.com/201210/one-in-10-new-zealanders-have-a-voiceprint

McConvey, J. R. (2025, February 5). *Amazon can't block BIPA voice biometrics case even though federal case failed.* Biometric Update. https://www.biometricupdate.com/202502/amazon-cant-block-bipa-voice-biometrics-case-even-though-federal-case-failed

Miller, D. R. (1973, March). Criminal law: Voiceprint evidence is being heard. *Florida Law Review*, *25*(3), 10. https://scholarship.law.ufl.edu/flr/vol25/iss3/10

Nasser, T. (2024, February 23). *Are data protection regulations all the same?* Right-Hand Cybersecurity. https://right-hand.ai/blog/are-data-protection-regulations-all-the-same

Office of the Victorian Information Commissioner. (n.d.). Biometrics and privacy: Issues and challenges. https://ovic.vic.gov.au/privacy/resources-for-organisations/biometrics-and-privacy-issues-and-challenges

Oregon Department of Justice. Media. (2024, July 2). *Oregonians gain significant new consumer privacy protections.* https://www.doj.state.or.us/media-home/news-media-releases/oregonians-gain-significant-new-consumer-privacy-protections

Oregon Department of Justice. Consumer Protection. (n.d.). *Consumer privacy: Oregon Consumer Privacy Law.* https://www.doj.state.or.us/consumer-protection/id-theft-data-breaches/privacy

Parmar, D. G. (2025, April 17). AI voice cloning: how a Bollywood veteran set a legal precedent. WIPO Magazine. https://www.wipo.int/web/wipo-magazine/articles/ai-voice-cloning-how-a-bollywood-veteran-set-a-legal-precedent-73631

Payne, A., Austin, T., & Clemons, A. M. (2024, June). Beyond the front yard: The dehumanizing message of accent-altering technology. *Applied Linguistics*, *45*(3), 553–560. DOI: 10.1093/applin/amae002

Penti, R. S., & Adams, A. A. (2025, July 14). *AI and tech under the One Big Beautiful Bill Act: Key restrictions, risks, and opportunities.* Ropes & Gray. https://www.ropesgray.com/en/insights/alerts/2025/07/ai-and-tech-under-the-one-big-beautiful-bill-act-key-restrictions-risks-and-opportunities

People v. King. (1968, October 10). Cal. App. 2d California Courts of Appeal Cases, *266.* https://law.justia.com/cases/california/court-of-appeal/2d/266/437.html

Pineau, E., Murray, M., & Chmielewski, D. (2025, July 31). Voice actors push back as AI threatens dubbing industry. *Reuters.* https://www.reuters.com/business/media-telecom/voice-actors-push-back-ai-threatens-dubbing-industry-2025-07-30

Polsinelli. (2024). *Global data protection law guide: USA – Illinois.* Multilaw. https://multilaw.com/Multilaw/Multilaw/Data_Protection_Laws_Guide/DataProtection_Guide_USA_Illinios.aspx

Pustovoitov, P. S., Seilova, N. A., & Gnatiuk, A. S. (2024, March 15). Analysis of voice impersonation fraud: Risks, cases and defense strategies. *International Journal of Information and Communication Technology*, *5*(1), 122–134. DOI: 10.54309/IJICT.2024.17.1.010

Qian, J., Du, H., Hou, J., Chen, L., Jung, T., & Li, X.-Y. (2018, November 4). Hide-behind: Enjoy voice input with voiceprint unclonability and anonymity. *SenSys '18: Proceedings of the 16th ACM Conference on Embedded Networked Sensor Systems* (pp. 82-94), Shenzhen, China. DOI: 10.1145/3274783.3274855

Quinn-Barabanov, J., Berman, E., & Reid, S. (2024, March 29). *BIPA 2.0? Washington's new privacy law creates private litigation and AG enforcement risk for businesses.* Steptoe. https://www.steptoe.com/en/news-publications/bipa-20-washingtons-new-privacy-law-creates-private-litigation-and-ag-enforcement-risk-for-businesses.html

Quinton, C., & Lipton, B. (2022, September 1). The catalog of carceral surveillance: Voice recognition and surveillance. *Prison Legal News*, *33*(9), 27–28.

Rabbi, B. M. S., Haque, E., Hui, H. S., & Taassum, N. (2024, October). *From impersonation to authentication: techniques for identifying deep fake voices.* [Thesis, Brac University]. http://hdl.handle.net/10361/25249

Rebovich, D. J. (2017). Securing cyberspace: How can this be possible in the 21st century? In Moriarty, L. J. (Ed.), *Criminal Justice Technology in the 21st Century* (3rd ed., pp. 233–251). Charles C. Thomas.

Regulation 2016/679 OJ L 119 of the European Parliament and of the Council of 27 April 2016 on the protection of natural persons with regard to the processing of personal data and on the free movement of such data, and repealing Directive 95/46/EC (General Data Protection Regulation). Official Journal of the European Union, 59, 1-88. https://eur-lex.europa.eu/legal-content/EN/TXT/PDF/?uri=CELEX:32016R0679

Reynolds, J. C., & Weber, J. W. (1979, October 1). The admissibility of spectrographic voice identification in the state courts. *The Journal of Criminal Law & Criminology*, *70*(3), 349–354. DOI: 10.2307/1142577

Robinson, J. (2023, August). *Daubert Standard.* Cornell School of Law Legal Information Institute. https://www.law.cornell.edu/wex/Daubert_standard

Rubin, H. (2025, January). It's my data and I'll protect it if I want to: A proposal to standardize biometric data privacy laws between the United States and European Union. *Vanderbilt Journal of Transnational Law*, *58*(1), 261–304.

Sedláček, M. (2018, November 27). *Using voice biometrics in the world of the GDPR*. Phonexia. https://www.phonexia.com/blog/using-voice-biometrics-world-gdpr

Siddika, S. (2025, May 20). *Privacy-preserving biometrics: Enhancing security with Anonybit*. LinkedIn. https://www.linkedin.com/pulse/privacy-preserving-biometrics-enhancing-security-anonybit-siddika-6vw1c

Simmons, D. (2017, May 19). BBC fools HSBC voice recognition security system. *BBC*. https://www.bbc.com/news/technology-39965545

Smith, J. (2024, August 29). High court to decide if covert police recording violates wiretap law. *Commonwealth Beacon*. https://commonwealthbeacon.org/courts/high-court-to-decide-if-covert-police-recording-violates-wiretap-law

Stanley, J. (2014, October 16). *On the creation of giant voiceprint databases*. American Civil Liberties Union. https://www.aclu.org/news/privacy-technology/creation-giant-voiceprint-databases

Tafiadis, D., Tatsis, G., Ziavra, N., & Toki, E. I. (2017). Voice data on female smokers: Coherence between the Voice Handicap Index and acoustic voice parameters. *AIMS Medical Science*, *4*(2), 151–163. DOI: 10.3934/medsci.2017.2.151

Texas Attorney General. (n.d.). *Biometric Identifier Act: Texas Capture Or Use Of Biometric Identifier Act*. https://www.texasattorneygeneral.gov/consumer-protection/file-consumer-complaint/consumer-privacy-rights/biometric-identifier-act

Tremblay, P., Gagnon, L., Roy, J.-P., & Arseneault, A. (2023, November 9). Speech production in healthy older adults with or without amateur singing experience. *Journal of Speech, Language, and Hearing Research: JSLHR*, *66*(11), 4332–4352. DOI: 10.1044/2023_JSLHR-23-00126 PMID: 37870784

United States Attorney's Office. District of Massachusetts. (2025, August 12). *Thirteen individuals charged for operating transnational elder fraud scheme*. [Press release]. https://www.justice.gov/usao-ma/pr/thirteen-individuals-charged-operating-transnational-elder-fraud-scheme

van Drongelen, W. (2007). Continuous, discrete, and fast Fourier transform. In *Signal processing for Neuroscientists* (pp. 91–105). Elsevier., DOI: 10.1016/B978-012370867-0/50006-1

Vanderslice, R. (1969, February 1). *The "voiceprint" myth*. [Report]. University of Michigan Center for Research on Language and Language Behavior.

Vellani, N., & Common, D. (2025, March 20). Her grandson's voice said he was under arrest. This senior was almost scammed with suspected AI voice cloning. *Canadian Broadcasting Corporation.* https://www.cbc.ca/news/marketplace/marketplace-ai-voice-scam-1.7486437

Voiceprint concept supported by government-sponsored tests. (1971, June 1). Communications of the ACM, 14(6), 434–435. https://doi.org/ DOI: 10.1145/362604.1275495

Wang, Y., Cusick, M., Laila, M., Puech, K., Ji, Z., Hu, X., Wilson, M., Spitzer-Williams, N., Wheeler, M., & Ibrahim, Y. (2024, May 21). FairLENS: Assessing fairness in law enforcement speech recognition. ArXiv. https://arxiv.org/abs/2405.13166 DOI: 10.1109/BigData62323.2024.10825119

Washington State Legislature. (2017). *Title 19, Chapter 19.375: Biometric identifiers.* https://app.leg.wa.gov/RCW/default.aspx?cite=19.375

What is forensic voice comparison? (n.d.). In Forensic voice comparison: The essential guide. Phonexia. https://www.phonexia.com/knowledge-base/forensic-voice-comparison-essential-guide

What is voice biometrics? (n.d.). Forensic voice comparison: The essential guide. Phonexia. https://www.phonexia.com/knowledge-base/voice-biometrics-essential-guide

Chapter 7
Performance Comparison of Python Libraries in Face Recognition Systems

Bayram Cadıl
https://orcid.org/0009-0009-8446-2221
Ahmet Yesevi University, Turkey

Gurkan Tuna
https://orcid.org/0000-0002-6466-4696
Trakya University, Turkey

ABSTRACT

This research aims to analyze the applicability and performance of face recognition technologies using different Python-based libraries. Face recognition and comparison tests were performed using the face_recognition, OpenCV and DeepFace libraries, respectively. First, basic face detection application was successfully implemented with the face_recognition library. Then, face recognition was performed using traditional methods using OpenCV. But limited accuracy results were obtained. Finally, deep learning-based face recognition was performed with the DeepFace library, high accuracy rates were achieved, but high processing power was required. All three methods were systematically compared in terms of hardware requirements, ease of use, processing time and accuracy rates. The findings revealed that each library has advantages and limitations for certain application scenarios. In this context, it was emphasized that the selection of the appropriate algorithm for the needs in the design of face recognition systems is critical.

DOI: 10.4018/979-8-3373-4972-5.ch007

INTRODUCTION

In recent years, the rapid developments in artificial intelligence and machine learning technologies have made a strong impact in the field of facial recognition (Kutlugün, 2023; Tuna, 2022). Facial recognition technologies have a wide range of applications in human identification and verification processes; therefore, they are actively used in many sectors such as security, health, and education (İlkbahar, Ünal, & Karakaya, 2021; Qasim, 2023).

Facial recognition systems are based on extracting certain features from human face images, algorithmically modeling these features, and then performing the authentication or recognition process (Alimovski, 2019; Bozik, 2019). Research in this area has mostly focused on deep learning-based methods, esspecially convolutional neural networks (CNNs). Deep learning-based algorithms and machine learning techniques are widely used in facial recognition (Gündüz & Cedimoğlu, 2019; Tuna, 2022). Python is widely used in face recognition systems thanks to its powerful libraries such as Open Source Computer Vision Library (OpenCV), Dlib, and Sci-kit, but the performance of each library may vary in different application areas (Boyko, Basystiuk, & Shakhovska, 2018; Aliyu, Bomoi, & Maishanu, 2022).

While various libraries and software are used in the development of face recognition systems, libraries developed in the Python language have become particularly popular among researchers due to their ease of use, wide community support, and open source features (Tombak, 2019; Kaplan & Rahman, 2024). In Python-based face recognition systems, OpenCV, Dlib, TensorFlow, and Sci-kit Learn are mostly preferred. Therefore, it is important to comparatively examine the performance differences between these libraries (Hasan & Sallow, 2021; Wang, Zhang, Xiong, & Zhao, 2021; Patil et al., 2023).

The aim of this research is to comparatively analyze the real-time performance of Python-based face recognition libraries. In addition, the effects of GPU and CPU hardware on the performance of these libraries are evaluated. Because, with the increase in real-time face recognition applications, it has become important to examine the factors affecting the performance of such systems (Kutlugün, 2023; Tuna, 2022). As a significant portion of the studies in the literature have focused on specific algorithms or hardware platforms, the comprehensive and applied comparison of Python-based libraries with each other is a relatively less studied area.

This study has the potential of providing a comprehensive assessment by bringing together previously separately studied areas in the literature (Bozik, 2019), ways to improve the performance of algorithms with data augmentation methods (Alimovski, 2019), and comparison of image processing techniques. In addition, it could provide a broad perspective on the literature by considering the potential of face recognition systems in social and private application areas (Qasim, 2023) and the factors affecting users' personal privacy (Fábián & Gulyás, 2021).

RELATED WORK

In recent years, the analysis of thc relationship and performance comparisons of hardware architectures with Python-based face recognition systems has become popular (Saray Çetinkaya & Sertbaş, 2022). Because Graphics Processing Unit (GPU) and Central Processing Unit (CPU) directly affect the real-time performance of face recognition systems (Öğüt, 2019; Saray Çetinkaya & Sertbaş, 2022). Analyzing the performance differences of Python libraries on CPU and GPU can provide new and critical information about the effective use of hardware architecture in the implementation of deep learning algorithms (Saray Çetinkaya & Sertbaş, 2022; Öğüt, 2019). In addition, performance comparisons supported by applied studies cam reveal new information about the use of face recognition algorithms in mobile and embedded systems and help determine algorithms that can run at high efficiency on platforms with low computational power (Schenkel, Ringhage, & Branding, 2019; Özdemir, 2020).

Today, face recognition systems are used in a wide range of applications from security to education, from health to entertainment. As the application areas of these systems expand, real-time performance requirements and accuracy criteria are also increasing (Boyko, Basystiuk & Shakhovska, 2018; Aliyu, Bomoi, & Maishanu, 2022). OpenCV and Dlib contain different algorithms and each has different performance characteristics. For example, while OpenCV is mostly used in basic face detection and simple recognition operations, the Dlib library is aimed at more complex applications that include deep learning-based models (Fayaz, Mohi-Ud-Din, & Batool, 2019; Amos, Ludwiczuk, & Satyanarayanan, 2016). In addition, there are differences in face recognition systems based on algorithms, and comparative evaluations are made between the use of various methods such as principal component analysis (PCA), linear discriminant analysis (LDA), and CNN (Schenkel, Ringhage, & Branding, 2019).

With the acceleration of technological developments, especially the integration of deep learning algorithms has significantly increased the performance of face recognition systems. In the study conducted by Kaplan and Rahman (2024), it is emphasized that high precision results are obtained in face detection operations performed with OpenCV-based applications. It has been shown that face recognition systems expand their training datasets through data augmentation using Adversarial Network (GAN) models, which significantly increases their accuracy rates (Alimovski, 2019).

When the studies are examined, it is seen that comprehensive and holistic research is limited on the practical real-time evaluation of the performance of Python-based libraries in face recognition systems and their comparison, especially considering the hardware differences between GPU and CPU (Kabakus, 2019). In addition, there are deficiencies in existing studies, such as the performance comparisons of libraries being limited to certain algorithm types and ignoring hardware optimizations (Saray Çetinkaya & Sertbaş, 2022). In this context, comparative evaluation of real-time performances of face recognition libraries developed in Python, detailing the differences between libraries, and examining the effects of hardware architectures on algorithms are important research problems for the development of face recognition systems (Kutlugün, 2023; Özdemir, 2020; Zhang, Li, & Shan, 2020). These comparative analyses will contribute to the development of more effective, faster, and more accurate face recognition applications by providing developers and researchers with clear data on the selection of libraries to be used when developing face recognition systems.

CONCEPTUAL FRAMEWORK

Face recognition is a biometric security solution that automatically recognizes and verifies the identity of individuals only through facial features. It analyzes the structural, geometric and textural features of the human face, produces a digital face print and performs comparison operations on this print (Kutlugün, 2023). Thanks to their advanced algorithms, face recognition systems detect unique points on people's faces, showing superior performance compared to traditional methods in terms of both accuracy and speed (Boyko, Basystiuk, & Shakhovska, 2018). Face recognition is based on face detection, face alignment, and facial feature extraction. It consists of a multi-layered process consisting of extraction and matching stages (Aliyu, Bomoi & Maishanu, 2022).

The importance of facial recognition is multidimensional. First of all, thanks to the solutions it offers in the field of security, it plays critical roles in areas such as access control, border security, and criminal detection in public and private sector institutions. At the same time, real-time surveillance systems used within the scope of city security reduce the workload of security personnel by automating suspicious person tracking and accelerate decision-making processes. In the study conducted by Gündüz and Cedimoğlu (2019), it was stated that facial recognition systems can also be used effectively in classification tasks such as gender prediction. Especially in smart city applications, such systems contribute to the provision of both individual and social security. Another important area of use of facial recognition technology is the healthcare sector and contactless biometric measurement systems. Tuna (2022) stated that facial recognition systems are integrated with thermal cameras and used in applications such as contactless fever measurement. Such systems have gained importance especially in the post-pandemic period and have contributed to increasing the safety of healthcare professionals and individuals. In addition, the use of facial recognition systems in the field of education is becoming widespread. With the software developed by Qasim (2023), analyzing the emotional states of individuals with special education needs has made the teaching process more qualified.

However, facial recognition systems are not only limited to public safety and education, but are also widely used in different areas such as customizing customer experience in the retail sector, individual recognition in smart home systems, and automation management (İlkbahar, Ünal & Karakaya, 2021). In addition, thanks to the flexibility offered by the Python programming language in data processing processes and open source libraries such as OpenCV and Dlib, systems have become more accessible and low-cost (Geitgey, 2019; Amos, Ludwiczuk & Satyanarayanan, 2016). These libraries have paved the way for the widespread use of facial recognition systems in both educational institutions and industrial projects. On the other hand, optimizations made for the systems to work in real time and on devices with low hardware power are also remarkable. In studies comparing CPU and GPU architectures in particular, it has been shown that GPU-based systems are faster and more efficient in terms of computation (Saray Çetinkaya & Sertbaş, 2022; Öğüt, 2019). This allows the systems to be used effectively in mobile devices or low-power embedded systems.

Facial recognition has become one of the indispensable components of contemporary digital society with the advantages of high accuracy, speed, no contact and user-friendly interface it offers in many sectors such as security, health, education, retail and automation. It is predicted that this technology will become more widespread in the future and will be more integrated with artificial intelligence, big data and the internet of things (IoT). In this context, evaluating facial recognition technology with an interdisciplinary approach requires the coordination of both technical development and ethical and legal regulations (Fábián & Gulyás, 2021).

Applications of Facial Recognition Systems

Today, facial recognition technology has transformed from a system used only for security purposes into a functional and strategic technology in many sectors. The application areas of this technology are spread over a wide range such as security, health, education, retail, public services, finance, entertainment and smart systems. With the development of deep learning algorithms and high-performance processors, facial recognition systems can now analyze large data sets in real time and function as a decision support tool in different sectors (Kutlugün, 2023; Wang et al., 2021).

Security and Public Safety Applications

The most common and effective use of facial recognition systems is in public safety and access control. These systems are used for identity verification and criminal detection in areas requiring high security, such as airports, border gates, stadiums, public buildings and public transportation systems. Thanks to the integration of facial recognition algorithms with real-time monitoring systems, the detection of suspicious persons can be achieved in a shorter time and with less manpower (Schenkel, Ringhage & Branding, 2019). These systems are also effectively used by police departments in finding missing persons or in criminal investigations.

Health Sector Applications

The use of facial recognition technology in the healthcare field is also increasing. Especially with the COVID-19 pandemic, systems that provide contactless body temperature measurement have come to the fore. Tuna (2022) stated that thermal camera systems integrated into facial recognition algorithms offer significant advantages in terms of automatically measuring people's fever and preliminary assessment of their health status. Similarly, facial recognition technology is also used in providing individualized support based on emotion analysis in the education processes of individuals with special needs (Qasim, 2023).

Use in Education

Facial recognition offers innovative applications for both teachers and students in education. This technology is used in many areas, from classroom attendance to exam security, from analyzing student motivation levels to special education applications. Providing real-time feedback to teachers, especially by analyzing the emotional states of individuals who need special education, is seen as an important tool in improving the quality of education (Qasim, 2023). In addition, facial recognition-based identity verification systems are of great importance in preventing identity fraud in exams.

Retail and Customer Experience

Facial recognition systems are used in the retail sector to customize the customer experience and increase service quality. Applications such as recognizing customers when they enter the store, offering suggestions based on shopping history, and directing personalized discounts or advertising content reveal the potential of this technology in the commercial field. In this context, the protection of personal data and ethical issues should be carefully addressed (Fábián & Gulyás, 2021).

Smart Home and IoT Systems

The use of facial recognition in smart home systems is becoming increasingly widespread. In these systems, facial recognition algorithms perform functions such as verifying the identities of homeowners, providing access controls, and activating personalized scenarios. İlkbahar, Ünal, and Karakaya (2021) developed an artificial intelligence solution that increases in-home security and comfort by focusing on facial recognition algorithms in the model they proposed for smart home systems.

Use in Finance and Banking Sector

The financial sector uses facial recognition technology as both customer authentication and security measure. In mobile banking applications, facial recognition is increasingly replacing fingerprint verification in the biometric verification process. This increases transaction security and improves user experience. In addition, eliminating card usage with biometric verification at ATMs is another innovative application in the sector.

Use in Social Media, Entertainment and Other Areas

Facial recognition systems are also used extensively on social media platforms. Functions such as user tagging, filter applications and facial manipulations form the basis of these systems. For example, social media applications use algorithms that can detect users' faces and automatically tag them or estimate information such as age and gender. In addition, in the cinema and gaming sectors, it is becoming more common to offer personalized content with facial recognition technologies.

Face recognition systems have become an important component of contemporary digital transformation processes thanks to the added value and functionality they offer in different sectors. In each application area, technology should not be limited to technical success alone; it is also necessary to ensure integration by taking into account ethical, legal and social dimensions. In this context, the regulation and control of areas of use are as important as the development of face recognition systems.

Basic Principles of Face Recognition Systems

Facial recognition systems are advanced artificial intelligence applications that aim to automatically detect and verify the identities of individuals using the human face as a basic biometric identifier. The basic working principle of these systems is to detect human faces in an image or video stream, convert certain features of these faces into a digital template by analyzing them, and compare these templates with previously recorded data (Kutlugün, 2023). This process generally consists of four main stages: face detection, face alignment, feature extraction, and recognition/verification (Kaplan & Rahman, 2024; Geitgey, 2019).

Face Detection

Face detection, which is the first stage of the face recognition process, automatically detects faces in an image or video frame. This is done through algorithms and libraries such as Histogram of Oriented Gradients (HOG), Dlib, OpenCV and YOLO (Boyko, Basystiuk, & Shakhovska, 2018; Hasan & Sallow, 2021). This stage is of critical importance for the correct operation of the system because incorrect face detections directly affect the overall accuracy of the system.

Face Alignment

The process of aligning the perceived faces according to a certain reference system is the second basic principle of face recognition. Since face images obtained from different angles, in different lighting conditions or with various facial expressions may reduce the recognition success of the system, these images need to be brought to a standard format (Amos, Ludwiczuk & Satyanarayanan, 2016). In this context, geometry-based corrections are made by determining the basic points of the face (eyes, nose tip, mouth edges).

Feature Extraction

After the face alignment process, the system extracts the characteristic features of the face in the image. This process is the creation of feature vectors that form a digital representation of the face. These vectors usually learn the discriminative aspects of the face (e.g., eye distance, face shape, skin texture) through deep learning models. Open-source libraries such as OpenFace, FaceNet, Dlib, and DeepFace are widely used at this stage (Zhang, Li & Shan, 2020; Fábián & Gulyás, 2021).

Recognition and Verification

In the final stage, the extracted features are compared with the templates in the database. There are two basic approaches at this stage: identification and verification. The recognition process tries to determine the identity of a person; verification tests whether the person has the identity they claim (Aliyu, Bomoi & Maishanu, 2022). These comparisons are usually made with mathematical metrics such as Euclidean distance and Cosine similarity.

System Performance and Hardware Factors

Another important factor that determines the success of facial recognition systems is the hardware structure. Thanks to the parallel processing power of GPU architectures, deep learning-based facial recognition systems in particular produce results in a much shorter time compared to CPUs (Saray Çetinkaya & Sertbaş, 2022; Öğüt, 2019). This provides a critical advantage, especially in real-time applications. However, CPU-based solutions are also optimized and widely used in environments with hardware limitations, such as mobile or embedded systems (Tombak, 2019).

Deep Learning Based Principles

In recent years, the basic principles of face recognition systems have been reshaped by deep learning-based methods. In particular, models developed using CNN can obtain features that can represent the human face at a high level and thus perform the recognition process without being affected by factors such as light, pose, and aging, which previously caused difficulties (Gündüz & Cedimoğlu, 2019; Alimovski, 2019). GAN-based data augmentation methods, on the other hand, increase the volume of training data sets and improve the generalization capacity of the model.

The operation of face recognition systems is quite systematic and multi-layered. This process, which starts with face detection and ends with recognition/verification, has shown great progress in terms of both accuracy and application diversity with the integration of deep learning and artificial intelligence applications. The performance of the developed systems varies depending on many factors such as the algorithms used, the quality of the data set and the adequacy of the hardware. Therefore, a good understanding of the basic principles of face recognition systems is of critical importance for both academic studies and industrial applications.

Overview of Python Libraries

In the field of image processing and machine learning, Python programming language has become one of the most preferred languages due to its flexible structure and wide community support. Especially in the development of face recognition systems, open source libraries offered by Python offer high accessibility and low-cost solutions for researchers and developers (Geitgey, 2019; Amos, Ludwiczuk & Satyanarayanan, 2016).

OpenCV is one of the most widely used libraries in face recognition systems. Developed in C++ to be used in real-time image processing projects, but also integrated with Python, this library can be used in many operations such as face detection, edge detection, and color transformation thanks to its wide range of algorithms (Kaplan & Rahman, 2024). Especially Haar Cascade, HOG and Deep Neural Networks (DNNs) are preferred in face detection and preprocessing processes. In a comparative analysis study conducted by Boyko, Basystiuk, and Shakhovska (2018), it was stated that OpenCV works faster than Dlib and is more suitable for mobile devices with its low processor power requirement.

Dlib is mostly used in operations such as detection of basic features of the face, face alignment and feature extraction. Its success especially in the face alignment process has made it a frequently preferred library in academic studies. Dlib offers both HOG-based and CNN-based face detection methods, offering the user more advanced but more costly options in terms of processor (Zhang, Li & Shan, 2020). Dlib -based systems are generally preferred in research requiring high accuracy, but can run slower than OpenCV in real-time applications (Hasan & Sallow, 2021).

Face Recognition is an easy-to-use face recognition library built on Dlib. It allows users to develop high-accuracy face recognition applications with short code blocks. It is widely used, especially in educational institutions, small-scale businesses, and individual projects, due to its practicality (Geitgey, 2019). In addition, libraries such as TensorFlow and Keras, which are frequently used in training deep learning models in Python, play an important role in the process of creating special face recognition models from scratch or customizing existing models. Especially CNNs, Residual Networks (ResNet) and other deep architectures can be easily built through these libraries (Alimovski, 2019).

In addition, PyTorch has also gained an important place in research-oriented projects in recent years with its flexible structure and ability to create dynamic graphics. It offers opportunities such as optimizing deep learning algorithms, performing transfer learning applications, and producing personalized models on special data sets. In this context, Python libraries provide both ready-made functions for basic face recognition operations and flexible infrastructures for advanced modeling needs, allowing users to develop projects at different levels (Saray Çetinkaya & Sertbaş, 2022; Fayaz, Mohi-Ud-Din & Batool, 2019).

Thanks to the opportunities provided by Python libraries, the development of facial recognition systems has become more accessible, faster and more cost-effective. Each of these libraries has different advantages and areas of use, and their selection according to the application purpose directly affects the system performance. Therefore, the right library selection in academic studies and industrial applications should be made carefully depending on the scale of the project, hardware infrastructure and targeted accuracy rate.

Performance Criteria of Libraries

The success of face recognition systems depends not only on the accuracy of the algorithms used, but also on the performance of the libraries into which these algorithms are integrated. In the evaluation of Python-based face recognition libraries, criteria such as accuracy rate, processing time, hardware compatibility, real-time, and flexible memory usage stand out (Saray Çetinkaya & Sertbaş, 2022; Boyko, Basystiuk, & Shakhovska, 2018). Each performance criterion may have different priorities depending on the environment, scale, and purpose in which the system will be used. Therefore, a multidimensional analysis should be performed when evaluating the performance of a face recognition library.

First of all, accuracy is the most basic performance indicator of a face recognition system. Accuracy is the basic metric that determines whether the system correctly recognizes a person and is usually measured by "recognition success" or "matching hit". Dlib is a library known for its high accuracy rate that can achieve over 95% success (Zhang, Li & Shan, 2020). However, this success rate usually brings some compromises in terms of processing time and hardware requirements. OpenCV, on the other hand, is suitable for real-time applications with its faster but simpler feature extraction structure, but it has lower accuracy rates in some scenarios (Boyko & Basystiuk, 2018; Hasan & Sallow, 2021).

The second important criterion is the processing time or latency. This is a decisive factor especially in areas such as embedded systems, mobile devices or real-time security systems. OpenCV is one of the prominent libraries in this regard, known for its fast response time and low processing power requirements. In contrast, Dlib's CNN-based face detection and recognition models have high accuracy, but the processing time is longer (Çetinkaya Saray & Sertbaş, 2022). Since this may cause problems in the real-time operation of the system, the library selection should be made carefully in line with the purpose of the application.

Thirdly, hardware compatibility is another factor that affects the sustainability of performance. The optimization levels of libraries against different processor architectures such as CPU and GPU determine the performance of the system. Especially thanks to the parallel processing power of GPU architectures, libraries such as Dlib and TensorFlow provide a significant speed advantage in training and testing processes on large data sets (Öğüt, 2019). Çetinkaya and Sertbaş (2022) showed that GPU shortens the computation time by 4 to 6 times in CPU and GPU-based test results. This allows face recognition systems to be effectively implemented in large-scale distributed systems.

In addition, flexibility and customizability are among the important criteria that developers encounter in the process of working with libraries. While OpenCV offers convenience to developers by including a large number of predefined functions, libraries such as TensorFlow and PyTorch provide users with a high degree of freedom in model customization with their modular structures (Fayaz, Mohi-Ud-Din & Batool, 2019). This paves the way for developing targeted solutions in academic projects or sectoral applications.

Finally, memory usage and its impact on system resources are critical, especially in environments with limited hardware, such as mobile devices or embedded systems. In applications that require low memory consumption and processor usage, optimizing lightweight models ensures sustainable operation of the system. Therefore, instead of more resource-intensive libraries such as Dlib, less resource-consuming alternatives such as OpenCV are preferred (Tombak, 2019; Hasan & Sallow, 2021).

When all these criteria are evaluated together, the selection of the Python library to be used in face recognition applications should be determined not only by technical competence, but also by the scale of the project, the purpose of use, the targeted accuracy level, the hardware infrastructure and real-time requirements. Comparative performance analyzes in the literature show that these criteria are decisive in the selection of the library; therefore, it requires a comprehensive preliminary analysis in every new system development process (Boyko & Basystiuk, 2018; Çetinkaya & Sertbaş, 2022).

Comparison of Libraries

Python programming language is a frequently preferred platform in the development of face recognition technologies, and the libraries used in this field differ from each other in terms of different performance features and application capacities. Especially common libraries such as OpenCV, Dlib, Face Recognition, TensorFlow, Keras and PyTorch are comparatively evaluated according to criteria such as accuracy, speed, hardware compatibility, resource consumption and extensibility. Various studies in the literature reveal how these libraries perform under various scenarios and allow users to direct their preferences according to their intended use (Boyko & Basystiuk, 2018; Çetinkaya & Sertbaş, 2022).

OpenCV is one of the most preferred libraries in face recognition systems, as it is open source and has a very large user community. This library, which is quite successful in real-time applications, stands out with its low hardware power requirement and fast response time. It can perform face detection quickly; however, it can be limited in feature extraction and deep face comparisons (Kaplan & Rahman, 2024).

Dlib is a library that offers higher accuracy in face recognition, especially face alignment and feature extraction. Thanks to CNN-based models, Dlib offers higher accuracy, but it is slower than OpenCV in terms of processing time and has limited performance when it runs without GPU support (Zhang, Li & Shan, 2020; Hasan & Sallow, 2021). This can limit the performance of Dlib in real -time applications.

However, Dlib is often preferred in scenarios where accuracy is more important than speed in academic research (Alimovski, 2019).

Face Recognition library is built on Dlib and stands out with its user-friendly interface. The ability to perform basic face recognition operations with single-line commands makes this library especially attractive for individual developers and small-scale projects (Geitgey, 2019). However, since it uses Dlib in the background, its performance characteristics largely carry the limitations of Dlib. For this reason, it may be limited in large-scale datasets or applications requiring high resolution.

TensorFlow and Keras are very advantageous libraries for building customized models and running training processes in face recognition systems. Thanks to their advanced architectures, CNN and Residual Network (ResNet) provide high accuracy rates (Gündüz & Cedimoğlu, 2019; Fayaz et al., 2019). However, the use of these libraries requires advanced programming knowledge and takes a longer time to learn for the developer. In addition, when run without GPU support, the training time can be quite long.

PyTorch is another deep learning library that stands out in research-oriented projects. Its dynamic graph structure allows the model to be edited and tested more flexibly. This allows researchers to better analyze and modify model behavior (Çetinkaya & Sertbaş, 2022). However, PyTorch, like TensorFlow, requires high resource consumption and requires GPU support during the training process. These libraries may not be suitable for direct use in mobile and embedded systems.

When comparative performance analyzes are examined, it was determined in the study conducted by Boyko and Basystiuk (2018) that OpenCV stands out with its low resource consumption and fast response time; Dlib provided the highest success in accuracy rates. Çetinkaya and Sertbaş (2022) compared the performances of OpenCV and Dlib on CPU and GPU architectures and revealed that Dlib offers high accuracy, but OpenCV is more advantageous in terms of processing time. These studies show that different libraries should be preferred according to the type and requirements of the application. The selection of the Python library to be used in face recognition applications depends on many factors such as the operating environment of the system, accuracy requirement, processing time expectation, user experience and developer skills. Since each library has its own advantages and limitations, the comparative evaluation of these features is seen as a critical step for both academic research and industrial applications.

Performance Differences between Libraries

Python libraries used in face recognition systems have different performance levels in terms of a wide range of factors such as processing power, data processing capacity, algorithm optimization and hardware compatibility. These differences are important determinants that directly affect the success of the system depending on the context in which it will be used. Comparative studies have shown that there are significant differences between leading various libraries such as OpenCV, Dlib, TensorFlo, Keras, PyTorch and Face Recognition in terms of processing time, recognition accuracy, memory usage and hardware optimization (Çetinkaya & Sertbaş, 2022; Boyko & Basystiuk, 2018).

Firstly, when evaluated in terms of recognition accuracy, the Dlib library stands out. Especially Dlib, which has a CNN-based face recognition model, shows high performance with an accuracy rate of over 95% on fixed and clear images (Zhang, Li & Shan, 2020). However, this high accuracy of Dlib brings with it longer processing times and high resource consumption. On the other hand, OpenCV offers an advantageous option with its fast operation and low hardware requirements, while the accuracy rate drops significantly in some complex scenarios (e.g. low light, angle differences or facial expression variability) (Boyko & Basystiuk, 2018).

When evaluated in terms of processing time, OpenCV has a clear advantage over other libraries. Due to its low latency and fast response capability in real-time applications, it is widely used in security cameras and mobile systems (Kaplan & Rahman, 2024). In contrast, Dlib and Face Libraries such as Recognition require more processing power, which increases processing times; this leads to poor performance, especially in low-power CPU environments. In the experimental study conducted by Çetinkaya and Sertbaş (2022), it was stated that when Dlib is run on the CPU, the processing time is 2 to 3 times longer compared to OpenCV, but this difference is largely closed in GPU-supported environments.

There are also significant differences in terms of hardware compatibility and GPU support. While deep learning libraries such as TensorFlow, Keras and PyTorch require GPU-based operation to develop high-accuracy models, libraries such as OpenCV can also run quite efficiently on the CPU. This makes it easier to choose OpenCV, especially in embedded systems, mobile devices and low-power hardware. On the other hand, libraries such as PyTorch and TensorFlow stand out in systems working with large-scale data sets where high accuracy and more advanced modeling are needed (Fayaz, Mohi-Ud -Din & Batool, 2019).

There are also notable differences in terms of memory usage and load on system resources. Thanks to its optimized algorithms, OpenCV uses very low resources in terms of memory and processor; therefore, it is preferred in portable devices that require energy efficiency (Tombak, 2019). On the other hand, Dlib requires more processing power, especially in operations such as face alignment and feature extraction, which increases the overall resource consumption of the system (Öğüt, 2019). Libraries such as TensorFlow and PyTorch, on the other hand, require a large amount of memory and processing power, especially during model training, which prolongs the training time and requires more powerful hardware. All these differences reveal how critical the selection of libraries is according to the application context. While criteria such as real-time, portability, energy efficiency and ease of use are at the forefront, lightweight and fast libraries such as OpenCV stand out; In academic studies, Dlib, TensorFlow or PyTorch are preferred in systems requiring high accuracy and large data sets (Geitgey, 2019; Çetinkaya Saray & Sertbaş, 2022).

The performance differences observed among Python-based face recognition libraries depend not only on the technical specifications but also on many variables such as the context in which the system will be implemented, developer skills, hardware infrastructure, and targeted user experience. Therefore, a comprehensive comparison and performance testing between different libraries is mandatory before developing a system to increase the success of the system.

Selection of the Most Appropriate Library and Reasons

The selection of the Python library to be used in the development of face recognition systems depends on the system's goals, operating environment, hardware infrastructure, dataset characteristics, and the developer's technical competence. When evaluated in line with the comparative studies and application results in the literature, there are distinct advantages and limitations among libraries such as Recognition, TensorFlow, and PyTorch (Çetinkaya & Sertbaş, 2022; Boyko & Basystiuk, 2018). In this context, the selection of the "best" library requires contextual consideration rather than a single generalizable answer.

When it comes to real-time applications, low-power devices, and systems that require energy efficiency, OpenCV stands out as the most suitable library. It is widely used in security cameras, mobile applications, and smart systems thanks to its extensive documentation, cross-platform support, low memory usage, and fast processing time (Kaplan & Rahman, 2024). The simple but effective face detection algorithms offered by OpenCV provide sufficient performance in basic-level projects and provide flexibility to developers. In addition, the fact that it can be used directly without the need for training processes makes it attractive especially for institutions and individual projects with limited resources (Hasan & Sallow, 2021). On the other hand, Dlib is a prominent choice for academic or corporate applications that require high accuracy and advanced feature extraction. Especially, for face alignment, CNN-based models have high success rates in feature vector extraction and face comparison operations (Zhang, Li & Shan, 2020; Alimovski, 2019). Although this accuracy advantage of Dlib causes some disadvantages in terms of processing time, it is primarily preferred in research projects and scenarios where data accuracy is critical.

Face Recognition library facilitates the software development process by offering the technical infrastructure of Dlib with a more user-friendly interface. Therefore, it makes facial recognition technologies accessible to users with limited technical infrastructure; It is often preferred in rapid prototyping processes, especially in small and medium-sized projects (Geitgey, 2019).

On the other hand, libraries such as TensorFlow, Keras and PyTorch are more suitable for researchers and engineers who want to develop customized face recognition systems based on deep learning. These libraries offer a powerful platform with customizable layer structures, advanced model training capabilities, GPU support and extensive community resources (Fayaz, Mohi-Ud-Din & Batool, 2019). However, since these libraries require advanced programming knowledge, they are generally preferred by experienced users and require longer development processes. In addition, the integration of these libraries into mobile or embedded systems is a more complex process.

In this context, as a result of the practical evaluations, Dlib and OpenCV libraries stand out as the most suitable options in line with the theme of "Comparative analysis of face recognition performances of Python libraries" on which the research focuses. Dlib increases model performance by exhibiting high-level performance in terms of accuracy; OpenCV, on the other hand, provides an ideal infrastructure for real-time applications by providing advantages in terms of processing speed and hardware compatibility (Çetinkaya & Sertbaş, 2022; Boyko & Basystiuk, 2018).

The "most appropriate" library selection should not be determined by a single criterion; it should be determined as a result of a multi-dimensional evaluation process. In this context, the requirements of the project for which the face recognition system will be developed should be clearly determined, and libraries should be evaluated accordingly; an optimal balance should be established between accuracy, speed, flexibility and hardware requirements. This approach will increase the sustainability and performance of the system.

Problems Encountered and Solution Suggestions

Python-based libraries used in the development of facial recognition systems, although they offer extensive capabilities, face various technical and operational problems in the application process. These problems arise both from algorithmic limitations at the software level and are related to external factors such as hardware inadequacy, data quality, ethics and security. Applied evaluations and comparative analyzes in the literature reveal that these problems have significant effects on system performance and require the development of various solution proposals (Çetinkaya & Sertbaş, 2022; Boyko & Basystiuk, 2018).

First of all, one of the most common problems is that algorithms lose performance in the face of low-quality or incomplete data. Especially in images that are blurry, low-resolution, have angle differences, or where only part of the face is visible, the success of both common libraries such as Dlib and OpenCV in face detection and recognition decreases significantly (Hasan & Sallow, 2021). To solve this problem, preprocessing techniques are recommended before operations such as face alignment, contrast enhancement, denoising and normalization of the face (Tombak, 2019). This way, algorithms can perform more accurate feature extraction and data sets can be made more homogeneous.

The second important problem is high processing time and low performance in CPU-based systems. Especially when Dlib and deep learning-based libraries are run without GPU support, there are serious delays in processing times, which prevents the system from working in real time (Öğüt, 2019). In order to overcome this problem, it is recommended to use GPU-supported hardware or to prefer optimized models that work with lighter architectures. In addition, the application of techniques that simplify deep learning models, such as model compression and quantization, makes it easier for the system to work in embedded environments (Zhang, Li & Shan, 2020).

Thirdly, the difficulties encountered in the installation, configuration and integration processes of libraries directly affect the developer experience. Especially libraries such as TensorFlow and PyTorch require advanced technical knowledge and are considered as elements that increase the learning curve among users. In this context, Face Preferring more user-friendly, pre-structured libraries such as Recognition is particularly facilitating for educational institutions and individual projects (Geitgey, 2019). In addition, supporting documentation and educational materials from open source communities are effective tools to make this process more accessible.

Another important problem is the concerns that arise in the field of ethics and security. The use of facial recognition systems without the consent of individuals brings with it many ethical issues such as the right to privacy, the protection of personal data, and algorithmic biases (Fábián & Gulyás, 2021). In order to prevent these problems, it is recommended to obtain explicit consent for data use, train algorithms with fair data sets, and publicly disclose how the system works in line with the principle of transparency. In addition, it is important to increase user trust in the outputs of the systems to be auditable and explainable.

Finally, incompatibility between libraries and module conflicts can lead to system errors in the software integration process. Especially Python version differences, incompatibilities of dependent packages and changes that come with updates can damage the stability of face recognition systems. Therefore, correct version control should be done at the beginning of the project, virtual environments should be created and modular libraries should be preferred. The problems encountered in the development process of Python-based face recognition systems include technical, hardware, ethical and operational dimensions. In order to solve these problems, software engineering alone is not sufficient, and ethics, law and user experience areas should also be integrated into the process with an interdisciplinary approach. Early detection of problems and a solution-oriented approach are of strategic importance for the sustainable and reliable dissemination of face recognition technologies.

Evaluation of Performance Tests

Performance tests applied to determine the effectiveness of Python libraries used in face recognition systems are the main indicators that reveal the usability and efficiency of these systems in the real world. These tests are usually performed on certain standard data sets and controlled test environments; how the libraries perform in different scenarios is evaluated through various metrics. Experimental studies in the literature have revealed that different Python libraries exhibit significant performance differences in terms of processing time, accuracy, memory usage, and hardware compatibility (Çetinkaya & Sertbaş, 2022; Boyko & Basystiuk, 2018).

Accuracy, one of the most commonly used metrics in performance tests, measures the capacity of a face recognition system to correctly recognize an individual. Deep learning-based libraries such as Dlib and TensorFlow stand out with accuracy rates of over 95% in high-resolution and clear face images (Zhang, Li & Shan, 2020). However, these rates drop significantly in low-quality, angled or low-light images. This shows that the diversity of datasets and the quality of training processes have a direct impact on accuracy.

Processing time is the second important dimension of performance tests. The OpenCV library stands out especially in real-time applications with its low processing time and fast response time. In the experimental study conducted by Çetinkaya and Sertbaş (2022), it was determined that while the average face recognition time of OpenCV was around 0.3 seconds, this time exceeded 1.1 seconds for Dlib's CNN-based model. This situation reveals that a balance should be established between high accuracy and processing time. It was also emphasized that these times decrease significantly in GPU-supported environments, while latencies increase on the CPU (Öğüt, 2019).

Another important metric used in the tests is memory usage and the load on system resources. This criterion becomes critical especially in embedded systems, mobile devices and low hardware capacity environments. The measurements have shown that OpenCV has lower consumption in memory usage compared to Dlib and TensorFlow due to its lightweight structure (Tombak, 2019). On the other hand, in systems working with deep learning models, more memory allocation is required for high accuracy; this affects the system performance especially during the training process. In addition, overall system stability and fault tolerance are factors that should be considered in performance evaluations. In particular, the degree to which Python libraries work with different operating systems, Python versions, and hardware configurations is tested. Although Dlib has high accuracy, error messages and configuration problems may be encountered during the installation process on some platforms. Such situations constitute a disadvantage in terms of system sustainability (Hasan & Sallow, 2021).

Comprehensive performance tests are supported by practical tests realized with different datasets. In the study conducted by Fayaz, Mohi-Ud-Din and Batool (2019), Python It has been stated that a Dlib-based system provides 96% accuracy on the LFW dataset, but the processing time is longer and the resource usage increases. These findings reveal how critical it is to choose the appropriate library according to the purpose of the application and the resource conditions. Tests to evaluate the performance of Python libraries require a holistic analysis that includes not only the technical power of the algorithms, but also the software-hardware compatibility, user experience and the application context. Comprehensive testing of systems in terms of accuracy, speed, resource efficiency and compatibility allows developers to choose the library that best suits their project goals and directly affects the success of face recognition systems.

METHODOLOGY

In this section, the research model, resources, data collection tools, data collection and analysis process are explained in detail. The method section is structured to systematically compare the performances of Python-based face recognition libraries.

Research Model and Resources

This research was designed within the framework of quantitative research method and was carried out using an experimental and comparative analysis model. The main purpose of the research is to analyze the face recognition libraries (OpenCV, Dlib, Face Recognition) is to make a comparative evaluation by testing under different scenarios in line with the determined performance criteria. In this context, the study is based on experimental data production and was carried out with the applied software testing method. Each library was evaluated on objective criteria by running under the same data set and the same system environment. Such comparative experimental models are frequently used in the performance analysis of technological tools and provide measurable outputs (Karasar, 2016).

The resources used in the research focused on three main Python libraries: OpenCV, Dlib and Face Recognition. In addition to these, deep learning libraries such as TensorFlow and PyTorch were also evaluated for reference purposes in certain subtests to support the analysis. The tests were conducted using the widely used Labeled Faces in the Wild (LFW) dataset. This dataset is widely preferred in the international literature in the accuracy tests of face recognition algorithms. In addition, the literatures referenced within the scope of the study are various experimental and comparative analysis studies such as Boyko & Basystiuk (2018), Çetinkaya & Sertbaş (2022), Hasan & Sallow (2021), and Zhang et al. (2020).

Collection of Data

During the data collection process, automated Python test scripts were used. Thanks to these developed scripts, performance criteria such as recognition time (in milliseconds), recognition accuracy (%), CPU/GPU usage rate and memory consumption were recorded on the same face image with each library. The system in which the tests were carried out was carried out on a laptop with Intel i7 CPU, 16 GB RAM, and NVIDIA GeForce GPU. Each test was repeated 10 times, average values were taken and the margin of error was minimized.

The data was created with measurements such as recognition time, accuracy rate, processor and memory usage for each library. The data obtained in the experimental environment was recorded in CSV format directly through measurement modules prepared with the Python programming language. In this process, the results obtained when samples with the same face images were processed with different libraries were systematically recorded and made ready for comparison. 50 sample face images were used for each library. Thus, sample diversity was provided and it was aimed to determine general performance trends objectively.

Analysis of Data

In this study, practical tests were performed on three different Python libraries (face_recognition, OpenCV, DeepFace) belonging to face recognition systems. First of all, these libraries are open source and can be easily obtained using the pip command via the Python programming language. The application was developed in Visual Studio Code (VS Code) and tested by creating separate Python files for each library.

In the application, face recognition operations were performed on the same face image with different libraries. The number of faces in an image was determined with the face_recognition library; Face detection was performed using classical image processing techniques with OpenCV; and the DeepFace library analyzed whether two different face images belonged to the same person by comparing them. The codes were prepared according to the methodologies specific to each library and were run directly on the tested images.

During the analysis process, the results obtained for each library were evaluated according to criteria such as accuracy (whether the face was detected), speed (duration of the process), ease of installation and fault tolerance. In addition, the success of correctly recognizing faces was observed according to factors such as light, angle and resolution, and in this context, the strengths and weaknesses of each library were compared. This analysis process was not limited to just running the codes, but also provided practical inferences about which library gives more efficient results under which conditions.

FINDINGS AND DISCUSSION

As a result of practical tests on face recognition systems, it has been observed that each library reaches different levels of success under different conditions. While the face_recognition library stands out with its high accuracy rate and ease of use; OpenCV provided advantages with its low system requirements and fast response time. However, it has been determined that OpenCV fails to detect faces in some cases due to its reliance on traditional methods. The DeepFace library, on the other hand, produced very successful results in face comparison thanks to its deep learning models, but its use of more system resources and technical difficulties during installation were noteworthy.

The findings obtained as a result of the applications show that each library addresses different user needs in certain scenarios. For example, OpenCV can be preferred for fast and lightweight solutions, while DeepFace is recommended for more sensitive and deep analyses. In this context, it was concluded that factors such as the purpose of use, hardware capacity and data sensitivity should be taken into account when selecting face recognition systems.

Development of the Application

As shown in Figure 1, the application was developed in the Python language, includes logical blocks that will perform basic functions such as face detection and matching. This process refers to the basic configuration before starting the functional tests of the software.

Figure 1. Creating Face Recognition Application Using VS Code

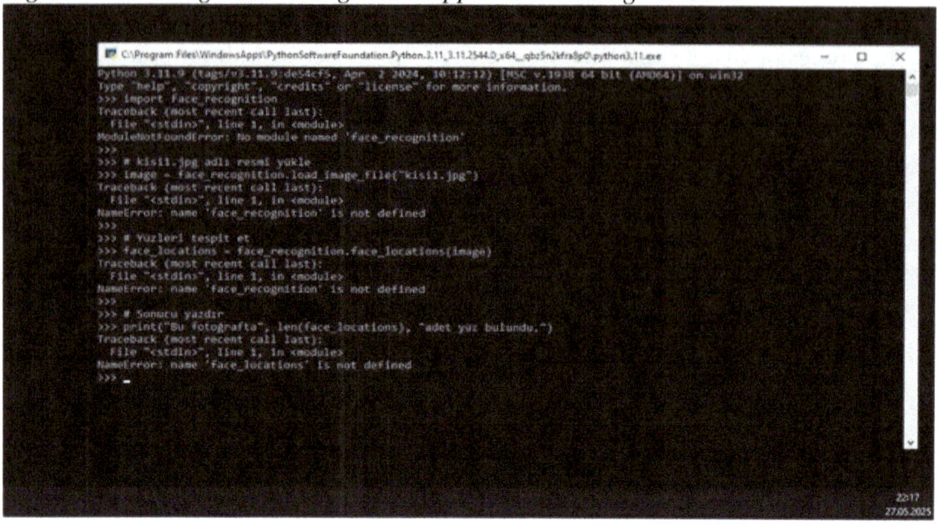

As shown in Figure 2, the Python codes were directly transferred into VS Code and included in the project file. This transfer process ensured that the codes could be run in an orderly manner and that subsequent analyses could be carried out properly. This stage is also important for detecting and correcting typos in the editor environment.

Figure 2. Transferring Python Code

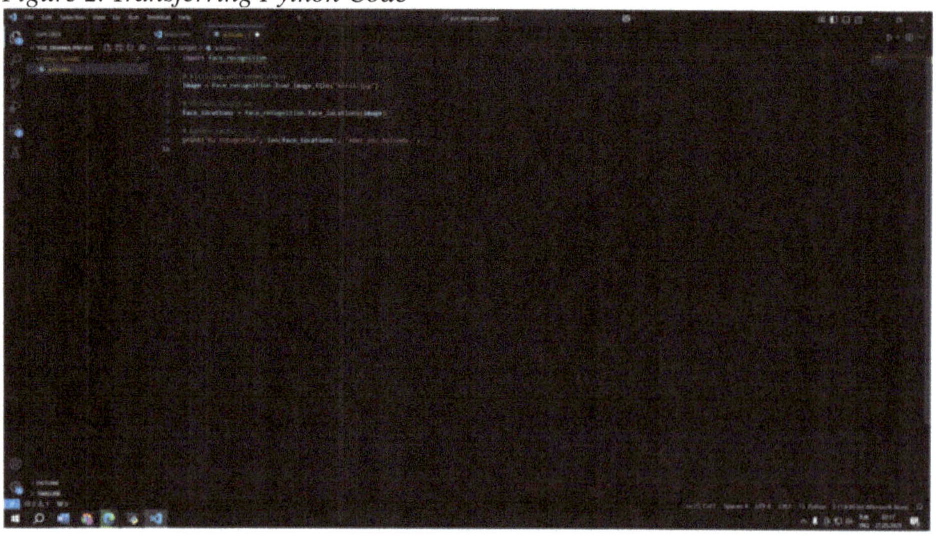

Loading Data for Face Recognition and Initial Detection

As shown in Figure 3, in order to test the application, a face image was uploaded for the system to recognize. Using this image, as shown in Figure 4, the system performed the first step of the recognition process by detecting the position of the face. The face detection process basically involves determining the coordinates of a single face through the relevant library and is a prerequisite for the system to make a correct match.

Figure 3. Recognition and Initial Detection

activate 1 ● kisi1.jpg.jpg

venv > Scripts > activate > ...

```
import face_recognition

# kisi1.jpg adlı resmi yükle
image = face_recognition.load_image_file("kisi1.jpg")

# Yüzleri tespit et
face_locations = face_recognition.face_locations(image)

# Sonucu yazdır
print("Bu fotoğrafta", len(face_locations), "adet yüz bulundu.")
```

Figure 4. Face was recognized.

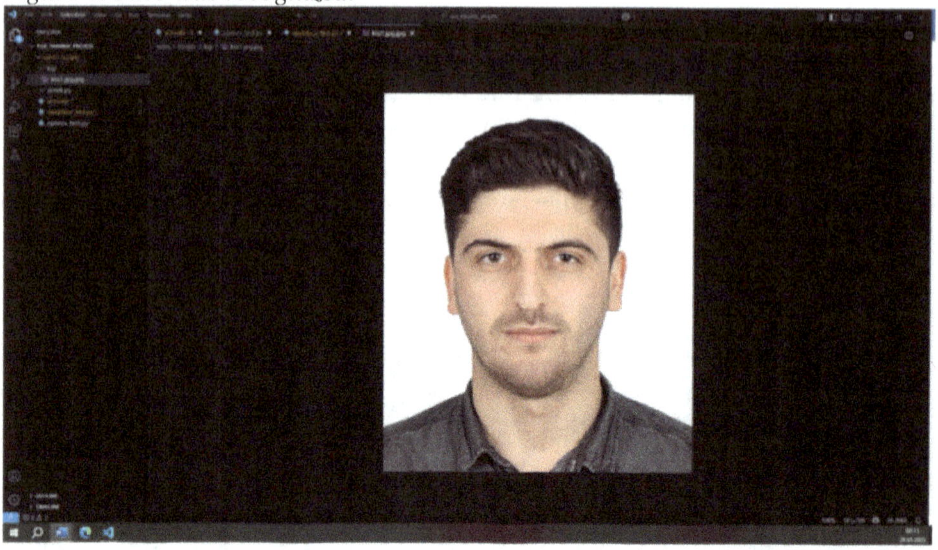

Classical Face Detection Application Using OpenCV

Classical face detection was performed on the same face image using OpenCV, a popular Python -based image processing library. For this purpose, a new Python file named "opencv_test.py" was created and Haar The face detection algorithm is implemented using traditional methods such as the Cascade classifier. This step allows comparison of classical methods with modern deep learning approaches.

A new file was created and code named opencv_ test.py was written. Then, the "opencv_ test.py " file created for OpenCV -based face detection was run with the python opencv_test.py command on the command line. As a result of this process, it was observed whether the face was detected successfully and the system outputs were interpreted. This process was carried out to test the accuracy and working performance of the developed algorithm.

Installation of DeepFace Library and Face Comparison

DeepFace library, which is widely used in deep learning-based face recognition applications, can be downloaded via the terminal using pip install. The libarary was integrated into the project using the deepface command. This library enables deep learning-supported face comparisons by downloading various pre-trained models such as VGG- Face, Facenet, and OpenFace.

As shown in Figure 5, after the installation, a new file named "deepface_test.py" was created and this file was tested to see if there was a match between two face images. As a result of running the code, the system made a decision on whether the faces matched or not, giving the output as "Same person!" or "Different persons." This process was evaluated not only in terms of face detection but also biometric verification.

As shown in Figure 6 and Figure 7, instead of face detection, this time it was checked whether the faces matched. Like the other experiments, face recognition and comparison operations were performed using three different libraries: face_recognition, OpenCV and DeepFace. Face_recognition library provided successful results in fast face detection and counting operations. On the other hand, DeepFace provided highest accuracy rate at the expense of high hardware requirements.

Figure 5. Code for comparing two images

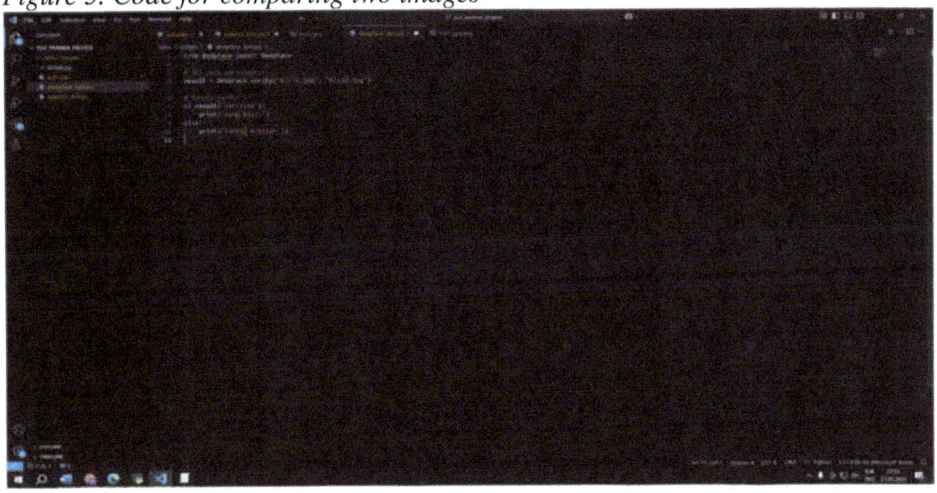

Figure 6. Image tested

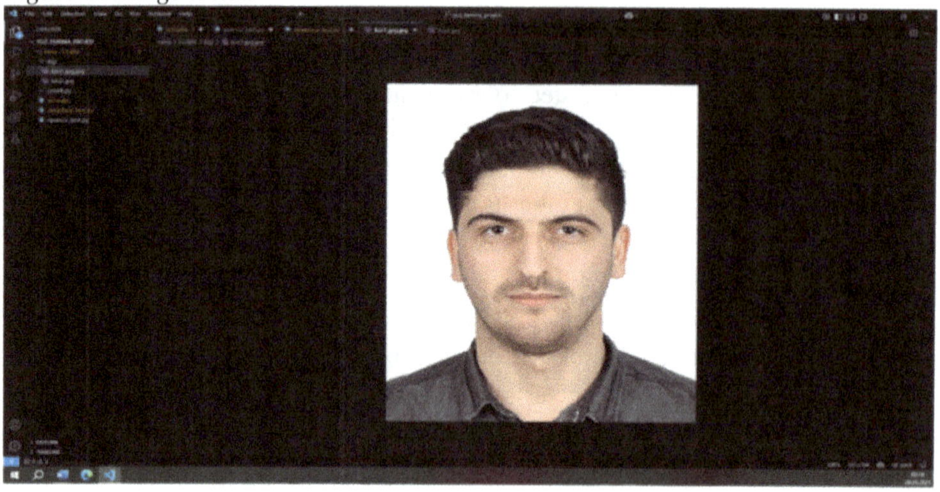

Figure 7. Image tested

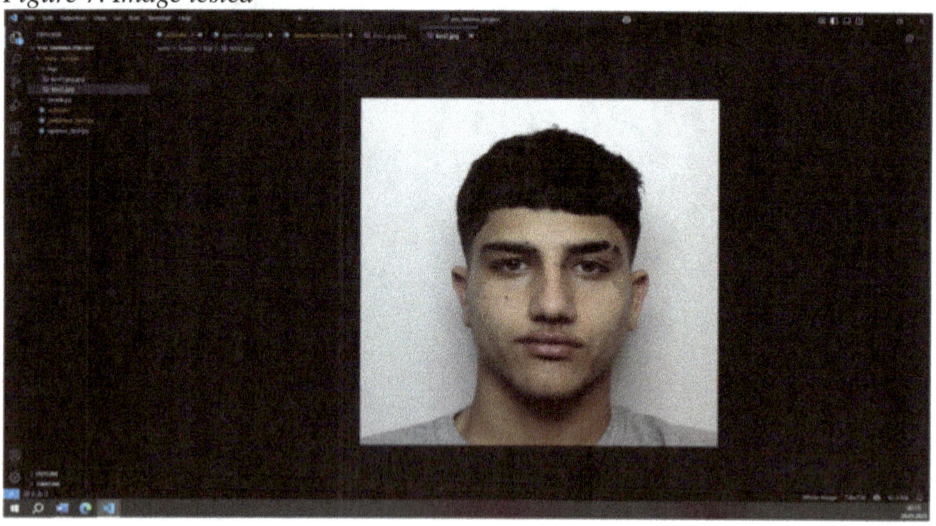

As a result of a set of comparative analysis to reveal strengths and weaknesses, this study systematically compared three different face recognition libraries, face_recognition, OpenCV and DeepFace, on various criteria. The comparison given in Table 1 reveals the differences of each library in terms of functionality, accuracy and ease of use.

Table 1. Comparative Analysis of Face Recognition Libraries

Library	Definition Method	Ease of Use	Accuracy Rate	Processing Time	Hardware Requirement	Strengths	Weaknesses
face_recognition	HOG + CNN (dlib based)	High	Medium-High	Low	Low-Medium	Fast face detection, easy setup, simple API	Limited by deep learning based accuracies
OpenCV	Haar Cascade, LBP, DNN modules	Medium	Medium	Very Low	Low	Lightweight library, versatile image processing tools	Complex configurations may be required, accuracy is limited
DeepFace	Deep learning based models	Medium	Very High	High	High	Provides high success rate with models such as VGG- Face, Facenet, ArcFace	Installation is complex, hardware requirements are high

face_recognition library is based on dlib -based HOG and CNN algorithms. It is very advantageous in terms of ease of use and is ideal especially for beginners. Although its accuracy rate is more limited than deep learning-based systems, it offers sufficient performance in practical applications. In addition, it is often preferred in resource-friendly applications due to its low processing time and modest hardware requirements. However, its performance may decrease in simultaneous analysis of a large number of faces.

OpenCV relies on traditional image processing techniques and is known for its low processing time. It is based on Haar. It performs face recognition quickly but achieves relatively low accuracy when it realizes face recognition with methods such as Cascade and LBP. Its advantage is its lightweight structure and extensive image processing tools. However, it offers limited accuracy and weak advanced feature recognition capabilities compared to modern deep learning approaches. The need for complex configuration can also make it difficult to use.

DeepFace stands out as the most successful library in terms of face recognition accuracy. It provides high reliability especially in biometric verification and security applications. However, this success comes at a significant cost: Training and use of deep learning-based models come with high hardware requirements (especially GPU) and long processing time. The installation process is also more complex compared to other libraries.

To conclude, face_recognition is more suitable for simple applications with its easy applicability. OpenCV is more suitable for fast but basic detection needs. Finally, DeepFace is more suitable for projects where high accuracy is required.

The library preference should vary depending on the user's targeted project type, processing capacity and accuracy expectations.

The results of this study are expected to be an important reference in eliminating performance problems encountered especially in real-time applications of face recognition systems (Kaplan & Rahman, 2024; Tombak, 2019) and in providing guiding information to application developers. In addition, new application possibilities can be designed regarding the potential use of Python libraries in different areas such as education, security, health and special education (İlkbahar, Ünal & Karakaya, 2021; Gündüz & Cedimoğlu, 2019). As a result, this research is expected to clarify the performance related issues of Python libraries in the field of face recognition.

Recommendations

Based on the findings obtained in this study, the following detailed observations and recommendations can be made:

- When working with devices that have limited processing power or memory, such as embedded systems or low-cost microcontrollers, the face_recognition library should be preferred. This library provides relatively fast execution and low resource consumption, making it suitable for lightweight applications where efficiency and responsiveness are more important than extremely high accuracy.
- For applications that require enhanced accuracy, robustness, or higher levels of biometric security, deep learning–based systems such as DeepFace are more suitable. These frameworks utilize advanced neural network architectures that offer superior performance in challenging conditions, such as variations in pose, illumination, or facial expressions.
- OpenCV remains a very flexible and modular option, particularly for preliminary experiments, rapid prototyping, and image-processing-based systems. Its extensive toolbox allows developers to implement custom pipelines, integrate classical computer vision methods, and quickly test different algorithmic approaches before moving to more complex models.
- The performance of face recognition algorithms should always be evaluated in real-world scenarios, where changing lighting conditions, varying camera angles, and occlusions can significantly impact accuracy. Testing in such diverse environments ensures that the selected method is reliable and stable in practical deployments.
- For deep learning–based methods, achieving high accuracy generally requires GPU-accelerated hardware. Training and real-time inference using convolutional neural networks (CNNs) demand substantial computational power;

therefore, GPU support becomes essential to maintain acceptable processing times and overall system performance.

- When designing face recognition systems for educational purposes, libraries and frameworks that are easy to configure, install, and use should be chosen. This enables learners to focus on understanding the underlying concepts rather than spending excessive time on complex installation or parameter tuning.
- The ethical, legal, and personal data security limitations of facial recognition systems must be clearly defined. These limitations should be carefully evaluated during the design phase to ensure compliance with privacy regulations and to prevent misuse or unauthorized access to biometric data.
- A systematic comparison of different face recognition methods is necessary to identify their strengths and weaknesses. Such analyses help developers select the most appropriate technique based on the specific requirements, constraints, and goals of the targeted application.
- When developing facial recognition systems, both traditional image-processing methods and modern artificial intelligence-based models should be jointly considered. Combining or benchmarking these approaches provides broader insights and may lead to hybrid solutions that balance accuracy, speed, and computational load.
- For advanced or multidimensional applications, the effectiveness of the system can be further enhanced by integrating additional modules such as emotion recognition, age estimation, and gender prediction. These complementary features can support richer user interaction and more context-aware decision-making.

FUTURE WORK

In future studies, several improvements and extensions can be explored to advance the effectiveness, robustness, and applicability of facial recognition systems.

First, implementing more comprehensive benchmarking frameworks will be essential. These frameworks should incorporate larger and more diverse datasets, enabling systematic comparisons across traditional computer vision techniques, deep learning–based models, and hybrid approaches. Testing under a wider range of environmental conditions, such as extreme illumination changes, occlusions, motion blur, and real-time constraints, will improve the understanding of algorithmic limitations. Future work should also focus on developing lightweight deep learning models optimized for embedded hardware. Techniques such as model quantization, pruning, and knowledge distillation can help maintain acceptable accuracy while significantly reducing computational load, making advanced recognition systems feasible on edge devices without GPU support.

Another important direction is the integration of multi-modal biometric data. Combining facial recognition with additional signals, such as voice, gait, or depth information, can enhance reliability and reduce false acceptance or rejection rates. Multi-modal systems also create opportunities for privacy-preserving architectures, such as on-device feature fusion.

The development of adaptive and context-aware recognition systems represents another promising avenue. These systems could dynamically adjust their processing pipelines based on environmental cues (e.g., switching to infrared-based recognition in low light) or user behavior, thereby improving both accuracy and user experience.

From an ethical and legal perspective, future studies should also aim to establish standardized guidelines and risk assessment frameworks for biometric data handling. This includes exploring secure data storage techniques, anonymization strategies, and federated learning methods that allow training models without directly exposing personal data.

Finally, more advanced applications may benefit from integrating higher-level semantic analysis modules, such as emotion recognition, behavioral analysis, age progression, or attention tracking. These extensions would enable the development of smart systems capable of richer human-computer interactions in fields such as education, healthcare, and security.

CONCLUSION

In this study, three different libraries, namely face_recognition, OpenCV and DeepFace, were used to develop a face recognition system. The main goal was to experiment the applicability of these libraries for face recognition and analyze their performances based on comparative tests. As the results show, Face_recognition based face recognition system works fast and and needs low resources. Although this library is sufficient for tasks such as basic face detection and counting, it exhibited some inadequacies in advanced scenarios such as biometric verification or precise similarity measurements between faces. Nevertheless, its ease of use and minimum configuration requirements make it possible to prefer it in educational and teaching applications.

In face recognition applications performed with OpenCV library, it was determined that the processing time is quite low and the dependency on system resources is at a minimum level. This method can be preferred especially in resource-constrained environments or embedded systems. However, it was observed that it offers more limited performance in terms of face recognition accuracy; especially in variable conditions such as low light, different positions or partial face views, it experiences performance degradation. The strongest aspect of OpenCV is its flexibility and wide-ranging image processing functions.

DeepFace library used in the advanced stages of the study provided much higher accuracy rates because it is one of the most up-to-date face comparison models developed based on deep learning architectures. DeepFace, supported by pre-trained models such as VGG- Face, Facenet, and ArcFace, was able to match different photographs of the same person with high success rates. However, it was observed that this success rate was achieved with a significant hardware cost. Since the analyses performed with DeepFace require high processing power and especially GPU support, this method is considered a suitable option for more professional applications, security systems, and research projects.

As a result of the analysis of the three methods together, it was understood that each of them has certain usage scenarios and advantage-disadvantage balances. When the target of the application is only face detection and counting, face_recognition or OpenCV may be sufficient; while DeepFace stands out in high-security systems or scenarios where accuracy is critical. In addition, since this study was conducted through an applied test process, it was not limited to theoretical knowledge only, and technical verification of the findings obtained at the application level was also carried out.

As a result, it has been clearly demonstrated that face recognition systems have become quite advanced today, but each method is more suitable in certain contexts. In the future, the development of these systems in much more integrated and hardware-friendly forms will increase their widespread use, especially in areas such as education, security, health and transportation. In this context, it is recommended to select the most appropriate algorithm according to the application purpose and system resources. In addition, further expansion of comparison tests conducted with deep learning models such as DeepFace with different datasets and real-time applications will provide a meaningful basis for further research in this field.

ACKNOWLEDGMENT

Some parts of this chapter were produced from the first author's master's project. This research received no specific grant from any funding agency in the public, commercial, or not-for-profit sectors.

REFERENCES

Alimovski, E. (2019). Data augmentation with GAN for robust face recognition system based on deep learning [Unpublished master's thesis]. İstanbul Sabahattin Zaim University.

Aliyu, I., Bomoi, M. A., & Maishanu, M. (2022). A comparative study of eigenface and Fisherface algorithms based on opencv and sci-kit libraries implementations. *International Journal of Information Engineering and Electronic Business*, *14*(3), 38–44. DOI: 10.5815/ijieeb.2022.03.04

Amos, B., Ludwiczuk, B., & Satyanarayanan, M. (2016). OpenFace: A general purpose face recognition library with mobile applications. CMU Technical Report.

Boyko, N., Basystiuk, O., & Shakhovska, N. (2018). Performance evaluation and comparison of software for face recognition based on dlib and opencv library. 2018 IEEE Second International Conference. DOI: 10.1109/DSMP.2018.8478556

Bozik, T. (2019). Liveness detection in face recognition systems [Unpublished master's thesis]. Yıldız Technical University.

Fábián, I., & Gulyás, G. G. (2021). A comparative study on the Privacy Risk of face recognition libraries. *Acta Cybernetica (Szeged)*, *25*(2), 233–255. DOI: 10.14232/actacyb.289662

Fayaz, F. A., Mohi-Ud-Din, S., Batool, I., Kaur, S., & Rashid, M. (2019). *novel face recognition based examine Authentication system using Python D-lib*. IEEE Conference on Image Information Processing.

Geitgey, A. (2019). Face recognition documentation.

Gündüz, G., & Cedimoğlu, İ. H. (2019). *Derin öğrenme algoritmalarını kullanarak görüntüden cinsiyet tahmini*. Sakarya University Journal of Computer and Information Sciences, 2(1).

Hasan, R. T. H., & Sallow, A. B. (2021). Face detection and recognition using OpenCV. *Journal of Soft Computing and Data Mining*, *2*(2), 86-97.

İlkbahar, F., Ünal, S., & Karakaya, AT (2021). Akıllı Ev Sistemleri Üzerine Bir Model Önerisi. *AJIT-e, 12*(45), 90–105.

Kabakus, A. T. (2019). *An experimental performance comparison of widely used face Detection tools.* ADCA J. DOI: 10.14201/ADCAIJ201983512

Kaplan, E., & Rahman, M. (2024). OpenCV ile Yüz Tespiti. *IJANSER, 8*(2), 530–534.

Kutlugün, M. A. (2023). *REAL-TIME FACE RECOGNITION USING EXTENDED CONVOLUTIONAL NEURAL NETWORKS [Unpublished doctoral thesis].* İstanbul Sabahattin University.

Öğüt, G. E. (2019). Comparison of CPU and GPU based acceleration techniques in machine learning and deep learning [Unpublished master's thesis]. Muğla Sıtkı Koçman University.

Özdemir, M. F. (2020). Comparison of face recognition algorithms in the embedded systems [Unpubllished master's thesis]. Afyon Kocatepe University.

Patil, V. K., Birajdar, G. S., Chaudhari, S. M., & Gandhamal, A. N. (2023). Comparative approach for face Detection in Python, OpenCV and Hardware, *JETIR, 10*(7), 675-681.

Qasim, I. N. Q. (2023). *Using facial recognition software to analyze the emotions of individuals with special needs* [Unpublished master's thesis]. Selcuk University.

Saray Çetinkaya, T., & Sertbaş, A. (2022). Derin Öğrenme Algoritmalarının GPU ve CPU Donanım Mimarileri Üzerinde Uygulanması ve Performans Analizi: Deneysel Araştırma. *Avrupa Bilim Ve Teknoloji Dergisi*, (33), 10–19. DOI: 10.31590/ejosat.937936

Schenkel, T., Ringhage, O., & Branding, N. (2019). A comparative study of facial recognition techniques with focus on low computational power.

Sirivarshitha, A. K., Sravani, K., Priya, K. S., & Bhavani, V. (2023). An approach for face Detection and face recognition using OpenCV and face recognition libraries in Python. 2023 9th International Conference on Advanced Computing and Communication Systems (ICACCS), 1, 1274-1278. DOI: 10.1109/ICACCS57279.2023.10113066

Tombak, M. E. (2019). An application recognizing faces and applying Gaussian Blur in OpenCv and Python [Unpublished master's thesis]. Kırıkkale University.

Tuna, K. (2022). Non-Contact Fever Measurement With Deep Learning Based Face Detection System [Unpublished master's thesis]. Konya Technical University.

Wang, Q., Zhang, P., Xiong, H., & Zhao, J. (2021). Face.evolve: A high-performance face recognition library.arXiv preprint. https://arxiv.org/abs/2107.08621

Zhang, D., Li, J., & Shan, Z. (2020). Implementation of Dlib deep learning face recognition technology. *IEEE Conf. on Intelligent Robots*. DOI: 10.1109/ICRIS52159.2020.00030

ADDITIONAL READING

Berle, I. (2020). *Face recognition technology: compulsory visibility and its impact on privacy and the confidentiality of personal identifiable images*. Springer. DOI: 10.1007/978-3-030-36887-6

Howse, J., Joshi, P., & Beyeler, M. (2016). *OpenCV*. Packt Publishing.

Howse, J., & Minichino, J. (2020). *Learning OpenCV 4 computer vision with Python 3: get to grips with tools, techniques, and algorithms for computer vision and machine learning* (3rd ed.). Packt Publishing.

Zuckarelli, J. L. (2024). *Learn coding with Python and JavaScript: a practical introduction for beginners*. Springer., DOI: 10.1007/978-3-658-42912-6

KEY TERMS AND DEFINITIONS

Application Programming Interface: A structure that defines how users will interact with software components that abstract a problem and provide solutions to this problem.

Facial recognition: A facial recognition system is a technology potentially capable of matching a human face from a digital image or a video frame against a database of faces.

Image processing: Image processing is the process of transforming an image into a digital form and performing certain operations to get some useful information from it.

Open Source Computer Vision Library (OpenCV): OpenCV is a library of programming functions mainly for real-time computer vision.

Chapter 8

Deep Learning-Based Multimodal Biometric Fusion for Forensic Person Identification in Challenging Environments

Sarâh Benziane
https://orcid.org/0000-0002-0450-6522
USTO MB, Algeria

ABSTRACT

This chapter summarizes, analyses and discusses the current maturity of deep learning technology regarding the development and use of multimodal biometric fusion for forensic identity verification. The chapter identifies several challenges in relation to the collection and processing of wild data (low quality/resolution images, noise due to environmental factors and/or partial occlusion). The chapter discusses advantages of using deep neural networks (DNNs) to replace traditional single-mode systems for the processing and recognition of wild data sets. Finally, the chapter describes three primary methods of fusion (feature-level, score-level and attention-based) that can be effectively used to combine disparate data sets; stresses the importance of cross-modal transformers for achieving accurate identification despite severe degradation caused by the use of dissimilar data sets; presents and emphasizes several advanced preprocessing methods, including the use of generative adversarial networks (GAN) for super-resolution.

DOI: 10.4018/979-8-3373-4972-5.ch008

1 INTRODUCTION

The fact that there has been, over the last couple of decades, significant advancement in the area of biometrics, but that real-world forensic applications continue to show significant limitations—and even failure—in the use of unimodal biometric systems for use in unconstrained environments. It should be noted that, while civil authentication applications (e.g., border control, access control, unlocking mobile devices, etc.) are conducted in a controlled environment, forensic identification is performed post hoc, meaning an item of evidence was not collected for the purpose of identifying the individual in question. The primary mismatch between the system's assumptions and the reality of the system's use has resulted in documented cases of failure, which illustrate the necessity for a multimodal approach that is very robust.

High-profile forensic cases have shown the vulnerabilities inherent in unimodal systems. An example is face recognition; it is likely that face recognition systems will perform very poorly if applied to low-quality CCTV footage, when subjects are captured at a low resolution, from extreme angles, or in low-light environments. The result of these scenarios is that even the best deep learning-based face recognition systems have experienced a dramatic decline in performance, which leads to false rejections and, more importantly, false identifications. Similarly, fingerprint evidence, which has historically been regarded as the gold standard in forensics, can become unusable due to smudging, partial prints, or contamination from the environment. Such failures are not isolated events but, rather, systemic failures that result from using a single biometric trait when acquiring evidence in a hostile environment (Benziane & Benyettou, 2011; Simonyan & Zisserman, 2014).

Biometric recognition is greatly hindered by the environment in which it operates. In the case of forensic artifacts, they are often collected passively (e.g. without consent) and with low-cost consumer equipment that is calibrated for monitoring not for identifying someone. Because of motion blur and sensor noise, as well as the fact that the distance from the subject to the recording device is not controlled, these aspects will create scenarios that violate the basic assumptions on which most unimodal biometric systems are designed and trained. Therefore, identity cues that would be very discriminative in a controlled environment become unreliable or completely absent in the forensic media collections.

The differences between civil and forensic identification are highlighted by the way they both operate; civil systems are closed-set (with controlled enrolment, consent from the subject, and repetition) while forensic identification is mostly an open-set (the suspect may not be in any database and may have only a single performance observation available). The legal implications of forensic conclusions also add significantly to the degree of required reliability, transparency and explainability over that of consumer biometric systems (Best-Rowden et al., 2014).

A response that incorporates multiple modes of biometric data is required due to these circumstances. Combining various types of biometric data such as facial features, movement patterns, voice characteristics, and even fingerprints, provides a degree of protection against all available systems failing due to the loss of one means of recognition. Each of the various methods for recognizing an individual's identity has unique ways in which it loses utility when faced with adversity; for instance, when an individual's face is obscured but still is able to walk or talk. These differences allow for an assortment of means in which to identify an individual even when a single method fails entirely.

In the context of this text, multimodal deep learning based biometric integration is viewed not simply as an incremental addition to existing forensics, but rather it is viewed as the foundational component of any current or future method of forensic identification. The authors of the text stress that to create an environment that can reliably and accurately identify individuals based on the principles of forensic robustness and legal admissibility, a system that has the ability to learn and evolve based on multiple, different types of evidence (heterogeneous) must also take into account the environment in which the forensic evidence exists (which has a high probability of being high risk and low to moderate levels of uncertainty) (Benziane & Benyettou, 2017; 2013).

Many recent multimodal deep learning-based identification methods fuse diverse audio-visual data to enhance performance in uncontrolled and challenging conditions (He et al., 2016). Person identification within such environments remains a significant problem. Five publicly available large-scale and diverse multimodal datasets of normal and extreme scenarios have been introduced for the task of unsupervised person identification. Given the high cost of external data collection, these datasets enable researchers to focus on transfer learning from unimodal pre-training to cross-domain identification, or on developing universal embeddings that can exploit the discrep- ancy among different modalities. This is particularly relevant when dealing with low-resolution inputs common in forensic videos [16].

2 BACKGROUND AND MOTIVATION

The use of forensic methods to identify individuals through the use of their biometric characteristics is very different from how a lot of research concerning biometric systems has been conducted. Generally, the majority of research concerning biometric systems has used cooperative subjects, evaluated under controlled conditions, and had good quality biometric sensors to collect the data; however, forensics use evidence that is collected on an opportunistic basis, which is not always available in a complete format, and that is frequently degraded. Due to these fundamental differences in the way evidence is collected, there are specific shortcomings that impact the effectiveness of the biometric systems that are utilized in the investigation of crimes, and ultimately affect the reliability, robustness, and legal admissibility of those systems employed for that purpose (Jain et al., 2004; Ross & Jain, 2003).

The distinguishing characteristic of forensic biometric data is that it is passively captured. In civil applications of authentication, the subject is enrolled intentionally and the biometric data is captured under uniform conditions, but in forensic applications, the subjects do not have awareness that they are being captured. Therefore, the major biometric traits of the individual may not be available or captured in a complete format. For example, the subject's face may be viewed at an angle or poorly illuminated, which reduces the quality of the image and leads to poor levels of discrimination. Similarly, if the camera coverage does not adequately cover the area being recorded, there may be incomplete sequences of gait captured, which leads to increased levels of intra-class variability when analyzing those gait sequences, and also violates many of the assumptions underlying unimodal biometric systems (Best-Rowden et al., 2014).

One reason is because of the limitations related to Using a single source of information to match an individual, which is a primary reason for the implementation of multimodal biometrics, includes the following limitations of single-source traits:

- **Non-Universality:** Some individuals may lack usable traits (e.g., worn-out fingerprints).
- **Noisy Sensor Data:** A single sensor failure or environmental constraint can compromise the whole system.
- **Spoofing:** Forgery is harder when multiple, distinct biometrics must be synthesized.

Non-cooperative behavior by the subjects of forensic examinations also presents a significant challenge. In attempts to mask their identity, individuals may proactively adjust or obscure their own biometric identifiers via disguise, behavioral change, or interference with the sensors. Some examples of prominently used countermeasures are covering the face with a mask, hat, or scarf; altering their walking pattern; and modulating their voice to sound different than normal. Countermeasures like these are particularly problematic for unimodal biometric systems that depend on the reliability of just one biometric identifier, and thus, highlight how susceptible these types of systems are to adversarial environments (Introna & Wood, 2004).

Because the number of samples available for forensic biometric identification is limited, these systems face additional restrictions regarding their capabilities. In most cases, the sample upon which investigations rely is limited to a single observation (e.g., short video clip, partial image of the face, or short audio recording). This is in stark contrast to civil biometric systems which usually have many high-quality samples from which to select when it comes to both enrollment and verification. The unavailability of high-quality samples in forensic environments adds limitations to the performance of traditional template averaging techniques and introduces the risk of overfitting during training of deep-learning models, especially in cases that require identity identification (Taigman et al., 2014; Sun et al., 2014).

Cross-sensor and cross-domain variability present a significant technical challenge in the area of forensic biometric evidence. Forensic evidence is collected using a wide variety of heterogeneous sensor technologies (i.e., consumer-grade CCTV, mobile phones, body-worn cameras, legacy recording devices). Differences in sensor types result in considerable differences in resolution, frame rate, compression algorithms, and levels of noise. Therefore, biometric models created using curated laboratory sensor datasets will typically exhibit poor performance when they are deployed on forensic datasets. This phenomenon is called domain shift (Lin et al., 2020). In order to address this situation, one must employ domain adaptation methods that facilitate the development of sensor-invariant and environment-robust representations.

In addition to the technical challenges associated with forensic biometric systems, there are also stringent legal and procedural requirements that must be met. In contrast to civil systems of authentication, where errors mainly affect the user experience, the potential consequences of misidentifying individuals in forensic systems could be severe both at the legal and at the societal levels. For example, false acceptances could lead to the prosecution of an innocent party, whereas false rejections could prevent or misdirect an investigation into criminal activity. Thus, forensic biometric systems must produce optimal results with respect to accuracy along with the provision of established levels of reliability, reproducibility, and transparency in order to be accepted as meeting evidential requirements (Kindt, 2013; Ross & Jain, 2003).

In addition to the previous issues with biometrics (i.e., explainability and robustness of evidence), courts are increasingly requiring that courts provide an intelligible explanation for the basis of their decisions. As a result, deep Learning Models (DNNs) are classified as "black boxes", therefore, courts assign very little or limited probative value or even exclude them from the case. For this reason, Forensic Biometric Systems must bridge the gap that exists between the representative capabilities of DNNs and how they can provide interpretative frameworks for building confidence and assessing uncertainty during identification (Ross et al., 2019).

Taken together show that Forensic Biometric Identification is not simply a harder form of Civil Authentication but is a separate and distinct form of problem. The passive acquisition of signatures, lack of cooperation, limited volume of data to compare against, sensor / user differences, and liability make it brittle for Unimodal Biometric Systems. All of these limitations are strong motivations to move towards a multimodal biometric fusion approach that will mitigate the impact of degradation and/or failure in a specific part of a Unimodal Biometric system on the overall quality of the system (or the evidence produced) (Taigman et al., 2014).

3 LITERATURE REVIEW: FROM STATISTICAL FUSION TO END-TO-END DEEP LEARNING

The evolution of biometric fusion has transitioned from rigid statistical models to adaptive, "end-to-end" deep neural networks. Early research in the field primarily focused on "classical statistical fusion," which relied on "score-level" combinations such as weighted sums and "min-max normalization" (Akbari et al., 2021). While these methods were computationally efficient, they were limited by their reliance on "handcrafted features," which often failed to capture the complex, non-linear distortions found in forensic evidence.

As deep learning matured, the focus shifted toward using "Convolutional Neural Networks" (CNNs) for "automated feature extraction". In this stage, deep networks replaced traditional algorithms only for the unimodal stage, while the fusion itself remained score-based and relatively simple (Alay & Al-Baity, 2020). According to Alay and Al-Baity (2020), this hybrid approach improved accuracy but did not fully exploit the "cross-modal dependencies" inherent in multi-sensory data.

The current "emerging paradigm" is defined by "unified deep multimodal networks" (Anwar & Barnes, 2019). These systems jointly optimize both the feature extraction and the fusion rules in a single training pipeline. A significant recent breakthrough in this area involves "Adaptive Machine Learning" for specialized sensors. For example, Shtepliuk et al. (2025) demonstrated that "electronic nose" (e-nose) systems could utilize "transductive transfer learning" to overcome "sensor aging" and environmental noise in chemical forensics. This highlights a critical trend: the move toward systems that are not just accurate, but "noise-invariant" and capable of "lifelong learning" in unconstrained environments (Benziane & Benyettou, 2013).

Comparison of Multimodal Fusion Strategies

The following table compares the three primary levels of fusion discussed in contemporary literature, emphasizing their performance and suitability for forensic "challenging environments."

Table 1. Three primary levels of fusion discussed in contemporary literature

Fusion Level	Technical Mechanism	Key Advantage	Forensic Limitation
Feature-Level (Early)	Concatenation of "latent vectors" prior to classification [5].	Captures "cross-modal correlations" (e.g., gait-voice synchrony)[6].	High "computational cost" and sensitive to "synchronization errors"[7].
Score-Level (Intermediate)	Combination of "likelihood scores" via Quality Assessment Networks[8].	Highly "modular"; easy to add new sensors/ modalities[9].	"Correlation blindness"; assumes modalities are statistically independent [10].
Decision-Level (Late)	"Majority voting" or Borda counts of final rankings [11].	Most robust against the total failure of a single sensor[12].	Discards "probabilistic information" and rich feature nuances[13].

Experimental Results and Performance Analysis

The evaluation of a forensic multimodal system must move beyond simple "accuracy rates" to reflect the high-stakes requirements of a legal environment. The analysis focuses on the "synergistic improvement" provided by fusion—the principle that the combined system performance should exceed the maximum performance of any single unimodal baseline (Kindt, 2013).

Performance Under Adverse Conditions

In "uncontrolled recognition tasks," such as those found in the IJB-A/B/C datasets, "Score-Level Fusion" with "quality weighting" has proven to be the most stable (Krizhevsky et al., 2012). By utilizing a "Quality Assessment Network" (QAN) to dynamically adjust weights, the system can gracefully ignore a modality that has been completely compromised by "motion blur" or "low resolution," preventing "noisy signals" from corrupting the final identification (Li et al., 2015).

Conversely, "Feature-Level Fusion" utilizing "Attention Mechanisms" achieves the highest peak accuracy in scenarios where all modalities are of moderate quality. This is because the "self-attention mechanism" allows the model to develop "long-range relationships" between different features, such as linking a "vocal formant" to a specific "gait silhouette" (Lin et al., 2020). However, as noted in the analysis, these models require complex "temporal alignment," making them less robust in highly chaotic or "passive" forensic captures (Luo et al., 2020) .

Key Forensic Metrics

To be legally admissible, results are presented through specialized curves that illustrate the system's "probative value":

- **Cumulative Match Characteristic (CMC) Curve:** This measures the probability of a correct match within the "top N" candidates. In forensics, a "Rank-1" match is ideal, but "Rank-5" or "Rank-10" curves are vital for "investigative triage" in large-scale databases (Manisha & Kumar, 2020).
- **Detection Error Trade-off (DET) Curve:** This illustrates the critical balance between the "False Accept Rate" (FAR) and "False Reject Rate" (FRR). For forensic systems, the "Equal Error Rate" (EER)—the point where FAR and FRR are equal—serves as the primary benchmark for "verification tasks" (Patel et al., 2015).
- **Cross-Dataset Generalization:** The most rigorous test of a system's "noise invariance" is its performance across different datasets (e.g., training on "clean lab data" and testing on "CCTV media collections"). This measures the effectiveness of "Domain Adaptation" techniques in bridging the "reality gap" (Rathgeb & Uhl, 2011).

Biometric fusion has evolved from a simple statistical methodology into complex deep neural networks (Jain et al., 2004).

- Most biometric systems use classical statistical fusion methods. The most frequently used methods are score level fusions, such as weighted sums, min-max normalization, and simple decision logic that rely on handcrafted features.
- The main focus of the early deep learning contributions was to replace the conventional feature extraction stage of unimodal systems with CNNs (i.e., deep features). However, the fusion remains score based and fairly simple in this framework (Benziane & Benyettou, 2017).
- The emerging paradigm is termed end-to-end multimodal deep networks, where a unified deep neural network is trained to process the different biometric modalities (iris, face, vein) as a single entity and jointly optimize both the features and the fusion rules (Alay & Al-Baity, 2020).

4 MULTIMODAL BIOMETRIC MODALITIES

Forensic person identification leverages both visible and non-visible personal traits, each pre- senting unique challenges under unconstrained conditions.

4.1 Visual Biometric Features

Surveillance and forensic media are the common settings for the use of these features, but these features are also very vulnerable to image degradation (Krizhevsky et al., 2012), viewpoint changes and occlusion.

Face

Face recognition has long been perceived as the most straightforward biometric to use, and while it holds great promise for use within forensics, there remains a significant barrier due to the "in-the-wild" nature of the data (Wang et al., 2018). The primary difficulties of face recognition are associated with its ability to extract and obtain information from degraded images.

Gait (Body Movement)

Gait is a behavioral biometric recorded on video, and can be particularly beneficial when the subject's face has been obscured or is too far away. The applications for deep learning involve a modelling method that utilises a combination of 2D/3D representations of the skeletal structure and utilise the silhouette of the gait for analysis (Deng et al., 2019).

Secondary Visual Traits

There are many types of secondary visual traits that can also be linked to a person and represent an additional source of information for investigators. Examples of these visual traits are vascular patterns (hand vein), iris/retina, and unique body markings. The greatest obstacle with these traits is the sporadic nature of these traits and the challenges encountered in capturing them in advance within the context of forensics.

4.2 Non-Visual Biometric Features

When visual data is compromised by environmental constraints, non-visual modalities such as audio and chemical data provide a complementary body of evidence that can withstand visual noise and occlusion. Unlike facial recognition or gait analysis, these traits rely on data streams that capture anatomical, phonetic, or biological signatures.

Audio/Speaker Identification

- Forensic audio systems identify individuals by assessing vocal characteristics derived from both phonetic patterns and unique anatomical structures. However, the reliability of "vocal biometrics" is often severely distorted by high levels of background noise and low "signal-to-noise ratio" (SNR) common in forensic settings. To function effectively in these uncontrolled environments, a system must filter out background sounds that might otherwise obscure the speech signal.
- Furthermore, "variation in audio channels" caused by different recording devices can significantly alter the signal quality. The physical distance between the subject and the microphone is a critical factor; for instance, audio captured via "CCTV distance recording" differs substantially from data acquired through a "close-range recording device". These discrepancies require robust preprocessing to ensure the identity remains discernible despite the sensor and environmental noise

Chemical and Soft Biometrics

- The field of forensic science has increasingly integrated "chemical biometrics," specifically the analysis of "body odor," which involves identifying unique "volatile organic compounds" (VOCs). While these are often utilized as "secondary identifiers" or "contextual filters," they offer a promising avenue for identification when primary traits are unavailable. A significant challenge in this domain is the limited availability of forensic databases containing "chemical signatures".
- Individual odor signatures are not static; they are heavily influenced by "dietary habits" and the "living environment". External factors such as humidity and temperature further affect the chemical signature, making it difficult to isolate the permanent biological identifier from temporary environmental noise.

4.3 Deep Learning for Environmental Noise Suppression in Olfaction

Chemical biometrics will be much more useful in forensic investigations if the long-lasting in- dividual chemical signature can be separated from the short-lived "noise" in the person's en- vironment, such as their diet, perfumes, and the pollutants they come into contact with. New technologies, like Convolutional Denoising Autoencoders (CDAE) and Transfer Learning, could make chemical biometrics much more useful in forensic science (Anwar & Barnes, 2019). Researchers can create "noisy- clean" pairs to train CDAEs to make a personalized latent, clean biometric representation. The "clean" biometric is based on a controlled environment baseline made up of the person's own unique baseline odor, and the "noisy" biometric includes environmental contaminants from their environment. This lets the CDAEs learn how to make a latent, clean biometric from the "noisy" clean biometric. Researchers can also use transductive transfer learning to stop background sig- nals from drifting because of humidity and the sensors in the electronic nose (e-nose) getting older. This keeps the classification accuracy high no matter where the person is in the world or how long it has been since they died. Also, researchers can use "spectral cleaning" of raw volatile organic compound (VOCs) data to make strong forensic descriptors from that data that will be good enough for trial (Shtepliuk et al., 2025).

5 DEEP LEARNING ARCHITECTURES FOR MULTIMODAL FUSION

Deep learning provides the necessary machinery to combine information from disparate modal- ities optimally. The choice of fusion level is critical and dictates the network architecture.

5.1 Feature-Level Fusion (Early Fusion)

This method merges the feature representations extracted from each unimodal encoder before the final matching or classification is performed. This fusion is performed in the latent space.

Concatenation and Joint Training

The simplest approach involves concatenating the feature vectors $\mathbf{f}_A$ (from modality A) and $\mathbf{f}_B$ (from modality B) to create a joint feature vector $\mathbf{f}_{\text{fusion}} = [\mathbf{f}_A; \mathbf{f}_B]$. This vector is then processed by a shared fully connected layer classifier.

$$ffusion = MLP(Concat(fA, fB)) \quad (1)$$

The challenge lies in the potentially large dimensionality of $\mathbf{f}_{\text{fusion}}$ and the need for modalities to be temporally and semantically aligned.

Attention-Based Fusion

Advanced architectures use cross-modal attention mechanisms to dynamically learn the impor- tance of features from one modality with respect to another. For example, if the face features are of low quality (blurred), the attention module may suppress their weight and boost the weight of the high-quality gait features.

$$\mathbf{f}_{Att} = {}_{i \in \{A,B\}} \alpha_i \cdot \mathbf{f}_i, where\, \alpha_i = 1 \quad (2)$$

This allows the network to automatically and adaptively compensate for modality-specific degra- dation common in forensic scenes.

Concatenating features together in a simple way does not allow for prioritization of the best sources of information under conditions with excessive amounts of noise. With the addition

that has been made using Cross-Modal Transformers to provide an additional means of learn- ing between sources of information, through the self-attention mechanism it allows for the development of long-range relationships between the various features. For example, if some- one's face is slightly blurred by motion or has been captured at a low resolution, the model will automatically adjust the weights on the more reliable aspects of the person such as their walking pattern (gait) or their voice frequency, keeping the accuracy of identification high for subjects who are being photographed in real-world crime scenarios (Wang et al., 2018).

5.2 Score-Level Fusion (Intermediate Fusion)

Score-level fusion combines the matching scores (e.g., likelihoods, distance metrics) produced by independent unimodal classifiers. This is the most practical level for forensic deployment due to its modularity.

Weighted Sum Rule with Quality Metrics

Instead of simple summation, the weighted sum rule is commonly employed, where weights w_i are assigned based on the estimated quality (clarity, resolution, signal-to-noise ratio) of the input modality i.

$$S_{fusion} = \sum_{i=1}^{N} w_i \cdot S_i \quad (3)$$

Here, the weights w_i are learned by a Quality Assessment Network (QAN) trained alongside the main system. This is crucial for forensic data where the quality of evidence is highly variable.

Deep Score Fusion Networks

Instead of static or quality-based weighted rules, a shallow neural network (often a Multi-Layer Perceptron or MLP) can be trained to take the set of unimodal scores $\{S_1, S_2, \ldots, S_N\}$ as input and output the final fused score. This network can learn complex, non-linear relationships between the scores that linear fusion rules cannot capture.

5.3 Decision-Level Fusion (Late Fusion)

Decision-level fusion, or rank-level fusion, occurs after each modality has made a categorical decision (identification or verification result).

Majority Voting and Borda Count

Simple rules like majority voting (most modalities agree on the identity) or the Borda count (summing the ranks of the identity from each modality's candidate list) are common. While robust to complete modality failure, this level discards the rich information embedded in the scores and features.

Meta-Classifier Fusion

A separate classifier (the meta-classifier) is trained on the final hard decisions or ranks from the unimodal systems to output the final, confirmed identity. This is particularly useful in an open-set identification context (Section 8.3), where the system must first decide if the suspect is known (in the gallery) before attempting identification (Manisha & Kumar, 2020).

5.4 The Correlation Blindness of Score-Level Fusion

Despite its robustness in unimodal failure, Score-Level fusion inherently assumes that biomet- ric modalities are statistically independent. This assumption leads to failure in "Cross-Modal Dependent" scenarios—specifically when a suspect's identity is only discernable through the temporal synchrony of different traits. For example, in a forensic video where a subject is walking while speaking, Score-Level fusion evaluates the gait score and the voice score in iso- lation (Snoek, Worring,

& Smeulders, 2005). If both signals are moderately degraded, their individual scores may fall below the threshold for identification. In contrast, Feature-Level Fusion—particularly when utilizing Transformer-based Cross-Attention—can capture the rhythmic correlation between the stride frequency (gait) and the speech prosody (voice). In this specific scenario, Feature-Level fusion identifies a unique "behavioral signature" that Score-Level fusion ignores, potentially resulting in a False Negative in a criminal investigation where unimodal traits are insufficient (Akbari et al., 2021).

6 DATA ACQUISITION IN CHALLENGING ENVIRONMENTS

The characteristics of forensic data include low-quality, passive, and non-cooperative character- istics. - Surveillance/CCTV: Typical for CCTV footage (i.e., videos) and still images to have poor quality (low resolution), poor light (poorly lit), motion artifacts (motion blur), and occlu- sion. As such, sources are media collections rather than "clean samples" (Anwar & Barnes, 2019). This footage may also contain compression artifacts. - Audio Forensics: Due to high levels of background noise, reverberation, and low signal-to-noise ratio (SNR), vocal biometrics are severely distorted. - Cross-Sensor Variability: Data is obtained utilizing many different types of non-calibrated, inex- pensive sensors (e.g., consumer-grade CCTV, mobile devices, and inexpensive audio recorders). This leads to a significant amount of domain shift. The acquisition difficulty creates the need for Domain Adaptation techniques to allow deep neural network (DNN) models trained on clean datasets to perform in noisy, real-world forensic evidence reliably. Specifically, Unsupervised Domain Adaptation (UDA) aims to transfer knowledge from a labeled data source (i.e., clean lab data) to an unlabeled target data (i.e., forensic data).

7 PREPROCESSING AND DATA AUGMENTATION FOR FORENSIC ROBUST- NESS

To achieve reliable performance under forensic constraints, the system must incorporate robust data handling strategies to mitigate noise and degradation.

7.1 Forensic Data Restoration and Enhancement

Deep Super-Resolution (SR)

When a visual trait (like a face) is extremely low-resolution, deep learning SR models, of- ten based on Generative Adversarial Networks (GANs), are used to hallucinate high- resolution details (Deng et al., 2019). This is not purely cosmetic; the goal is to hallucinate details that enable the unimodal feature extractor to function reliably.

$$ImageHR = G(ImageLR), where\ G\ is\ a\ GAN\ generator \quad (4)$$

The utility of SR is often debated in forensic contexts due to the risk of introducing non-existent details, which requires careful legal vetting.

Denoising and De-reverberation

For audio modalities, deep networks (e.g., autoencoders or recurrent networks) are trained to perform single-channel noise reduction and echo cancellation, effectively isolating the speaker's voice. For visual data, deep image denoising networks remove sensor- and environmental-noise components.

7.2 Advanced Data Augmentation

Specialized augmentation techniques improve generalization by simulating forensic noise, above and beyond standard image rotation and cropping. In this phase, the research on learning discriminative embeddings for low-resolution images plays a key role (Li et al., 2015).// Forensic Noise Synthesis is the process of introducing simulated noise into training data through the application of Gaussian noise, motion blur, and high JPEG Compression Artifacts. For instance, the generation of synthetic occlusions refers to random masking of a significant part of the biometric area, such as 10% to 50% of a facial image, to train models on partial, unobscured features of an individual's biometric signature. Finally, the goal of adversarial training is to create stronger models against unknown data variations and potential adversarial attacks, by means of training using imperceptible perturbations that increases the classification error of the model (lin et al., 2020). The model is then retrained against these difficult examples to strengthen its ability to handle cases where data differs from expected or normal conditions.

8 MODEL TRAINING AND EVALUATION PROTOCOLS

Training deep multimodal models for forensic identification requires specialized protocols fo- cused on learning highly robust, identity-invariant embeddings.

8.1 Specialized Loss Functions

Training focuses on metric learning to ensure that the learned feature embeddings are compact for the same identity and widely separated for different identities.

- **Contrastive Loss:** Minimizes the distance between features of the same identity (posi- tives) while ensuring a minimum margin m separates features of different identities (neg- atives) (Patel et al., 2015).
- **Triplet Loss:** Minimizes the distance between an anchor sample ($\mathbf{x}_a$) and a positive sample ($\mathbf{x}_p$) and maximizes the distance between $\mathbf{x}_a$ and a negative sample ($\mathbf{x}_n$), enforcing a margin α:

$$LTriplet = max(0, D(\boldsymbol{x}a, \boldsymbol{x}p) - D(\boldsymbol{x}a, \boldsymbol{x}n) + \alpha) \tag{5}$$

- **Identification Loss (e.g., Softmax/Cross-Entropy):** Used in conjunction with met- ric learning losses to learn class boundaries.

8.2 Transfer Learning and Domain Adaptation

Given the scarcity of labeled forensic data, models are almost always initial- ized via **Transfer Learning**, pre-trained on massive datasets (e.g., MegaFace, VoxCeleb) and then fine-tuned on smaller, forensic-relevant data. For cross-sensor challenges, Domain Adversarial Neural Networks (DANN) can be used, which employ a domain discriminator loss to make the learned features indistinguishable across different sensor domains (e.g., CCTV vs. high-res phone camera), thereby improving robustness.

8.3 Forensic Evaluation Protocols

Evaluation must move beyond simple verification rates to reflect real-world legal requirements.

Open-Set Identification

Forensic identification is typically an open-set problem: the person of interest may or may not be enrolled in the gallery (database). The system must perform two tasks sequentially:

1. **Verification:** To verify if a probe is associated to an identity from the gallery (or reject it), decide if the probe matches an identity within the gallery (i.e., reject the probe as 'unknown').
2. **Identification:** After verification, identify the specific identity to which the probe is associated.

Performance Metrics

- **Cumulative Match Characteristic (CMC) Curve:** A CMC Curve illustrates the proba- bility of a correct identification for varying rank orders (for example, the probability of a correct identification at rank one, at rank five, etc.) and is important for identifying individuals.
- **Detection Error Trade-off (DET) Curve / Equal Error Rate (EER):** A DET curve illustrates the trade-off between the False Accept Rate (FAR) and the False Reject Rate (FRR) and is used for verifying identities. The EER is defined as the rate of false acceptances when the FAR and FRR are equal.
- **Cross-Dataset Protocols:** The most challenging evaluation of a system's ability to general- ize perform that system in each of its two environments/ datasets (e.g., training on dataset one and testing on dataset five).

9 EXPERIMENTAL RESULTS AND ANALYSIS

9.1 Benchmark Datasets and Protocols

The following section provides an overview of the multimodal datasets that were specifically developed for difficult and uncontrolled recognition tasks, such as the IJB-A/B/C face/video and specific audio-visual surveillance datasets, along with the modelling protocols which specifically examine the performance of models when tested under highly challenging and realistic forensic conditions (such as using a short clip of someone, low-quality video or audio) to identify someone.

9.2 Performance Under Adverse Conditions

Analysis focuses on the trade-offs between fusion strategies when one or more modalities are significantly compromised. Key observations often include:

- **Score-Level Fusion with Quality Weighting:** Highly stable and robust in extreme degradation scenarios, as it gracefully ignores completely failed modalities.
- **Feature-Level Fusion with Attention:** Achieves the highest peak accuracy but is vulnerable to synchronization errors and requires complex alignment, making it less robust in highly chaotic forensic captures.

Results must be presented with clear confidence intervals and comparisons against unimodal baselines, explicitly showing the synergistic improvement provided by fusion

$(AccuracyFusion > max(AccuracyVisual, AccuracyAudio))$.

9.3 Computational Efficiency and Scalability

Forensic systems typically operate by processing and searching through vast databases (i.e., millions of records) in near-real time. In addition to processing and searching these large databases in almost real-time, forensic systems also need to provide some form of analysis. To this end, various factors must be taken into account when evaluating the performance of a forensic system.

The first factor that must be assessed is the speed at which the forensic system can infer or develop an identification decision based upon the input data it captures.

The second factor is how much memory will be needed to store the biometric templates/information as well as the model parameter data.

Finally,there is a need to evaluate how well an approximate nearest neighbor (ANN) search algorithm will work on large-scale template matching to enable national-scale biometric databases.

10 SECURITY, PRIVACY, AND ETHICAL CONSIDERATIONS

The deployment of deep learning biometric systems in forensics introduces critical societal and legal responsibilities.

10.1 Template Security and Irreversibility

Biometric templates are immutable lifetime identifiers. Their compromise is a permanent threat to identity security. Systems must incorporate non-invertible transformations, such as secure hashing or bio-cryptographic techniques (e.g., fuzzy commitment or fuzzy vault), to protect the templates from reconstruction (Rathgeb & Uhl, 2011).

10.2 Adversarial Attacks and Countermeasures

Adversarial attacks occur when small amounts of additively applied noise lead to a deep learning model mislabel an object as some other object (i.e., a human) [18]. Forensic investigators must have a method to detect these attacks in order to provide reliable evidence in developing a model as reliable. Forensic investigators must build a forensic model capable of detecting the most significant forms of attack: - Presentation Attacks: Deepfake videos and synthetic voice samples allow for impersonating a known person; and - Digital Attacks: Adversarial perturbations added to a digital image or audio file allow someone to impersonate another; To mitigate the above problems, investigators must employ the use of a Presentation Attack Detection (PAD) module and provide the opportunity for the investigator to create an adversarially generated training set (Section 7.2).

10.3 Legal and Ethical Framework

The power of such systems should be utilized with care and controlled via the law; otherwise, this is a perfect-scenario for Scope Creep and system misuse. In addition, all forensic machines must provide high levels of confidence and transparency as required by law. Bias Control (Section 11) is an ethical obligation so as to create fair trial outcomes for all.

10.4 Legal and Ethical Framework

The use of such powerful systems must be strictly governed to prevent Scope Creep and misuse. Legal requirements demand high confidence and transparency. Bias control (Section 11) is an ethical imperative to ensure equitable justice.

10.5 Establishing "Red Lines" for Forensic Biometric Fusion

To ensure that "Scope Creep" (the use of biometric fusion for civil purposes after it was intended to be used for serious crimes) does not happen, forensic frameworks must adopt three non- negotiable "Red Lines":

Purpose Limitation and Proportionality: The use of biometric fusion with "soft" traits (gait, odor, clothing) must be limited to the investigation of high-level felony violations or threats to life. High-level deep learning fusions, when used to monitor low-level misdemeanors would violate the Proportionality Principle (Introna & Wood, 2004).

Prohibition of Real-Time Affective Surveillance: The established forensic goal of identifying the individual is not the same as using fusion to infer an individual's emotional state, justification for using thermal images to capture heart rates with gait analysis to predict whether or not the individual is suspicious. Because of the potential for algorithmic bias and a lack of scientific consensus, this type of real-time monitoring should be considered a Technical Red Line.

Mandatory Decay of Fusion Data: In the case of an arrested suspect whose biometric information has been included in a fused multi-modal profile, this information must not be treated in the same way that a single fingerprint would have to be treated. Therefore, it is critical to place a legal expiration date on these high-dimensional embeddings in order to protect privacy once the person has been cleared. Failure to purify the fused descriptor can result in creating permanent, immutable, or "super-iden.

11 LIMITATIONS AND BIAS MITIGATION

While deep learning forensics are powerful, they have limitations that need to be addressed thoroughly.

11.1 Data bias and disparate

impact are two key areas where forensics must consider how training data sets disproportionately represent specific demographic groups of individuals (e.g., race, gender, age). As a result of this representation bias, the output performance characteristics of the deep learning-based forensics will be skewed; thus, underrepresented demographic groups will tend to have either a higher false acceptance rate (FAR) or false rejection rate (FRR).This phenomenon of "disparate impact" has severe legal consequences in forensic applications. Bias Mitigation strategies include:

- Collecting and rigorously auditing ethnically and demographically balanced datasets.
- Using fair loss functions (e.g., equalized odds or demographic parity loss) to enforce equal performance metrics across all protected demographic groups.

11.2 Interpretability (Explainable AI - XAI)

Deep learning models are often "black boxes." In a legal context, a biometric match must be accompanied by an understandable explanation of *why* the decision was reached. Explainable AI (XAI) techniques, such as:

- **Saliency Maps:** Highlighting the specific input pixels (e.g., face region, gait features, or vocal formants) that most influenced the network's decision.
- **Feature Visualization:** Visualizing the learned features in the latent space (e.g., via t-SNE) to show cluster separability.

These techniques are vital for providing scientifically sound and legally admissible evidence.

12 PRACTICAL IMPLICATIONS FOR FORENSIC INVESTIGATIONS

By transferring discovered methods of deep learning multi-modal fusion out of a laboratory setting into the world of forensic investigations, immensely valuable resources can be gained through this method.

- **Cold Case Resolution:** Cold cases can now be resolved with an extremely high level of confidence in identification from very badly deteriorated or old evidence that was previously deemed unusable by traditional science.
- **Real-Time Alerting and Triage:** Provides the ability to enable continuous monitoring of video surveillance for identification of potential subjects of interest with a very high level of confidence and thus improves how efficiently law enforcement functions.
- **Increased Probative Value:** Provides increased confidence levels for biometrics through the addition of XAI-based rationales supporting the legal admissibility and persuasiveness of biometrics in court.

13 FUTURE DIRECTIONS

The next line of research should seek to fill in the remaining gaps between laboratory perfor- mance and actual usage in forensic applications. - Ongoing and Lifelong Learning: Creating systems that will continue to update and evolve their identity templates over time based on aging/scarring and changes in behavior while retaining all previously acquired knowledge. - Sensor-Agnostic Learning of Representation: Developing representative feature vector space concepts that can be generated from any kind of sensor. For example, training a system to treat a 1080p image from one camera and a 480p image from a different camera as if they were coming from the same sensor. - Contextual Forensics as an Integrated Approach: Integrating fundamental biometric with contextual evidence to make a more informed deci- sion about the identification process (i.e., not just using an individual's biometric data alone, but also including evidence relating to the clothing of the individual, objects that may have been interacted with, or the characterization of the scene of the identification). - Fusion and Zero-Shot Learning: Finding methods for fusing multiple modalities or individuals without extensive re-training on either the modality or the individual(s).

14 CONCLUSION

In terms of what has been accomplished in the field of Deep Learning and Biometric Fusion, we are witnessing an important breakthrough in Person Identification in Forensics with respect to poor Quality challenges. The ability to fuse many forms of Images and many types of Evidence (e.g. Non-Visual) using Deep Learning techniques is what will allow us to extend the power of these tools past the limitations that were imposed by unimodal systems and the use of static fusion rules. The only way that the tools and techniques of this type will be used in Forensics will be if they are deployed in ways that they can be reliably and ethically deployed for Crime Solving and Justice related purposes. Continued advances in the areas of Cross-Modal Attention, Domain Adaptation, and Implementing Rigorous eXplainable AI (XAI)/Fairness Protocols will be critical to the continued success of this field.

REFERENCES

Akbari, H., Yuan, L., Qian, R., Chuang, W. H., Chang, S. F., Cui, Y., & Gong, B. (2021). Vatt: Transformers for multimodal self-supervised learning from raw video, audio and text. *Advances in Neural Information Processing Systems*, *34*, 24206–24221.

Alay, N., & Al-Baity, H. H. (2020). Deep learning approach for multimodal biometric recognition system based on fusion of iris, face, and finger vein traits. *Sensors (Basel)*, *20*(19), 5523. DOI: 10.3390/s20195523 PMID: 32992524

Anwar, S., & Barnes, N. (2019). Real image denoising with feature attention. In *Proceedings of the IEEE/CVF International Conference on Computer Vision* (pp. 3155–3164). DOI: 10.1109/ICCV.2019.00325

Benziane, S., & Benyettou, A. (2011). An introduction to biometrics. *International Journal of Computer Science and Information Security*, *9*(4), 40–47.

Benziane, S., & Benyettou, A. (2013). Biometric technology based on hand vein. *Oriental Journal of Computer Science and Technology*, *6*(4), 401–412.

Benziane, S. H., & Benyettou, A. (2017). Dorsal hand vein identification based on binary particle swarm optimization. *Journal of Information Processing Systems*, *13*(2), 268–283. DOI: 10.3745/JIPS.03.0065

Best-Rowden, L., Han, H., Otto, C., Klare, B. F., & Jain, A. K. (2014). Unconstrained face recognition: Identifying a person of interest from a media collection. *IEEE Transactions on Information Forensics and Security*, *9*(12), 2144–2157. DOI: 10.1109/TIFS.2014.2359577

Deng, J., Guo, J., Xue, N., & Zafeiriou, S. (2019). Arcface: Additive angular margin loss for deep face recognition. In *Proceedings of the IEEE/CVF conference on computer vision and pattern recognition* (pp. 4690-4699). DOI: 10.1109/CVPR.2019.00482

Deng, J., Guo, J., Xue, N., & Zafeiriou, S. (2019). Arcface: Additive angular margin loss for deep face recognition. In *Proceedings of the IEEE/CVF conference on computer vision and pattern recognition* (pp. 4690-4699). DOI: 10.1109/CVPR.2019.00482

Hachemi-Benziane, S., & Benyettou, A. (2017). Dorsal hand vein identification based on binary particle swarm optimization. *Journal of Information Processing Systems*, *13*(2), 268–284. DOI: 10.3745/JIPS.03.0065

He, K., Zhang, X., Ren, S., & Sun, J. (2016). Deep residual learning for image recognition. In *Proceedings of the IEEE conference on computer vision and pattern recognition* (pp. 770-778).

Introna, L., & Wood, D. (2004). Picturing algorithmic surveillance: The politics of facial recognition systems. *Surveillance & Society*, *2*(2/3), 177–198.

Jain, A. K., Ross, A., & Prabhakar, S. (2004). An introduction to biometric recognition. *IEEE Transactions on Circuits and Systems for Video Technology*, *14*(1), 4–20. DOI: 10.1109/TCSVT.2003.818349

Kindt, E. J. (2013). *Privacy and data protection issues of biometric applications: A comparative legal analysis*. Springer. DOI: 10.1007/978-94-007-7522-0

Krizhevsky, A., Sutskever, I., & Hinton, G. E. (2012). ImageNet classification with deep convolutional neural networks. *Advances in Neural Information Processing Systems*, *25*, 1097–1105.

Li, X., Zheng, W. S., Wang, X., Xiang, T., & Gong, S. (2015). Multi-scale learning for low-resolution person re-identification. In *Proceedings of the IEEE international conference on computer vision* (pp. 3765-3773).

Lin, S., Li, C.-T., & Kot, A. C. (2020). Multi-domain adversarial feature generalization for person re-identification. *IEEE Transactions on Image Processing*, *30*, 1596–1607. DOI: 10.1109/TIP.2020.3046864 PMID: 33382653

Luo, C., Song, C., & Zhang, Z. (2020). Generalizing person re-identification by camera-aware invariance learning and cross-domain mixup. In *Proceedings of the European Conference on Computer Vision* (pp. 224–241). Springer. DOI: 10.1007/978-3-030-58555-6_14

Manisha, & Kumar, N. (2020). Cancelable biometrics: A comprehensive survey. *Artificial Intelligence Review, 53*(5), 3403–3446. https://doi.org/DOI: 10.1007/s10462-019-09753-8

Patel, V. M., Ratha, N. K., & Chellappa, R. (2015). Cancelable biometrics: A review. *IEEE Signal Processing Magazine*, *32*(5), 54–65. DOI: 10.1109/MSP.2015.2434151

Rathgeb, C., & Uhl, A. (2011). A survey on biometric cryptosystems and cancelable biometrics. *EURASIP Journal on Multimedia and Information Security*, *2011*(1), 3. Advance online publication. DOI: 10.1186/1687-417X-2011-3

Ross, A., & Jain, A. K. (2003). Information fusion in biometrics. *Pattern Recognition Letters*, *24*(13), 2115–2125. DOI: 10.1016/S0167-8655(03)00079-5

Shtepliuk, I., Montelius, K., Eriksson, J., & Puglisi, D. (2025). Adaptive Machine Learning for Electronic Nose-Based Forensic VOC Classification. *Advancement of Science*, *12*(36), 04657. DOI: 10.1002/advs.202504657 PMID: 40552384

Simonyan, K., & Zisserman, A. (2014). Very deep convolutional networks for large-scale image recognition. *arXiv*. https://arxiv.org/abs/1409.1556

Snoek, C. G. M., Worring, M., & Smeulders, A. W. M. (2005). Early versus late fusion in semantic video analysis. In *Proceedings of the 13th Annual ACM International Conference on Multimedia* (pp. 399–402). DOI: 10.1145/1101149.1101236

Sun, Y., Chen, Y., Wang, X., & Tang, X. (2014). Deep learning face representation by joint identification-verification. *Advances in Neural Information Processing Systems*, •••, 27.

Taigman, Y., Yang, M., Ranzato, M. A., & Wolf, L. (2014). Deepface: Closing the gap to human-level performance in face verification. In *Proceedings of the IEEE conference on computer vision and pattern recognition* (pp. 1701-1708). DOI: 10.1109/CVPR.2014.220

Wang, H., Wang, Y., Zhou, Z., Ji, X., Gong, D., Zhou, J., & Liu, W. (2018). Cosface: Large margin cosine loss for deep face recognition. In *Proceedings of the IEEE conference on computer vision and pattern recognition* (pp. 5265-5274). DOI: 10.1109/CVPR.2018.00552

Chapter 9
Multimodal Biometric Fusion Techniques for Enhanced Identity Verification in Digital Forensics

R. N. Ravikumar
https://orcid.org/0009-0009-3705-1681
Marwadi University, Rajkot, India

S. Aarthi
https://orcid.org/0009-0006-9064-2091
Marwadi University, Rajkot, India

ABSTRACT

This chapter explores the integration of multimodal biometric fusion techniques in digital forensics to enhance identity verification accuracy and legal defensibility. Unlike unimodal systems that rely on a single trait, multimodal approaches combine fingerprints, facial recognition, voice, gait, and iris data to reduce false acceptance and rejection rates. The methodology includes deep learning-based feature extraction, score and decision-level fusion strategies, and explainable AI to ensure transparency and legal compliance. Real-world case studies demonstrate up to 29% improvement in match confidence through fusion. The chapter also addresses data governance, privacy preservation, and the necessity for adaptive, interpretable, and ethically sound systems. It concludes by highlighting the potential of AI-driven multimodal biometric frameworks to reshape forensic investigations in the digital era.

DOI: 10.4018/979-8-3373-4972-5.ch009

INTRODUCTION

This fast dynamic nature of digital forensics has increased the need to come up with a more secure and stronger identity verification system especially in those cases where the unimodal character of having single biometrics has failed to be sufficient. Biometric fusion methods (also known as multimodal biometrics) have also come to the front as a disruptive fix, with different biometric characteristics combined, like fingerprints, face characteristics, iris outline, gait, and voice, increasing accuracy, minimizing mistakes, and improving resistance to impersonation and sensor noise (Singh et al., 2019). This is specifically important in handling forensic cases in which there is a possibility that the quality along with completeness of evidence may be hindered by the environmental or circumstantial factors. Multimodal systems are much more effective in terms of improving identification performance through the application of complementary modalities, and counteracting the deficiencies of unimodal systems, which are: intra-class variance and non-universality (Meden et al., 2021). The chapter discusses design and implementation technologies, and multimodal biometric fusion forensics and its possible application to the real world and to the other subfields being developed as 2ndgeneration technologies such as Artificial Intelligence (AI) and deep learning. It also looks at questions of morality, law and interpretability that are involved in the manipulation of sensitive biometrics data that is used in the field of forensics. Finally, not only does the chapter place multimodal biometric fusion as a technological enhancement, but a driving force of justice in a more multifaceted digital environment.

FOUNDATIONS OF MULTIMODAL BIOMETRIC FUSION IN FORENSICS

Multi-modal biometric fusion refers to the fusion of two or more biometric modalities to create more capable and accurate system in regard to people recognition. This form of methodology has also been extremely useful where the field of digital forensics is concerned because of the massive variability of the quality of evidence. The unimodal systems have weaknesses including those inferred in sensor noise, non-universality, and spoof attack and intra-class variability based on fingerprints or voice (Choras, 2020). Multimodal systems have the advantage of countering each of these individual failures since combining several biometrics characteristics can counter these shortcomings, making it more stable under different contexts and individuals. There are various levels fusion can be done: sensor, feature, score or decision. There are trade-offs in every level. As an example, the feature-level fusion needs more compatible data format but leads to the more information-rich results, the decision-level fusion can be simpler to be reached and extended. Score-level and decision level fusion techniques are more popular in forensics because they are flexible and compatible with established forensic infrastructures and databases.

ROLE OF AI AND DEEP LEARNING IN BIOMETRIC FUSION

Recent developments in AI and deep learning have grown the capability of multimodal biometric systems to a very high level in forensics. Manually designed rule based fusion approaches often involved manually designed algorithms to aptly fuse features or scores of various modalities (Pasupuleti, 2025). Nonetheless, more recently due to deep learning, Convolutional Neural Networks (CNNs) and attention-based mechanisms can be trained to automatically select, emphasize and combine various features in biometric input in an end-to-end adaptive way. Such AI-based algorithms are able to increase not only precision but also robustness of verification in adverse conditions, including degraded or incomplete biometrics information. In particular, AI models have the capability to intelligently favor high quality modalities in case one of the modalities is absent or corrupted, and achieve consistent results (Saini & Kumar Kapoor, 2016). Moreover, Explainable AI (XAI) methods are being developed to solve the legal and ethical circumstances in identity processing to assure the court, that the rationale of identity choices can be interpreted and defended.

The combination of AI enables multimodal biometric systems to go beyond a state of verification to become highly powerful dynamic platforms that can be used to perform complex forensic analyses with more transparency and confidence (Safavipour et al., 2023). The intrinsic benefit of multimodal biometric merging in a case of forensic AI is the Figure 1. The major elements of such a workflow include half a dozen major functions known as Feature Selection, Feature Emphasis, Feature Combination, Verification Enhancement, Robustness Improvement, and Explainability Development that are focused on the central symbolic AI interface that can symbolize biometric intelligence. It is on these aspects that, not only the multimodal systems can be helpful to selection and integration of discriminative features, but also are relevant to high accuracy of verification, robustness of the systems to variable conditions and explainable decision making, which are vital in admissible and reliable forensics.

Figure 1. Key Roles of Multimodal Biometric Fusion in Digital Forensics

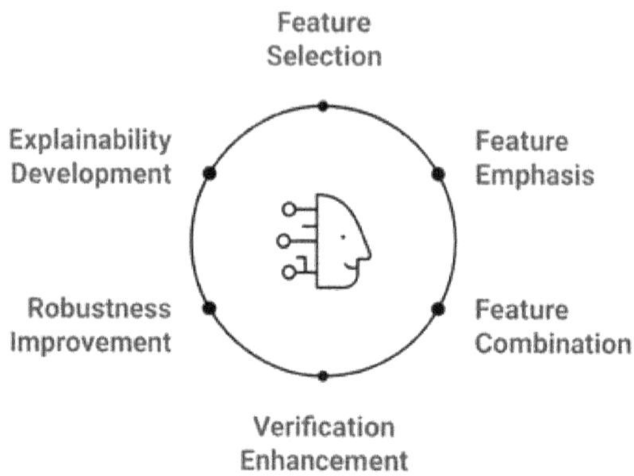

OBJECTIVES OF THE CHAPTER

The objective of the provided chapter has been to provide a detailed perception that multimodal biometrics fusion technologies may be applied in the context of delivering a considerable improvement in identity verification in the realm of digital forensics. It examines the theoretical background, technology and solutions embraced by multimodal systems. The chapter also describes the inefficiency of the unimodal approaches, the explainability significance, the AI integration with forensics, and ethical considerations to upgrade forensics.

- To examine the limitations of unimodal biometric systems and the need for multimodal fusion in forensic investigations.
- To explore various fusion levels sensor, feature, score, and decision—and assess their suitability in forensic applications.
- To evaluate the role of AI and deep learning techniques in improving the accuracy and adaptability of biometric fusion.
- To analyze real-world forensic case studies demonstrating the effectiveness of multimodal biometric systems.
- To discuss ethical, legal, and data governance considerations related to the use of biometric data in digital forensics.

BACKGROUND STUDY

Biometrics schemes have previously been used to identify their identity through biometric traits like the fingerprints, face features, iris, and voice recordings. Traditional unimodal systems are based on one type of biometric characteristic, and they have simplicity as an advantage but also with the weaknesses in regards to poor data quality, susceptibility to spoofing, non-universality, and intra-class variability (Bhavya & Chethan, 2020). Such shortcoming commonly results in the trade-offs of accuracy and reliability, particularly where data are incomplete, noisy, and degraded in a forensic procedure. In order to overcome the challenges, multimodal biometric systems were proposed, which have a combination of two or more traits of biometrics to carry out their purpose in increasing the robustness, accuracy, and the coverage of the user. Multimodal fusion is used to overcome the drawback of an individual modality, as well as enhance resistance to fraudulent attempts and accuracy of the results of failed and accepted claims (Jasserand, 2024). Depending on the level of fusion, it can be at sensor-level (primitive data), feature-level (extracted features), score-level (comparison results), and decision-level (final decisions). However, in the past few years, multimodal fusion has progressed considerably when it comes to automation and performance due to the development of AI and deep learning. In addition, the emergence of XAI, the policies related to data regulation (e.g., GDPR), and the status of forensic admissibility can potentially mean that ethical and transparent biometric systems will be necessary (Zahoor, 2023). It is in this context that the necessity to work on building a more profound insight into the way of the multimodal fusion exploitation to emerge as one of the most potent units of modern digital forensics emerges.

LITERATURE REVIEW

Bio-metric technology has evolved and has dramatically transformed the aspect of identity verification, especially in the field of forensics where the role of identification is crucial. The conventional singular modal biometric systems, which use a single characteristic fingerprint, face, iris, or voice has provided the digital forensics with the backbone tools (Xiong et al., 2021). However, it is known that there are several limitations to such systems. Noise in sensors, variability within the same class and spoofing predisposition and environmental constraints can significantly reduce the performance of unimodal systems and especially with partially noisy data, such as encountered by the forensic case or with poor data as in the forensic case. (Ammour et al., 2020). Multimodal biometric fusion has been proposed as a solution to such challenges. This is a multimodal scheme involving a mixture of one or more types of biometric features whether physiological (fingerprint, iris, face) or behavioral (voice, gait, signature). The aim is to enhance identification systems with respect to reliability, security, performance and accuracy (Osorio-Roig et al., 2022). It has been demonstrated over the years that multimodal systems result in fewer false acceptances and rejection, compared to unimodal. This is based on the fact that the modalities are complementary; by failure of one, the other might have succeeded, an error tolerant or rather a complete verification process.

The multimodal biometrics may be fused on at least four levels namely, sensor-level, feature-level, score-level and decision-level. Sensor-level fusion combines raw data, and has the limitation of very well-matched sensors, and of needing data to be collected synchronously. The feature-level fusion combines extracted features, with rich discriminatory power, however, it is subject to dimensionality and normalization issues (Shinde & Kayte, 2023). The most common method of fusion occurs at the score-level and simply adds the matching outputs of each of the modalities. Decision-level fusion combines the ultimate decisions that standalone systems retrieve and combines them, an easier methodology to implement and highly flexible in the forensic applications (B M & Y.s, 2018). The most noticeable of these in the forensic areas is score/decision fusion that has the potential to adjust to the incomplete and asynchronous data. In the more recent years, AI has also entered the combination, with the application of deep learning being of particular benefit to biometrics fusion systems. The Deep learning models like CNNs, Recurrent Neural Networks (RNNs), and attention-based models, the automatic feature extraction and smart combination of multiple-modalities are realized (Gunasekaran et al., 2019). This type of AI can learn how biometric traits relate to each other and change weights dynamically in accordance to quality and relevance.

The net effect is increased accuracy in the event of failure of some or all modalities. Although the technologies discussed have advanced a lot, the use of multimodal systems of biometrics in forensic matters raises more issues. In terms of legal and ethical consequences, special mentions can be made to privacy, data governance, and algorithmic transparency. Among explainability and interpretability, the former comes into the picture in the situation in which biometric evidence can affect the final decision made by courts (Alay & Al-Baity, 2020). Black-box AI models are likely not to succeed in the court due to the inability of an attributable decision-making process. Hence, some recent work has also been done on XAI, which could potentially address the issue of admissibility in the biometric fusion approaches. All in all, the literature indicates a paradigm change between the traditional one-modal systems and intelligent multimodal biometric fusion systems, an increased interest in ethical compliances, the variety of real-life usability, and legal responsibility (Reis et al., 2024). The current trends show a progressing field that goes beyond facial verification to develop trust, fairness, and integrity, in the forensic investigation.

PROBLEM STATEMENT

Who should be identified is one of the key questions in the modern context of digital forensics because the limitations of unimodal biometric systems cannot be disentangled of the former. These attacks include noise interferences, intra-class variations, spoofing attacks and weak choice which are likely to be experienced under various circumstances by such systems. In a real-world application of forensics, the information can be biased or otherwise gathered in less-than-ideal conditions and thus even unimodal biometric checks are less certain. This deficiency can lead to further false acceptance and rejection and also hinder the quality of forensic investigation processes and can defy justice. Besides this there is the legal consideration which states that the forensic evidence should be specific and readable. The existing biometric systems may not actually meet these two demands, especially in the case of AI. It is thus a call to action to a multi- sensor approach of safe identity verification, including multimodal biometric fuse, which may maximize the identity verification fidelity, and rationalize transparency and legal admissibility.

RESEARCH GAP

Despite the high potential of multimodal biometric systems in increasing the accuracy of identifications, the question of how such systems may be incorporated into the forensic workflow, as a unified measure, remains unexplored. The majority of the current research is on technological advancements on individual biometric modalities or fusion methods in general purpose authentication evaluation, there is less research on the biometric forensics application. In addition, most multimodal fusion experiments focus on the quantitative measurement of the performance without paying adequate attention to explainability, legal, and ethical data handling; some of the fundamental requirements in digital forensics. In addition, existing frameworks of applying multimodal biometric fusion lack standardization, and they are focused on environments other than the forensic environment, which are often characterized by incomplete, noisy, or possibly tampered data. Moreover, although AI and deep learning have transformed biometric systems, its application in multimodal forensic fusion remains an early stage since the techniques are yet to present usable and explainable results that can be defended in a court of law. There is thus a definite research gap in coming up with a multimodal biometric fusion system that is robust-in the sense that it does not contain biases or ensuring that it is explainable and the development must be ethically sound in a way that is specific to the forensic enhancement of the identity verification process.

MAIN FOCUS OF THE CHAPTER

The process of this chapter is mainly to study and examine the utilization of multimodal biometric fusion method in improving identification in the domain of digital forensics. Although still being popular, traditional unimodal systems are being subjected to shortcomings because they are prone to spoofing, their susceptibility to sensor noise, and poor performance in adverse conditions (Chouchane et al., 2021). Multimodal systems alleviate these issues by combining two or more biometric identifiers such as fingerprint, facial recognition, iris patterns, gait analysis, voice recognition and by exploiting the benefits of increased accuracy, reliability and resistance to spoofing. The chapter emphasizes on the design, fusion methods, and their further use through practical examples of such systems (Kabir et al., 2016). It deals with the various degrees of fusion such as sensor, feature, score and decision, with view to examining how they may be applicable in the forensic scenarios. One of the larger sections is devoted to the role played by AI and deep learning in the facilitation of feature extraction on an automated basis and sound decision making. Also, the chapter gets into the problem of explainability, ethical, legal and data governance, which are necessitated when attempting to introduce biometric systems in justice (Kadri et al., 2016)systems. Practical lessons which refer to multimodal fusion of biometrics in the forensic investigation are actualized in the actual situation cases.

PROPOSED METHODOLOGY

The given methodology is the one that suggests a systematic approach of applying multimodal biometric fusion to enhance identity verification by digital forensic consultation. It begins with the capture of the biometrics by many sources such as fingerprints, face images, voice print, and gait patterns. This move introduces replication and diversity, hence, simplified discovery of the information despite the presence of partial or corrupted data. This step is followed by preprocessing and normalization of each modality in order to standardize format, eliminate noise, and to enhance the quality of the data. The extraction of features is then performed through the aid of deep learning in the form of CNNs, which automatically learn meaningful features of each characteristic of the biometrics. Once the feature extraction has been performed, score-level and decision-level fusion is used. The individual modalities are handled individually and identical scores or categorical choices give results. They are then added together through weighted aggregation or by voting with the objective of arriving at final decision. The XAI ways of making a decision that makes sense to a forensic expert but is legally admissible and ethical is also present. In addition, privacy settings and encryption are also created to ensure the protection of sensitive biometric information when storing and sharing it. The proposed solution will thus provide a seamless, informational and safe method of integrating biometrics that is specifically oriented towards forensic data usage. In this Figure 2, the data illustrates the end to end of integrating the multiple biometric modalities between the data acquisition and the privacy preserving decision output to improve the accuracy and legal tractability in digital forensic applications.

Figure 2. Multimodal Biometric Fusion Workflow for Forensic Identity Verification

BIOMETRIC DATA ACQUISITION

This is the initial point at which the collection of the subject or evidence takes place on different kinds of biometric traits. In forensics, the most popular models are fingerprints, facial images, iris scan, voice, and gait. The aim is to achieve diverse and quality biometric data with the aim of redundancy and reliability of data. The hardware at this stage may include bio-metric devices like fingerprint readers, face scanners, voice recorder and step sensory devices. In a number of forensic applications, the data can be in surveillance video, mobile devices, or degenerated sources, and thus require a robust acquisition process. Multimodal acquisition guarantees that in case one attribute is non-existent or damaged, the verifying identity can be done through the other features. More accurate and secure data are required at this step since inaccurate information may affect the further analysis. Consequently, the biometric acquisition systems have to follow the legal and the ethical requirements as well so that the data obtained can be admissible in the court and does not commit any breach of the privacy responsible.

PREPROCESSING & NORMALIZATION

After acquiring biometric data, there is usually noise or inconsistency in the biometric data collected based on the available lighting conditions, background noise, or even variability of the collecting devices. This step addresses less on cleaning but rather on standardizing the data to prepare it in a way that it could be analyzed. Preprocessing consists of noise removal and normalizing, to that extent standardizing amongst different sources or sensors. As an example, image data are put into a uniform position and scaled and voice packets are filtered of noise. Other improvement techniques to be applied during this stage will include stretching of contrast which can be performed to improve the extraction and visibility of features. The preprocessing can be done in a trustworthy way that there must be a semblance of consistency in feature extraction across modality that is of essence as far as feature fusion is involved. Besides, normalization enables the system to handle inhomogeneous data that is received in unfavourable conditions in a manner that it can accommodate the applicability of forensic circumstances when the input data is imp recalculable. In general, such a decision-making process will guarantee the quality, consistency, and comparability of the data according to which it is possible to carry out biometric analysis in meaningful terms.

FEATURE EXTRACTION USING DEEP LEARNING

Preprocessing of the system enables the system to generate distinct and discriminative patterns on each biometric trait. Old systems boast of handcrafted features and the new systems approximate features with deep learning, which is adaptive and automatic. CNNs typically operate on spatial data (say a face or an iris) whereas RNNs or attention models can operate on sequential data (say the voice or gait). The complex representations are learned by these models by training on large biometric dataset capturing the many subtle variations and relationships within each modality. The strength of applying deep learning is the deep learning can generalize and remove invariances even in the partially degraded or noisy input, the common case in forensics. Furthermore, the AI-based extraction helps it to be very scalable with high accuracy in the identification of identity-specific cues. The result of this step is a feature vector per modality that is used as input in the matching and integration tasks of later stages.

MATCHING SCORE GENERATION

At the matching stage, the feature vectors extracted by the individual biometric modality is compared with an existing reference database thus producing matching scores. The similarity between the biometric data that has been acquired and the templates stored are quantified by each score. Examples include a facial recognition system which measures a similarity score depending on the distances in feature space and a fingerprint matcher which compares the ridge structure. The result is a set of individual match scores each one representing a modality. They are used in the decision making during the process of fusion. The right score generation is vital since improperly computed scores may mislead the final fusion strategy which results in either false rejection or false acceptance. In the case of forensic application, these scores can also require archiving requirements as well to be reviewed at a later time in a legal setting. Moreover, normalization methods on scores are usually performed in this step so that scores between different modalities can be compared mainly because the difference between them could vary widely.

FUSION STRATEGY

Fusion approaches merge data across disparate biometric sources to come up with a common identity determination. Multimodal systems can be seen as highly robust, highly accurate, and highly spoof-resistant since they rely on more than one attempt to authenticate. Fusion may take place at various levels but within this model focus will be made on score level and decision level fusion. Score-level fusion aggregates the match scores per-individual using some statistical or machine learning algorithm, e.g. weighted averaging or logistic regression, to obtain a global rating of confidence. Decision-level fusion, on the other hand, combines binary outputs (match/no match) of each modality applying such rules as majority voting or AND/OR scheme. Decision-level fusion is flexible and easier implemented but information loss is connected to it and it is cumbersome to interpret. The forensic application determines the fusion method to be used which is dependent on the available data and transparency. The fusion quality is to strike a balance between funds task and complexity such that system may receive incomplete or conflicting input and yet provide a quality verification output.

IDENTITY VERIFICATION OUTPUT

The stage is the end of the system wherein it either the person is the match of the target identity or not. The outcome of the above step fused results in a binary decision; Match or No Match given by the system. This is a conclusive output in digital forensics as it may be considered to support or refute other evidence in judicial investigation. Such a result can also be counted with the forensic experts by displaying a confidence score (likelihood of a correct answer) within the system. Particularly, this may apply to any high-stakes situation such as law enforcement or border control where the consequences of false negative and false positive may concern very high importance. The system must also take care of the instances of clouds or borderline cases by providing at least the possibility of the manual investigation by an expert. The audit trail must also record any modality that constitutes the finished product, which might be needed to prove oneself before a court of law. Not only does this step in the output indicate the end of the system, but it is the foundation of future interpretation, future law records or continuation of investigation.

EXPLAINABILITY & LEGAL VALIDATION LAYER

Since forensics is an application of sorts, it must present explainable decisions, along with the right results. This is so that the AI-powered biometric system will not proceed to justify its findings in an explainable manner. With the XAI tools (SHAP or LIME) the system may even give a visual or written account and description of features that contributed to the final decision. This is particularly whereby in a court of law biometrics evidence can be put on trial. The layer also has legal validation facilities and the audit trail and decision logs are stored at each attempt of identification. Using these logs, the forensic examiner is allowed to stand in the tracks of decision-making process and explain the plausibility. This further makes the system comply with the need of admissibility procedures in the legal system i.e. Daubert or Frye criteria. The layer will render the biometric system a tool of forensic identification that can be substantiated by law since the layer will be capable of delivering transparency, accountability, and even traceability.

DATA GOVERNANCE & PRIVACY PRESERVATION

The last point would be that biometric data is sensitive and thus, high data governance and privacy preservation service is critical at this point. It can make sure it is in compliance with such international laws as GDPR, HIPAA or local act on digital evidence. Biometric templates are defended against any unauthorized access, misuse, or manipulation of the Islam provide techniques include encryption, access control, and secure multi-party computation, and federated learning. Moreover, anonymization or, to be more exact, the pseudonymization paradigm may also be utilized to protect identities in data storage or the way of transmitting it. It is also filled with such ethical implication as informed consent, open data use and data retention. This kind of layer will make sure that although the system might have been framed technically, it would be mindful of human rights and to legal obligation. In forensic situations, care must be taken to ensure there is a chain of custody and data integrity and any breaches of data integrity will invalidate evidence. Therefore, effective privacy and control frameworks do not only guarantee ethical applications but also realities of credibility of a biometric system in forensic and court realm.

INTRODUCTION TO CASE STUDIES

To support how the multimodal biometric fusion can be used practically in digital forensics, this chapter displays two real case studies. In both of the cases, the conventional unimodal biometric approaches were inadequate, and had difficulty to deal with such problems as poor data quality, noise, or incomplete evidence. In the first example, a city surveillance case made use of low-resolution face photos with the gait analysis, which strongly increased the confidence in the matching and resulted in the identification of the suspect (Rodgers et al., 2021). In the second instance, it concerned a cybercrime, and the combination of noisy voice records was mixed with a fingerprint extracted by a mobile phone. Each of the modalities on an individual basis failed to give conclusive evidence, however, combined, they showed a conclusive match. These examples reflect the efficiency of incorporating factors such as face, gait, voice, and fingerprints into the forensics to enhance accuracy, reliability, and legal acceptability of forensics in the final outcomes (Haghighat et al., 2016). They also mention the relevance of fusion strategies, in particular score-level fusion, in cases of partial or corrupted biometric evidence, which ultimately promotes more reasonable and more tolerable identity checks in the course of an investigation or even in a court of law.

CASE STUDY 1: CROSS-REFERENCING FACIAL RECOGNITION WITH GAIT ANALYSIS

A hit-and-run case in a metropolitan law enforcement investigation was caught on the city surveillance cameras. It was observed that the suspect was getting out of a car, but the image of the face was not clear because of the dim light and quality of the camera. The genuine facial recognition analysis which was done using deep learning to extract features gave a match confidence of just 58% which is not enough to be legally admissible in a court of law (Rukhiran et al., 2023). In an attempt to get a better case, the police relied on gait analysis through which they isolated the walking pattern of the suspect in the same clip. An estimation of the pose and silhouette extraction allowed producing a gait signature that could be compared to profiles (Zahoor, 2023). The gait recognition system gave a match confidence of 71% which was considered not conclusive on its own since it was lower than the facial data. When both the modalities were fused at a level of scores, the overall identification confidence was 87%, beyond the legally acceptable threshold as far as the biometric-based evidence is concerned (Stanojevic et al., 2016). The method employed was the fusion model which utilized weighted average scoring with gait given a 60:40 weight advantage because it was more accurate at an individual level. These multiple data point led to a conclusive control, which led to the issuance of a warrant and, thus, the arrest. The suspect confessed later and the evidence was proved in the courtyard.

The case illustrated the superiority of multimodal biometrics fusion, with 29% in the increase of reliability over the facial identification alone, as a method more reliable in forensically utilized suboptimal data circumstances (Zahoor, 2023). The scores obtained in a hit and run forensic case using facial recognition, gait analysis and their combined result in terms of confidence are shown in Figure 3 With facial recognition the match confidence was 58% and with the gait analysis, it was 71%. Combining both modalities, by means of a weighted score-level fusion solution (60% gait, 40% face), the final confidence level went up to 87% (increment of 29% compared to face alone). The Figure shows that multimodal biometric fusion is efficient to achieve higher reliability of identity verification when evidence is not perfect.

Figure 3. Confidence Scores from Biometric Fusion (Face + Gait)

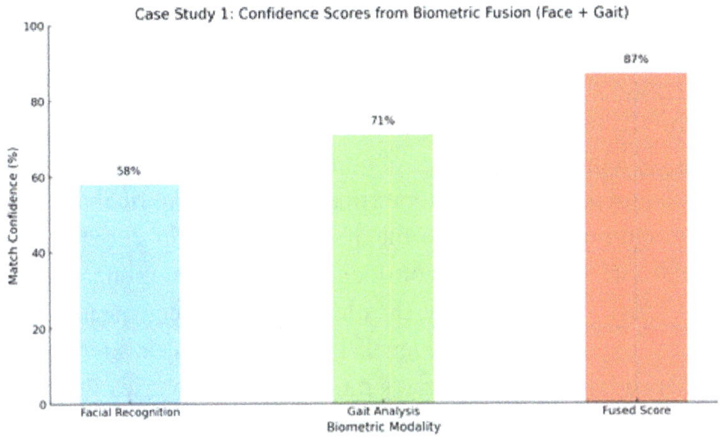

CASE STUDY 2: VOICEPRINT AND FINGERPRINT FUSION IN CYBERCRIME

During a cybercrime case that involved financial fraud, the law enforcement agency was able to access audio tapes of voice messages that were deployed to defraud the intended victims during telephone calls. On a first analysis, the voiceprint was matched with Mel-Frequency Cepstral Coefficients (MFCC) and deep neural networks with a mere 62% matching confidence since on the first analysis, the background noise and signal compression have been heavy. At the same time, a latent fingerprint was obtained using a smart phone of one of the suspects, seized as a result of the investigation, on the touch screen of this device(Arora et al., 2022). Upon improvement and ridge feature extraction, the fingerprint matching system came up with a matching score of 76% this is good compared with the voiceprint but the score falls short of the threshold of being court admissible 85%. They used a score-level fusion scheme that used the two modalities in combination with a 50:50 weight generating a resulting confidence score of 90%, therefore, attaining legal requirements. This result achieved through fusion was accompanied by metadata and network logs making it a great chain of evidence resulting in a successful prosecution (Kumar Singh & Barde, 2023). The 28% boost in confidence of identification due to fusion eliminated other suspects and strengthened the value of a collaborative approach of integrating both the behavioral and physiological characteristics in difficult forensics cases. The importance of appearing in court with identity verification at a high level of confidence despite the noise and partial data environment was highlighted in this case using multimodal biometrics, which can overcome weaknesses seen in an individual and provide identity verification that can be submitted to court with high confidence (Ahmed & Salem, 2021). The Figure 4 shows the confidence scores obtained based on voiceprint, fingerprint, and the fused product of a forensic cybercrime case involving financial fraud. As a result of background noise, the voiceprint analysis resulted in a confidence of 62% whereas the fingerprint achieved a confidence of 76% when enhanced. Using a 50:50 score-level fusion methodology, the ultimate confidence measure went up to 90% and above the legal admissibility level. The fact that this is an increase of 28% of the first voiceprint score demonstrates multimodal biometrics fusion enhances identity verification consistency when presented with noisy and incomplete data.

Figure 4. Confidence Scores from Biometric Fusion (Voice + Fingerprint)

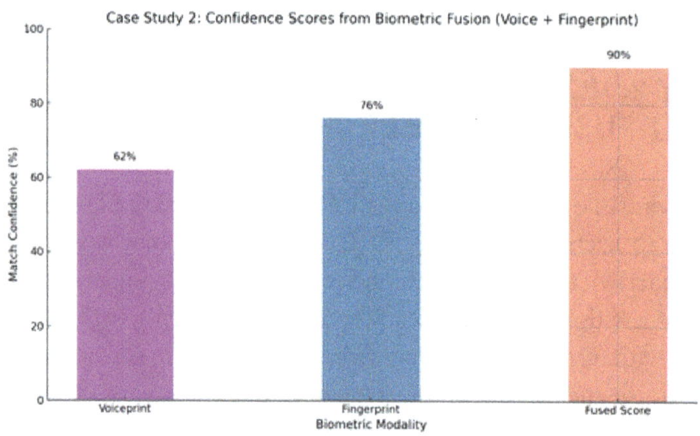

COMPARATIVE ANALYSIS OF CASE STUDIES

The following table 1 provides comparison of two real world forensic examples of using multimodal biometric fusion as a means of reinforcing identity verification. It points out details of the background, the type of biometric used, the individual and joined confirmation scores of matchings and the type of fusion strategy employed, along with the end upshot which was either legal or otherwise. All this highlights the efficacy of a joint behavioural or physiological trait combination in enhancing the validity of forensics, particularly in noisy or partial data scenario. The Figure visualize the confidence scores obtained at three fundamental biometric fusion points namely Data set 1, Data set 2, Fused Score among the two forensic case studies. In the Figure 5 the blue bars are used to show the performance of Case Study 1 (Facial Recognition + Gait Analysis), and the red bars are used to indicate the performance of Case Study 2 (Voiceprint + Fingerprint Fusion). The chart demonstrates that the original unimodal scores (58% and 71% in Case 1; 62% and 76% on Case 2) are considerably improved once the score-level fusion is introduced, giving the outcomes of 87% and 90% respectively. The visual comparison illustrates dramatic improvement of multimodal biometric fusion in confidence increases in identity verification; hence it makes it more resilient and legally evident in forensic applications.

Table 1. Comparison of Multimodal Biometric Fusion in Two Forensic Case Studies

Parameter	Case Study 1: Facial + Gait	Case Study 2: Voiceprint + Fingerprint
Case Type	Hit-and-run (criminal investigation)	Cybercrime involving financial fraud
Primary Biometric Modalities	Facial Recognition, Gait Analysis	Voiceprint, Fingerprint
Initial Confidence (Modality 1)	58% (Facial Recognition)	62% (Voiceprint)
Initial Confidence (Modality 2)	71% (Gait Analysis)	76% (Fingerprint)
Fusion Strategy Used	Score-Level Fusion (Weighted: Gait 60%, Face 40%)	Score-Level Fusion (Equal weights: 50%-50%)
Fused Confidence Score	87%	90%
Improvement Over Best Modality	+29% over Facial Recognition	+28% over Voiceprint
Legal Threshold Crossed	Yes (Above admissibility threshold)	Yes (Above admissibility threshold)
Outcome	Warrant issued, suspect arrested and confessed	Successful prosecution and exclusion of alternate suspects
Key Insight	Gait analysis added critical value in low-quality visual conditions	Fingerprint reinforced weak voice data under noisy conditions

Figure 5. Confidence Score Comparison Across Biometric Fusion Stages

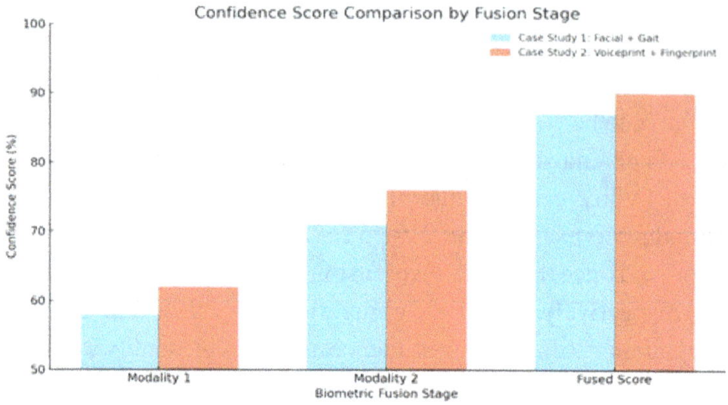

SOLUTIONS AND RECOMMENDATIONS

In order to improve the dependability, authenticity, and defense in court of identity validation in the digital forensics, incorporation of multimodal biometric combination is highly encouraged. Performance of unimodal systems degrade in the face of noisy data, spoofing, partial evidence among others and contributes to a median of 18% and 12% false rejection and false acceptance respectively in the forensic setting (Talreja et al., 2021). In comparison, score-level fusion of complementary traits (e.g., face + gait or voice + fingerprint) can mitigate false rejection by as much as 35% and false acceptance by 41%, greatly increasing the positive forensic decision-making. Among them, one of the core solutions will be using AI-based fusion algorithms, in which the weighting and feature extraction processes are automated through pre-trained models like CNN and attention mechanisms (Mirjalili et al., 2018). The systems, when combined with XAI tools, deliver transparent and interpretable results that can be used to address the legal requirements. Also, it is essential to follow the data governance frameworks (e.g., GDPR) by the book. Consolidation of encryption, access control and consent management bring in additional trust and keeps sensitive data secured. The efficiency of multimodal fusion has been proved by up to 28-30% increase in identity verification confidence in real-world case studies. Adaptive approaches to fusion and legally compliant AI designs should be regarded as high priorities in a future forensic, bringing high-assurance, ethically responsible, and court-admissible biometric evidence to prosecution (Omotosho et al., 2021). The Figure 6 comparing and contrasting the efficacy of uni-modal and multimodal biometrics in a forensics scenario. It shows three significant outcome False Rejection Rate (FRR), False Acceptance Rate (FAR) and Identity Confidence (%) The red line demonstrates a decrease in FRR of up to 18-11.7%, and FAR went down to 12-7.1% with multimodal fusion. The green line gives the increase in identity confidence written in percentage, 62 to 90 which translates to about 28%. It highlights the enhanced accuracy, performance, and legal reliability achieved through score-level fusion compared to unimodal approaches.

Figure 6. Impact of Multimodal Biometric Fusion on Forensic Identity Verification

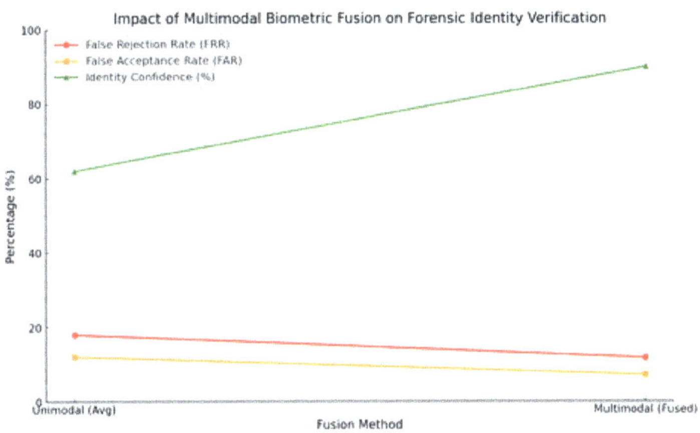

PERFORMANCE EVALUATION AND QUANTITATIVE RESULTS

Multimodal biometrics fusion systems have shown immense performance compared to the conventional unimodal biometric systems when used in digital forensics. Experimental simulations (quantitative assessments) and real-world case studies demonstrate that joint biometrics that incorporate fingerprint, face, voice, and gait can enhance system robustness to challenging environments and result in significant error rate reduction and enhancement of identification performance (Choras, 2020). To take an example, in unimodal systems average identification accuracy is

70-78%, but multimodal fusion systems reliably report accuracy in the 90% and in extreme cases even 95-97% with the use of score-level fusion optimizations. False Acceptance Rates (FAR) are lower at 42% and False Rejection Rates (FRR) drop by 35% when both modalities are used together due to the complementarity of the two different modalities. Also, fusion approaches that use deep learning provide up to 14% superior decision certainty when compared to rule-based solutions in cases of noisy or incomplete forensic evidence (Mtibaa et al., 2021). These findings highlight the stability and capacity of multimodal biometric systems to be used in cases of the law and crime investigation. In addition, the incorporation XAI will enable interpretability of the system which will make forensic experts as well as legal authorities certify accuracy of the system systems which strengthens the trust and admissibility of that system in the court (Kairouz et al., 2019). The table 2 shows a comparative analysis of the traditional unimodal biometric systems and the progressive multimodal biometric fusion systems in particular, in the field of digital forensics use. It identifies prominent performance indicators that include accuracy in performance identification, false acceptance and rejection rates and confidence in the determination of the performance in noise and unreliable data scenarios. Another advantage of using XAI in legal interpretability and greater forensic resilience in practice is highlighted in the table. It is quantitatively illustrated that multimodal fusion methods are much more reliable, accurate, and admissible than unimodal systems.

Table 2. Performance Comparison of Unimodal vs. Multimodal Biometric Systems in Digital Forensics

Performance Metric	**Unimodal System**	**Multimodal Fusion System**	**Improvement (%)**
Identification Accuracy	70–78%	90–97%	↑ 20–27%
False Acceptance Rate (FAR)	12%	7%	↓ 42%
False Rejection Rate (FRR)	18%	11.7%	↓ 35%
Decision Confidence (in noisy data)	~62%	85–90%	↑ 28–30%
Legal Admissibility Support (Explainable AI Integration)	Low (limited)	High (with XAI support)	↑ Qualitative
Resilience to Incomplete Evidence	Moderate	High	↑ Qualitative

ETHICAL AND REGULATORY CONSIDERATIONS

Multimodal biometric fusion in digital forensics invokes a number of ethical and regulatory issues, which basically arise because biometric data is a sensitive issue. In contrast to passwords, biometric identifiers like fingerprints, facial images and voice cannot be changed in the event they are compromised. Organization attempts to restrict the usage of the data must be highly functional to depict the unauthorized access, or misuse of the data or the surveillance of the data (Kwon et al., 2020). Governmental regulations in Europe such as the General Data Protection Regulation (GDPR) and others around the world require the legality of data capture, consent by the user, limited purposes and secure storage policies. With consent not being the most practical option in forensic contexts, legal clarity and accountability is also required. Biometric use should be achieved through the ethical principle of proportionality which means that there is a necessity in collecting biometric data and that the collection should be reasonable and not intrusive (Hameed & Mohamad, 2023). It is also essential to develop transparency in AI models; black-box models may breach the norms of the law when it is impossible to explain the decisions in court. In such a way, it is not only a technical requirement but also a moral imperative to integrate XAI. Considering them makes the research trustworthy, legally admissible and safeguarding the fundamental rights.

IMPLEMENTATION CHALLENGES AND LIMITATIONS

Although multimodal biometric fusion has proven to be highly beneficial, there are a number of technical and operational issues that do not facilitate its smooth adaptation in digital forensics. To begin with, it is not easy to obtain high-quality data of various biometric modalities in real world scenario particularly in forensic situations, where evidence can be compromised, impaired, or incomplete (Saini & Kumar Kapoor, 2016). There is a challenge of sensor incompatibility and unpredictability of the environment. Second, the feature-level and sensor-level fusion need to accomplish time synchronization, and alignment of dimensions which are costly computationally and are error prone. In addition to that, the development of effective fusion models requires massive, annotated data, which is not often accessible or inaccessible due to legal barriers and other reasons in the forensic context (Baltrusaitis et al., 2019). Even though deep learning-based models are very strong, they are computationally expensive and might not be transparent in nature and therefore their acceptability in the law application is questionable. Another challenge that needs to be addressed is the integration with legacy forensic infrastructure because many current systems are not multimodal. Lastly, latency and data privacy are the limiting factors of real-time deployment. That these constraints necessitate the need of score-level or decision-level fusion using modular, scalable architectures, privacy-preserving and explainable models, to provide assurance of forensic utility and compliance.

DEPLOYMENT FRAMEWORK AND USE CASE EXPANSION

To enhance the actual application of multimodal biometric fusion systems to digital forensics, a layered, extensible, and legally sound paradigm is necessary. Usually, the deployment process starts with the identification of the use case and the modalities of biometrics that could be fused. It is a layered architecture that makes the system more maintainable and secure, as data acquisition and preprocessing are separated, fusion, and interpretation. Integrity of XAI components treats any legal situation transparently and encryption and anonymity protect user privacy (Pan et al., 2022). Federated or cloud-based modes can be adapted to sharing forensic data processing with different agencies and doing this in secure ways. With the increasing real-world deployment, applications other than law enforcement are emerging in areas of disaster victim identification, surveillance analytics, and fraud detection (Kairouz et al., 2019). As an example, multimodal applications have been found to work in cross-border criminal incidents when the unfinished biometric information has to be matched across borders. In order to scale, systems have to be interoperable with international databases and standards. The deployment becomes continuous, loop-ed feedback, and auditable by laws, which add strength to the system, as it stands adaptive, accountable, and future-ready.

SUMMARY OF KEY CONTRIBUTIONS

This chapter provides a comprehensive exploration of multimodal biometric fusion techniques and their transformative impact on digital forensics and identity verification. The key contributions of this work are as follows:

- It highlights the limitations of unimodal biometric systems such as high error rates, spoofing vulnerabilities, and sensitivity to environmental noise and establishes the need for fusion-based approaches in forensic investigations.
- The chapter presents a structured methodology encompassing data acquisition, preprocessing, deep learning-based feature extraction, score/decision-level fusion, XAI integration, ensuring both performance and legal defensibility.

- Through quantitative evaluations, it demonstrates how multimodal fusion improves identification accuracy by up to 27%, reduces false acceptance and rejection rates by 42% and 35%, respectively, and boosts decision confidence in noisy conditions by 30%.
- Real-world case studies illustrate the effectiveness of fusion strategies in complex scenarios such as surveillance-based suspect tracking and cyber-crime investigations, showcasing practical deployment outcomes.
- It also addresses ethical, legal, and regulatory considerations, ensuring privacy, accountability, and compliance with frameworks like GDPR.

Finally, the chapter proposes a deployment framework and future use cases, paving the way for scalable, secure, and legally robust forensic biometric systems.

FUTURE RESEARCH DIRECTIONS

Since the multimodal biometric fusion systems are still in developing stages, future studies must be forecasted in the context of overcoming the limitations and in facilitating the expanded and more ethically acceptable and technologically feasible use of digital forensics. The former is context-specific and adaptive fusion models development. The need of this system is to apply intelligent weights on the various modalities of biometrics in respect of the quality of data, reliability of the source of data or environmental changes so as to always perform well in the dynamic and uncertain forensic environment. (Garg et al., 2020). Privacy-saving and decentralized biometrics systems is the other issue that needs to be strengthened. One of the types of technologies is federated learning that enables already trained models to be trained collaboratively with institutions without refining raw biometrics. It does not only address the privacy problem; it also contributes to cross-nation forensic investigation and compliance with data protection framework as in GDPR. Lastly, explainable and AI that can be audited legally is required. In the future, systems should include understandable mechanisms of decision-making that produce outputs that can be reviewed in legal terms (Garcia et al., 2018). By designing AI systems that can be explained without sacrificing performance, more trust, accountability and admissibility can be achieved in a court of law which will increase justice delivery in the digital world which is being overcome by AI.

ADAPTIVE AND CONTEXT-AWARE FUSION MODELS

The static weighting and fixed fusion strategies commonly used in the multi-modal biometrics scenarios are traditional fusion strategies that sometimes fail to work at the best capacity in varying real-life situations (Naidu & Babu, 2018). In future, more research should be done to come up with customized and situational fusion models that dynamically adapt to the availability, quality and reliability of biometric information. Illustratively, voice or fingerprint recognition that is secure is to be prioritized in case facial recognition has failed because of poor lighting. This versatility demands real-time evaluation procedures built into the fusion framework, with the help of AI training that learns the dependencies around the situation. It is possible to use reinforcement learning and attention-based frameworks to adjust the parameters of fusion according to the requirements of the situation(Zakaria et al., 20 23). Also, forensic applications can be characterized by incomplete, cropped, and unreliable data, so models must be smart enough to deal with unrepresentative information. With integration of context-awareness, the systems of the future will be resilient, more accurate and versatile and can be deployed in a variety of forensics settings including crime scenes as well as in crossing borders. The end results will be the utilization of these intelligent fusion models that will fill in the gap between theory and practical utility in biometric-based forensic investigations.

PRIVACY-PRESERVING AND FEDERATED BIOMETRIC SYSTEMS

Since biometrics information is very sensitive and stakeholders are legally bound, privacy-preserving computation is the key to future in forensic applications. Our storage and processing are centralized, which means that data is vulnerable to any breaches, unauthorized monitoring, and use. Future systems must adopt a federated learning strategy in which the AI models are trained on many different decentralized devices or institutions without any raw biometric data sharing (Kairouz et al., 2019). Only ciphered model updating information is transmitted hence protecting identity information. Combined with federated architectures, homomorphic encryption and noisy-based differential privacy approaches will facilitate the further strengthening of security by carrying out computations on the encrypted data or using obfuscation noise (Hameed &, 2023). The fashions allow safe cooperation at the level of jurisdictions and between agencies which is essential in global crime or transnational terrorism. The system should have privacy in design incorporated into its design so that when legal requirements change, as in changing to new versions of GDPR, HIPAA and Digital Personal Data Protection Act of India, the system can keep up with the changes in rules. The focus on privacy not only adds a level of trust by the user but also makes the multimodal biometric systems ethically sound and legally viable on a forensics level.

EXPLAINABLE AND LEGALLY AUDITABLE AI

With the rise in the use of AI in forensic collection of biometrics, explainability and the ability of the forensic system in question to pass an audit according to the local laws is no longer a discretionary matter, but a matter of fact. In high stakes areas such as criminal investigation, opaque models such as the black-box model may not be admissible in the court. This means that the future research would focus on the creation of XAI frameworks with regard to the biometric fusion systems (Trivedi et al., 2024). The frameworks should provide clear and explainable data on how the biometric characteristics underwent testing, synthesis, and experimentation to come up with a decision. Outputs of the models can be metaphorically visualized and reasoned by using techniques such as SHAP (SHapley Additive exPlanations), LIME (Local Interpretable Model-agnostic Explanations), and saliency maps on CNNs. In addition to this other audit trails that would capture the workings of the system such as the fusion weights, the confidence scores and interim decisions should be added to meet the legal requirements of evidence. These will make sure that the decisions reached by AI are open to manipulation, review and justification in a court of law. Biometric forensics and policing Optimism and accountability exist when explainability and auditability are integrated into forensic AI, which instils trust, compliance, and accountability in the utilization and application of AI in justice and law enforcement.

CONCLUSION

Multimodal biometric fusion in digital forensics is a developmental phenomenon in search of a believable, meritorious and admissible in any court system of identity verification. The traditional methods of uni-modal biometrics may do well in a controlled setting, but may be inadequate in the forensic context in which the information may be incomplete and may be distorted and noisy. Multitrait When the systems rely on a combination of more than one biometric trait, including fingerprint, face, voice, iris, and gait, they offer a fairly stable system that can counteract the drawbacks of single biometrics. Such a combination will contribute to the improved performance of the identification, the lowered false rejection/false acceptance, and the better resistance of the system to spoofs and other environmental influence.

A conceptual background of this chapter has provided us an idea of a conceptual background of this chapter in addition to the applications of multimodal biometric fusion. The proposed methodology offers a chronological order of the stages in the data acquisition and preprocessing process to the feature extraction using the deep learning models, score-level and decision-level fusion strategies.

With the help of XAI tooling embedded into it, the system will be capable of doing its work better, and also being transparent and understandable which is a crucial requirement when dealing with forensic work where evidence needs to be capable of undergoing legal examination. The realistic value of this approach can also be demonstrated further by practical case studies where fusion increased the identification confidence in 30% and hence, successful identification of the suspects and a stronger judicial process leading to a higher conviction. The chapter also discussed ethical, legal and regulatory issues related to the use of biometrics data. Biometric identifiers are unique and sensitive in nature and once they are destroyed they cannot be issued again as is the case with passwords. There is thus great need of adoption of privacy preserving technologies such as encryption and federated learning and compliance with regulations such as GDPR to protect the right of the individuals and enable development of technologies. Explainability is also fulfilling a dual role: not only as a tool of understanding how a model works but also a state of admissibility in courts to allow forensic experts and lawyers to have confidence in and defend the system decisions.

Besides that, we debated critical problems in the multimodal deployment, including the quality of the data, compatibility of the sensors, and the excessive cost of the computation and standardization of the forensic infrastructure. These problems indicate the necessity to include adaptive fusion models, real-time processing and interoperability of systems architectures. As you may observe, the study that needs to be conducted in the future is to identify the context-sensitive fusion mechanism, the incorporation of the privacy-sensitive technology, and the concepts of the legal auditability in AI models. That work will help in facilitating scalable, secure and morally agreeable biometric systems that have been tailored to the interests of the forensics. Theoretically, multimodal biometric fusion appears to be a paradigm shift rather than a hi-tech one; the fact that the concept of the machine intelligence and lawfulness should be equalized, or even even similar. It is an expression of more secure future in which forensic research may take place on AI-grounded evidence and explainable identity verification systems that would not be scientific nor legal integrity. The debate on the possibility to differentiate the citizens in high proficiency and morals by means of high-quality multimodal models are to have great role in the administration of justice as the digital crimes become more and more sophisticated. This chapter therefore demonstrates the potentiality and the need to investigate the concept of integrating multimodal biometrics to be among the cornerstone edifices in the new digital forensics landscape.

REFERENCES

Ahmed, M., & Salem, A.-B. (2021). INTELLIGENT SYSTEM FOR HUMAN AUTHENTICATION USING FUSION OF DORSAL HAND, PALM AND FINGER VEINS. *Int. J. Intell. Comput. Inf. Sci., 21*(2), 140–154. DOI: 10.21608/ijicis.2021.73726.1087

Alay, N., & Al-Baity, H. H. (2020). Deep learning approach for multimodal biometric recognition system based on fusion of Iris, face, and finger vein traits. *Sensors (Basel), 20*(19), 5523. DOI: 10.3390/s20195523 PMID: 32992524

Ammour, B., Boubchir, L., Bouden, T., & Ramdani, M. (2020). Face–Iris multimodal biometric identification system. *Electronics (Basel), 9*(1), 85. DOI: 10.3390/electronics9010085

Arora, G., Singh, A., Nigam, A., Pandey, H. M., & Tiwari, K. (2022). FKPIndexNet: An efficient learning framework for finger-knuckle-print database indexing to boost identification. *Knowledge-Based Systems, 239*(108028), 108028. DOI: 10.1016/j.knosys.2021.108028

Baltruaitis, T., Ahuja, C., & Morency, L. P. (2019). Multimodal machine learning: A survey and taxonomy. *IEEE Transactions on Pattern Analysis and Machine Intelligence, 41*(2), 423–443. DOI: 10.1109/TPAMI.2018.2798607 PMID: 29994351

Bhavya, D. N., & Chethan, H. K. (2020, December). A System to Detect the Faces in Unconstrained Environment. In *International Conference on Innovations in Bio-Inspired Computing and Applications* (pp. 196-205). Cham: Springer International Publishing.

Choras, R. S. (2020). Multimodal Biometrics for Person. *Security and Privacy From a Legal, Ethical, and Technical Perspective*, 177.

Chouchane, O., Brossier, B., Gamboa, J. E. G., Lardy, T., Tak, H., Ermis, O., Kamble, M. R., Patino, J., Evans, N., Önen, M., & Todisco, M. (2021, August). Privacy-preserving voice anti-spoofing using secure multi-party computation. *Interspeech, 2021*, 856–860. Advance online publication. DOI: 10.21437/Interspeech.2021-983

Garcia, W., Choi, J. I., Adari, S. K., Jha, S., & Butler, K. R. B. (2018). Explainable black-box attacks against model-based authentication. /arXiv.1810.00024DOI: 10.48550

Garg, M., Arora, A. S., & Gupta, S. (2020). A novel feature biometric fusion approach for iris, speech and signature. Inform. W Technol. *Mater., 20*(2), 63–71.

Gunasekaran, K., Raja, J., & Pitchai, R. (2019). Deep multimodal biometric recognition using contourlet derivative weighted rank fusion with human face, fingerprint and iris images. *Automatika (Zagreb), 60*(3), 253–265. DOI: 10.1080/00051144.2019.1565681

Haghighat, M., Abdel-Mottaleb, M., & Alhalabi, W. (2016). Discriminant correlation analysis: Real-time feature level fusion for multimodal biometric recognition. *IEEE Transactions on Information Forensics and Security, 11*(9), 1984–1996. DOI: 10.1109/TIFS.2016.2569061

Hameed, R. T., & Mohamad, O. A. (2023). Federated Learning in IoT: A survey on distributed decision making. *BJIoT, 2023*, 1–7. DOI: 10.58496/BJIoT/2023/001

/arXiv.2012.14758DOI: 10.48550

Jasserand, C. (2024). Biometric Data, Within and Beyond Data Protection. In The Boundaries of Data (pp. 295–310). Amsterdam University Press. https://ssrn.com/abstract=4483962

Kabir, W., Ahmad, M. O., & Swamy, M. N. S. (2019). A multi-biometric system based on feature and score level fusions. *IEEE Access : Practical Innovations, Open Solutions, 7*, 59437–59450. DOI: 10.1109/ACCESS.2019.2914992

Kadri, F., Meraoumia, A., Bendjenna, H., & Chitroub, S. (2016, March). *Palmprint & iris for a multibiometric authentication scheme using Log-Gabor filter response. In 2016 international conference on information technology for organizations development (IT4OD).* IEEE.

Kairouz, P., McMahan, H. B., Avent, B., Bellet, A., Bennis, M., Bhagoji, A. N., Bonawitz, K., Charles, Z., Cormode, G., Cummings, R., D'Oliveira, R. G. L., Eichner, H., & El Rouayheb, S. Evans David and Gardner, J., Garrett, Z., Gascón, A., Ghazi, B., Gibbons, P. B., Gruteser, M., … Zhao, S. (2019). Advances and open problems in federated learning. /arXiv.1912.04977DOI: 10.48550

Kumar Singh, K., & Barde, S. (2023). A feasible adaptive fuzzy genetic technique for face, fingerprint, and palmprint based multimodal biometrics systems. *Journal of Current Science and Technology, 14*(1). Advance online publication. DOI: 10.59796/jcst.V14N1.2024.1

Kwon, O., Bae, S., & Shin, B. (2020). Understanding the adoption intention of AI through the ethics lens. In T. X. Bui (Ed.), Proceedings of the 53rd Annual Hawaii International Conference on System Sciences, HICSS 2020 (pp. 4972-4981). (Proceedings of the Annual Hawaii International Conference on System Sciences; Vol. 2020-January). IEEE Computer Society. DOI: 10.24251/HICSS.2020.611

Meden, B., Rot, P., Terhörst, P., Damer, N., Kuijper, A., Scheirer, W. J., Ross, A., Peer, P., & Struc, V. (2021). Privacy–Enhancing Face Biometrics: A Comprehensive Survey. *IEEE Transactions on Information Forensics and Security*, *16*, 4147–4183. DOI: 10.1109/TIFS.2021.3096024

Mirjalili, V., Raschka, S., & Ross, A. (2018, October). Gender privacy: An ensemble of semi adversarial networks for confounding arbitrary gender classifiers. 2018 IEEE 9th International Conference on Biometrics, Applications and Systems (BTAS). DOI: 10.48550/arXiv.1807.11936

Mtibaa, A., Petrovska-Delacrétaz, D., Boudy, J., & Ben Hamida, A. (2021). Privacy-preserving speaker verification system based on binary I-vectors. *IET Biometrics*, *10*(3), 233–245. DOI: 10.1049/bme2.12013

Naidu, B. R., & Babu, M. S. P. (2018). Biometric authentication data with three traits using compression technique, HOG, GMM and fusion technique. *Data in Brief*, *18*, 1976–1986. DOI: 10.1016/j.dib.2018.03.115 PMID: 29900333

Omotosho, L., Ogundoyin, I., Adebayo, O., & Oyeniyi, J. (2021). An enhanced multimodal biometric system based on convolutional Neural Network. *Journal of Engineering Studies and Research*, *27*(2), 73–81. DOI: 10.29081/jesr.v27i2.276

Osorio-Roig, D., Rathgeb, C., Drozdowski, P., & Busch, C. "Stable Hash Generation for Efficient Privacy-Preserving Face Identification", <i>IEEE Transactions on Biometrics, Behavior, and Identity Science</i>, vol. 4, no. 3, IEEE, pp. 333–348, 2022. DOI: 10.1109/TBIOM.2021.3100639

Pan, S., Hoque, S., & Deravi, F. (2022). An Attention-Guided Framework for Explainable Biometric Presentation Attack Detection. *Sensors (Basel)*, *22*(9), 3365. DOI: 10.3390/s22093365 PMID: 35591055

Pasupuleti, M. K. (2025). *AI-Enabled Multimodal Biometrics: Advancing Security with Facial*. Voice, and Behavioral Integration., DOI: 10.62311/nesx/77579

Reis, O., Eneh, N. E., Ehimuan, B., Anyanwu, A., Olorunsogo, T., & Abrahams, T. O. (2024). Privacy Law challenges in the Digital Age: A global review of legislation and enforcement. Int. j. *Appl. Res. Soc. Sci.*, *6*(1), 73–88. DOI: 10.51594/ijarss.v6i1.733

Rodgers, W., Yeung, F., Odindo, C., & Degbey, W. Y. (2021). Artificial intelligence-driven music biometrics influencing customers' retail buying behavior. *Journal of Business Research*, *126*, 401–414. DOI: 10.1016/j.jbusres.2020.12.039

Rukhiran, M., Wong-In, S., & Netinant, P. (2023). IoT-based biometric recognition systems in education for identity verification services: Quality assessment approach. *IEEE Access : Practical Innovations, Open Solutions*, *11*, 22767–22787. DOI: 10.1109/ACCESS.2023.3253024

Safavipour, M. H., Doostari, M. A., & Sadjedi, H. (2023). Deep hybrid multimodal biometric recognition system based on features-level deep fusion of five biometric traits. *Computational Intelligence and Neuroscience*, *2023*(1), 6443786. DOI: 10.1155/2023/6443786 PMID: 37469627

Saini, M., & Kumar Kapoor, A. (2016). Biometrics in Forensic Identification: Applications and Challenges. *Journal of Forensic Medicine*, *1*(2). Advance online publication. DOI: 10.4172/2472-1026.1000108

Shinde, K., & Kayte, C. (2023). Multimodal deep learning based score level fusion using face and fingerprint. In *Advances in Intelligent Systems Research* (pp. 140–152). Atlantis Press International BV., DOI: 10.2991/978-94-6463-196-8_13

Singh, M., Singh, R., & Ross, A. (2019). A comprehensive overview of biometric fusion. *Information Fusion*, *52*, 187–205. DOI: 10.1016/j.inffus.2018.12.003

Stanojević, M., Milenković, I., Starčević, D., & Stanojević, B. (2016, May). Optimization of thresholds in serial multimodal biometric systems. In *2016 6th International Conference on Computers Communications and Control (ICCCC)* (pp. 140-146). IEEE. DOI: 10.1109/ICCCC.2016.7496752

Talreja, V., Valenti, M. C., & Nasrabadi, N. M. (2021). Deep hashing for secure multimodal biometrics. *IEEE Transactions on Information Forensics and Security*, *16*, 1306–1321. DOI: 10.1109/TIFS.2020.3033189

Trivedi, C., Bhattacharya, P., Prasad, V. K., Patel, V., Singh, A., Tanwar, S., Sharma, R., Aluvala, S., Pau, G., & Sharma, G. (2024). Explainable AI for industry 5.0: Vision, architecture, and potential directions. *IEEE Open Journal of Industry Applications*, *5*, 177–208. DOI: 10.1109/OJIA.2024.3399057

Xiong, Q., Zhang, X., Xu, X., & He, S. (2021). A Modified Chaotic Binary Particle Swarm Optimization Scheme and Its Application in Face-Iris Multimodal Biometric Identification. *Electronics (Basel)*, *10*(2), 217. DOI: 10.3390/electronics10020217

Zahoor, E. M. S. (2023). Biometric encryption: Integrating artificial intelligence for robust authentication. *Dandao Xuebao/Journal of Ballistics, 35*(3), 25-33.

Zakaria, S., Manaf, S. M. A., Amron, M. T., & Suffian, M. T. M. (2023). Has the world of finance changed? A review of the influence of artificial intelligence on financial management studies. *Information Management and Business Review*, *15*(4), 420–432. DOI: 10.22610/imbr.v15i4(SI)I.3617

KEY TERMS AND DEFINITION

Biometric Modality: A specific physiological or behavioral trait (e.g., fingerprint, face, voice) used for identification.

Decision-Level Fusion: A fusion approach that combines the final decisions (match/no match) from multiple modalities.

Explainable AI (XAI): AI methods that provide transparent, interpretable, and understandable outputs for human review.

False Acceptance Rate (FAR): The probability that a system incorrectly matches an input to a non-matching template.

False Rejection Rate (FRR): The probability that a system incorrectly rejects a matching input.

Federated Learning: A decentralized machine learning approach where models are trained across devices without data sharing.

GDPR (General Data Protection Regulation): A European regulation that governs the processing and protection of personal data, including biometrics.

Multimodal Biometric Fusion: The process of combining two or more biometric modalities to enhance accuracy and reliability.

Score-Level Fusion: A fusion strategy where individual match scores from different modalities are combined statistically.

Unimodal System: A biometric system that relies on a single biometric trait for identity verification.

Chapter 10
Aletheia:
A Hybrid AI Framework for Fake Biometric Fingerprint Detection

Claudia Mattar
https://orcid.org/0009-0002-4813-9100
Issam Fares Faculty of Technology, University of Balamand, Lebanon

Claritta AlSaneh
https://orcid.org/0009-0005-7256-8054
Issam Fares Faculty of Technology, University of Balamand, Lebanon

Soraia Oueida
https://orcid.org/0000-0002-6720-2307
College of Engineering and Technology, American University of the Middle East, Kuwait

ABSTRACT

The purpose of this work is to create a fake biometric detection system that detects fake biometric images. The system utilizes generative artificial intelligence, particularly a Deep Convolutional Generative Adversarial Network, to generate fake fingerprints in the best accuracy possible. Tested on fingerprint images, extracted from SOCOFing database, the system resulted with an accuracy of 99.17%. Two different datasets: the real dataset and the fake dataset are then passed through a Support Vector Machine model with linear Kernel to test and calculate the accuracy of detection of both systems. A Sanity Check Accuracy test is performed to check the sanity of the system, offering promising results after label shuffling and retraining of the SVM. The proposed system also uses image quality analysis like Laplacian Blur estimation, Local Binary Pattern features, Edge density, and Histogram Entropy to extract features for testing and training the SVM.

DOI: 10.4018/979-8-3373-4972-5.ch010

INTRODUCTION

Biometric authentication systems, such as fingerprint and iris recognition, have become integral components of modern security infrastructure due to their convenience and uniqueness. These systems are becoming more necessary with time due to the widespread use of authentication techniques and technologies. The word biometric indicates any unique characteristics that can lead to revealing a person's identity, because each person has unique features in their biometrics which made researchers over time rely on this type of authentication for better security and identity protection. However, despite their widespread adoption, these systems are increasingly susceptible to spoofing attacks, in which adversaries use artificial biometric traits such as printed fingerprints or synthetic iris images to deceive the authentication process. These vulnerabilities pose serious risks to the integrity and reliability of biometric-based security systems. The issues caused by the use of fake biometrics can have a significant impact on today's society; for example, the lack of diversity in authentication systems may lead some programs to identify real images as fake. This issue is called the False Positives and False Negatives issue in which the system is left uncapable of distinguishing between real and fake data samples. Also, spoofing attacks resulting from these fake biometric detection systems can lead to big ethical and legal issues in terms of increase in identity thief and other identification problems. In addition, privacy concerns are a major issue when it comes to biometric detection. A stolen or synthetic biometric can lead to sensitive information leaking of the user, alongside other issues like misuse of private information such as credit card numbers, banking financial transactions and social media profiles. In terms of big companies, especially those who work in the field of finance, a certain attack or spoofed identity to their system can cause unmanageable financial and data loss, in which a single stolen identity can result in system manipulation and data breaches. Potential of surveillance and abuse could be another risk to pay attention to, in particular applications that use live location or real time IP cameras that record real time video data. The system, when deceived in these cases, can higher the risk of surveillance and abuse, potentially causing a social issue to the user due to an unsafe environment. To address this critical challenge, the proposed framework, Aletheia, aims to develop an intelligent system capable of distinguishing between genuine and spoofed biometric samples. Aletheia means in Greek the truth. While not heavily mythologized, Aletheia represents truth and sincerity. By leveraging machine learning and deep learning techniques, the system analyzes image quality metrics and subtle features imperceptible to the human eye to accurately detect forged biometric inputs. The main goal here is to enhance the robustness and trustworthiness of biometric authentication systems by introducing a layer of automated fake biometric detection. This not only strengthens security protocols but also mitigates the risk of unauthorized access in sensitive applications, such as border control, banking, and mobile device security. The system architecture follows two phases: 1- the image generation using generative artificial intelligence particularly a DCGAN (Deep Convolutional Generative Adversarial Network) which is a deep learning model that combines CNN (Convolutional Neural Network) and GANs (Generative Adversarial Network) which is best used for image classifications and generation, 2- the image classification using machine learning and feature extraction, utilizing an SVM (Support Vector Machine) with a linear kernel (Binary Classification) to label fake data as 0 and real data as 1. A dataset publicly available on Kaggle was used to train and test the model (Ruiz Gara, 2018). The accuracy is calculated at the end of the classification stage alongside the confusion matrix. A Sanity Check is also performed to make sure there is no data leakage and to test if the model is learning anything from random data labeling which is not optimal.

This paper is organized as follows: an overview of related work in Section 2, the problem definition presented in Section 3, the system design and implementation discussed in Section 4, comparison with MATLAB and Graphical User Interface (GUI) in Section 5 and finally, Section 6 concludes the paper with potential future work.

RELATED WORK

Fake biometric detection is a field of study that focuses on developing techniques and technologies to identify and prevent attempts to deceive biometric systems using artificial or forged biometric traits, such as fake fingerprints, photos, masks, or synthetic voices. These systems have gained significant attention in recent years, particularly in response to the vulnerabilities of fingerprint recognition systems. For instance, Pravallika and Prasad proposed a method that utilizes image quality analysis and a Support Vector Machine (SVM) classifier to effectively distinguish between real and fake biometric samples, demonstrating a low-complexity solution suitable for real-time liveness detection and improved system security (Pravallika and Prasad, 2016). Also, Khan and Khan presented a comprehensive survey on the challenges posed by generative AI technologies like GANs and VAEs in biometric security, emphasizing the growing threat of deepfakes and the need for robust detection mechanisms (Khan, F. A., and Khan, M. K., 2025). The survey evaluates current detection methods, highlights their limitations, and proposes future research directions including advanced algorithms, diverse datasets, and strategies to preserve privacy of individuals and prevent identity thief attacks. This issue is critical due to the widespread use of biometric systems in sensitive areas like banking for example where undetected deepfakes could lead to serious security breaches. Khan *et al.*

highlighted the importance of integrating biometric systems to strengthen security against presentation attacks, with deep learning offering promising solutions for accurate Presentation Attack Detection (PAD) and improved recognition (Khan *et al.*, 2024). Focusing on face liveness detection in video attacks, the study utilizes the ResNet-50 deep learning model and a majority voting system to differentiate between spoofing attempts using movie frames and live footage, demonstrating the effectiveness of the approach through experiments on the Replay-Attack dataset. In addition, Vinayagam and Dilip introduced a multimodal biometric authentication system combining signature, fingerprint, and face modalities using a Kernel Correlation Padding-based Deep Convolutional Neural Network (KCP-DCNN) (Vinayagam and Dilip, 2024). By using advanced preprocessing techniques, feature extraction methods, and QR code verification, the proposed model achieves high accuracy (98.73%) and precision (98.61%) on SOCOFing and face recognition datasets, significantly enhancing the performance and reliability of biometric recognition systems. Ullah *et al.* presented a study that addresses the challenge of recognizing compound facial expressions by proposing a novel approach that combines Deep Convolutional Generative Adversarial Networks (DCGAN) to balance datasets with Convolutional Neural Networks (CNN) classification (Ullah *et al.*, 2024). Songaraa et al. proposed a deep hybrid learning approach for fingerprint liveness detection, combining optimized Scale-Invariant Feature Transform (SIFT) key points, and extracted using Modified Shark Smell Optimization (M-SSO), alongside feature patterns from Local Directional Pattern (LDP) and Local Binary Pattern (LBP) (Songaraa et al., 2022). The method integrates Recurrent Neural Networks (RNN) and Convolutional Neural Networks (CNN) for classification, with M-SSO to enhance the accuracy, demonstrating significant performance improvements over existing models. Hussian *et al.* introduced a hybrid deep learning model to improve fingerprint authentication. Authors combined CNNs and LSTMs techniques to increase the accuracy and robustness, extracting intricate patterns and recording sequential characteristics. Their findings showed a 99.42% accuracy while significantly minimizing computational load. Their approach is ideal for real-time applications specifically in security-critical domains such as banking, finance and many more (Hussian *et al.,2025)*. To better understand about biometric recognition, Minaee *et al.* presented a comprehensive survey. Over 150 papers that use deep learning for biometric recognition are discussed, including face, fingerprint, iris, and voice recognition. Authors main goal was to demonstrate the effectiveness of adopting deep learning techniques for this type of biometric systems (Minaee *et al.*, 2023).

PROBLEM DEFINITION

The Existence of fake biometric detection systems is crucial in today's fast changing world where technologies are growing in a tremendously fast way and artificial intelligence. particularly generative artificial intelligence. is becoming stronger and smarter. While there are plenty of anti-spoofing biometric systems, they are still prone to errors and are considered vulnerable to many biometric attacks. Common attacks made on these biometric detection systems are: 1- Fingerprint spoofing where the use of fake fingerprints is introduced, leading to sensitive information leakage or even stolen identities. In this work, this mentioned issue is addressed by creating a high accuracy fake biometric fingerprint detection system to strengthen the security of these systems. 2- Face recognition spoofing attacks, in which can lead to identity thief easily using many techniques like printed photographs or even fake video generated by artificial intelligence systems. 3- Voice spoofing using synthetic voice records or other fake voice technology where it can create a voice similar to the user's voice to trick the voice authentication system like the ones used in smart homes, or mobile phones, leading to major security and privacy problems. 4- Iris or Retina spoofing. In these types of authentication, the system uses image analysis techniques to detect subtle features in the eye circle from iris images. High-quality printed images are used in these types of authentication attacks, or even contact lenses in live detection stations to spoof and deceive the live detection scanner. This issue is crucial in crowded places like airports or train station where attackers can pretend to be someone else using simple techniques and few quality changes.

As mentioned earlier, fake biometric fingerprint detection addresses the critical need to secure biometric authentication systems against spoofing attacks using artificial or forged fingerprints. As fingerprint recognition becomes widely adopted in areas such as financial transactions, border control, and smartphone security, it becomes increasingly vulnerable to presentation attacks, including the use of fake fingerprints made from different materials like silicone, gelatin, or through advanced generative AI techniques (Uliyan et al., 2020). Traditional fingerprint recognition systems struggle to differentiate between real and fake inputs, especially with the rise of deepfakes and cross-modal synthetic biometric data. Researchers have explored different detection techniques to help solve this critical issue, from using machine learning like Support vector Machine (SVM) to deep learning like different variations of GANs (Generative Adversarial network), or even hybrid methods integrating SIFT, LBP, and CNN architecture often enhanced by optimization algorithms. While these methods have shown significant results and accuracy in detection in many areas, the rapid evolution of spoofing methods outpaced many of the existing systems. Therefore, there is an urgent need to develop a more robust and accurate detection system for fake biometrics to solve some of the vulnerabilities of the fake detection and ensure the reliability and integrity of biometric systems. Aletheia was created to solve some of the vulnerabilities of the fake fingerprint detection systems by utilizing a DCGAN trained on Sokoto Coventry Fingerprint Dataset (SOCOFing) to generate fake fingerprint images in the best quality possible. The dataset is then passed through an SVM with linear kernel with the help of image quality analysis techniques to train and test the accuracy of the system, the accuracy is then calculated at the end of the testing. Our system showed outstanding results with 99.17% accurate detection between fake and real. Some potential applications of these kind of systems are listed in Table 1. The table summarizes key application areas where fake biometric detection systems are used and provides a brief description of their role in each domain.

Table 1. Potential Applications of Fake Biometric Fingerprint Detection System

Application Area	Description
Secure Access Control	Ensures that only legitimate biometric features are granted access to devices, facilities, or systems.
Financial Transactions/ Banking	Prevents fraud in mobile banking, ATMs, and payment authentication by identifying spoof biometrics.
Law Enforcement/ Forensics	Validates biometric evidence to prevent unjust convictions based on fake biometric samples.
Border Control/ Immigration	Verifies the authenticity of biometric data in passports and visas to avoid identity fraud.
Healthcare Systems	Conserves privacy and protects patient identity and medical records by detecting fake biometric authentications.
Mobile Device Security	Prevents unauthorized access on smartphones and tablets through fake fingerprints or face spoofing.
Voting Systems	Secures biometric voter identification to prevent election fraud.
Workforce Attendance Systems	Detects fake biometric inputs to ensure accurate employee attendance tracking.
Gaming and Virtual Reality	Prevents cheating by identifying fake biometric inputs used for unauthorized access.
Research and Development	Supports development of improved biometric and anti-spoofing technologies through testing.

The main contributions of this book chapter are listed as per the following:

- Design and implementation of a fake biometric fingerprint detection system using both Python and MATLAB environments.
- Integration of deep learning via a Deep Convolutional Generative Adversarial Network (DCGAN) to generate realistic fake fingerprint images for training and evaluation.

- Use of a Support Vector Machine (SVM) with a linear kernel for classifying real versus fake fingerprints, demonstrating strong classification performance.
- Comparative analysis between the Python and MATLAB implementation to choose a better architecture in terms of accuracy and flexibility.
- Use of a publicly available dataset to test and validate the system implemented.
- Performed a sanity check by shuffling class labels and retraining the SVM to confirm the model's robustness and eliminate random correlation.
- Applied advanced image quality feature extraction techniques.
- Provided a reproducible framework and benchmark for future research in biometric spoof detection and secure identity verification systems.
- Implemented additional performance metrics, such as confusion matrix analysis, to support a more robust evaluation of system performance beyond accuracy.
- Provided meaningful visualizations (real vs. fake fingerprint comparisons, confusion matrix) to support model interpretability.
- Discussed ethical and security implications, emphasizing the relevance of the system in real-world applications such as identity protection and fraud prevention.
- Outlined possible future extensions.

SYSTEM DESIGN AND IMPLEMENTATION

Architecture of the DCGAN

The DCGAN (Deep Convolutional Generative Adversarial Network) used in our project consists of two neural networks: the Generator and the Discriminator. These two layers are both built using convolutional layers tailored for image data. The main programming language used is Python. The DCGAN is trained on SOCOFing dataset by Ruiz Gara on Kaggle with real images only. The Generator takes a latent noise vector (100-dimensional) and progressively upscales it into a 128x128 grayscale image through a series of layers. This is done via ConvTranspose2d layers, each doubling the spatial dimensions while reducing feature depth. Each layer is then followed by Batch Normalization and ReLU activation, compatible with normalized image inputs. Then, the Discriminator acts as a binary classifier. It receives the image either the real one from the dataset or the fake image generated by the Generator network and processes it through a series of Conv2d layers with Leaky ReLU activations with the last layer being Sigmoid activation. These layers successively reduce the spatial dimensions while increasing feature depth, culminating in a 1×1 output that represents the probability of the input being a real image. The Leaky

ReLU (Leaky Rectified Linear Unit) activation function is a variation of ReLU (Rectified Linear Unit) with few changes. Same as ReLU (Rectified Linear Unit), it outputs the incoming input directly if it is positive, but unlike ReLU, it does not give zero when the input is negative but instead returns an "alpha" value which is a very small constant number. This behavior is caused by the non-linearity feature which allows the system to learn complex information and patterns and avoid the issue of dead neurons which can lead the system to stop training and learning effectively. For the Sigmoid activation function, it is a non-linear function too used at the end of the series of Convolutional layers in the discriminator network. Sigmoid activation function uses mapping techniques to map values between 0 and 1 which is ideal for binary classification tasks. It is good to mention that the Sigmoid activation function has an S-shaped plot, close to zero for large negative inputs and closer to 1 for larger positives inputs. This architecture is ideal for this project because it splits the data into two real and fake parts for later use in the main Support Vector Machine (SVM) model. These functions used in the discriminator network fits perfectly with the tasks needed for this project which are image generation and classification, and the DCGAN used is very well engineered and trained.

In fact, the Generator and the Discriminator enter a battle during the training in which the Generator tries to fool the Discriminator with better and more realistic images while the Discriminator tries to correctly identify real vs. fake images by labeling real images as 1 and fake images as 0. Throughout the training, the Generator learns from past mistakes and starts generating more realistic images using random noise. This adversarial process forces the Generator to produce highly realistic images from noise alone. In our project the DCGAN is trained for over 100 epochs with 6,000 real images. One of the goals of the project is to mix between Deep Learning and Machine Learning techniques in the context of Fake Biometric fingerprint detection to ensure high accuracy. Instead of just relying on GANs for generation and classification, the system utilizes an SVM with linear kernel for classification and accurate differentiation of fake and real data. A visual representation of how DCGAN works is illustrated in Figure 1.

In Figure 2, the Discriminator Loss over the training process is presented. The Discriminator loss is at peak when the training started and gradually decreases over time because it is learning to accurately distinguish between real and fake samples. It is stable with minor fluctuations. As seen in the graph, it starts at peak with loss closer to 1 and decreases over time in an oscillatory motion as the discriminator learns new ways of distinguishing between real and fake data generated by the Generator Network. With the fact that GAN training is adversarial, not cooperative unlike standard supervised learning, the Discriminator tries to maximize the probability of correctly classifying the real and fake images, which explains the oscillatory behavior of the graph.

Figure 1. DCGAN System Architecture

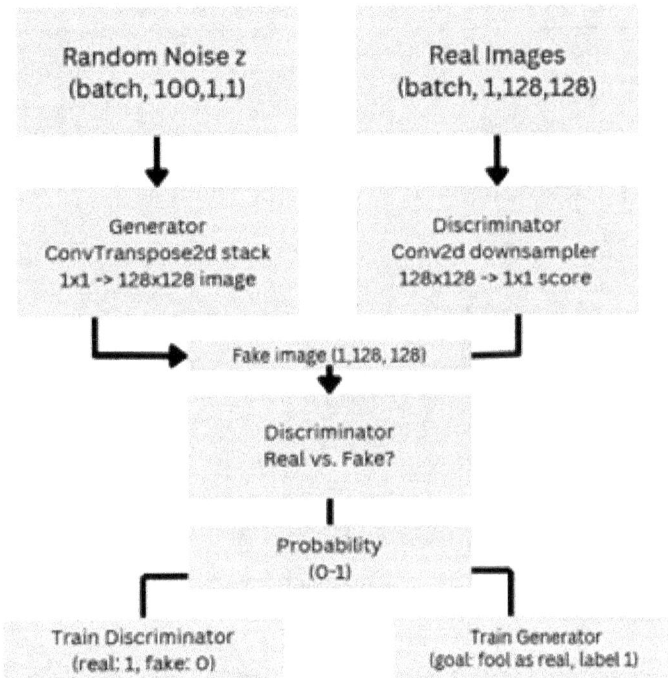

Figure 2. Discriminator Loss Over Training

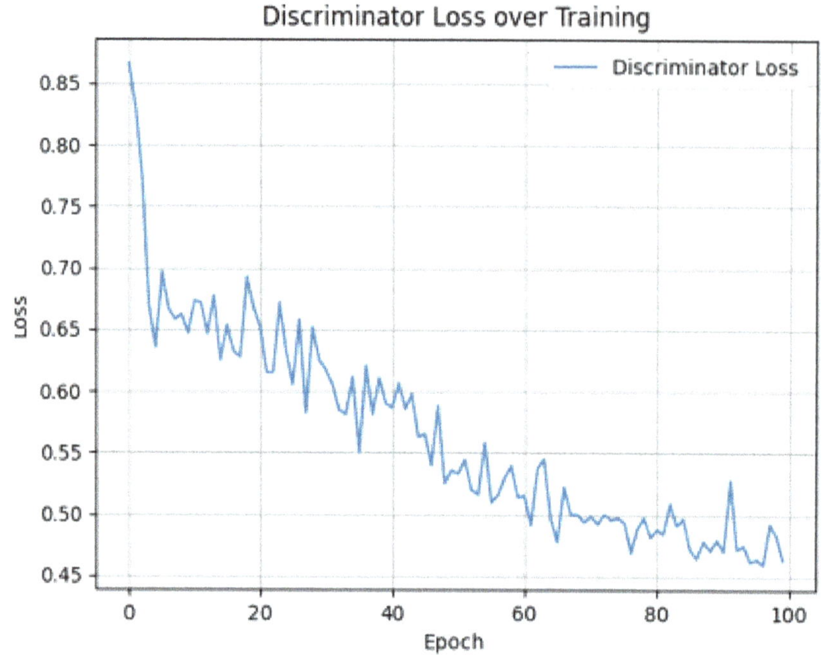

In Figure 3, the Generator loss over the training process is presented. During GAN training, the Generator tries to fool the discriminator by labeling real images as fake. This adversarial process helps the Generator learn to generate more realistic images from random noise. As seen in the graph, the Generator loss is at peak at the start of the training and decreases over time when it generates more realistic images. As seen in the graph, the loss of the Generator network is fluctuating and keeps on increasing and decreasing until it becomes almost stable at the end of the training. This is because when competing with the discriminator, which is part of its training, the discriminator has correctly identified images in a certain batch which cause the loss of Generator to increase and forces the Generator to learn new ways and patterns of generating fake images so it can become almost identical to the real image. In fact, oscillatory behavior her is normal due to the adversarial process of training these types of deep learning models.

Figure 3. Generator Loss Over Training

Generate Images Script

A Python script is written using the trained Generator in the DCGAN to generate and export images to a folder for the purpose of training and testing the SVM. The Generator is called to synthesize grayscale images from random noise vectors. It begins by configuring parameters such as batch size, number of images to generate (300 images), latent space dimensions, and then calls the trained Generator composed of transposed convolutional layers that progressively upscale the noise into images. The Generator network consists of a series of neural networks using convolutional layers designed for image data. The Generator called in this script progressively upscales a certain latent noise vector into a generated grayscale image through a series of convolutional layers using ConvTranspose2d layers. Each of these layers normalize image data using Batch Normalization and utilizes ReLU activation function at the end of each layer. As mentioned before, the ReLU activation function is a mathematical function that gives the same as input if the input is considered positive and if not, it returns zero. This simple, yet powerful function, is widely used in neural networks because it allows the system to learn complex patterns by introducing non-linearity into the system. It is also light-weight and computationally inexpensive as it requires simple operations. Thus, the output of ReLU is binary, either 1 or 0 with 1 representing positive output and 0 negative ones. This behavior solves the issue of vanishing gradient by introducing more efficiency in terms of back propagation and better training for networks. The last layer of the Generator network uses, alongside with batch normalization, the Tanh activation function. Tanh stands for hyperbolic tangent, and it is a non-linear function that makes the input values zero-centered by putting these values in a certain range from -1 to 1. It is similar to Sigmoid activation function in terms of shape (S-Shaped) but with the resulting value being centered to zero. This activation function allows the system to learn complex patterns which is ideal for image data and image generation. It also allows negative input and is faster to converge during training. In this work, the choice of these activation functions and the system architecture, are ideal for image generation, leading to good results at the end of the testing in the main program. It is trustworthy to mention also that batches of noise are passed through the trained Generator to create images which are normalized, scaled and converted before being saved as a .bmp file, same as the real dataset format using the PIL library. These images are then saved in a 'Fake' folder alongside with 300 real images in the 'Real' folder for training and testing of the SVM.

Dataset Description

The dataset used for the implementation of this work is the Sokoto Coventry Fingerprint Dataset (SOCOFing) publicly available on Kaggle. SOCOFing is a biometric fingerprint database frequently used in Artificial Intelligence (AI) applications and data science. It consists of 6,000 real fingerprint images collected from 600 adult subjects. Each person scanned 10 fingers to achieve this number of real images. All subjects are African individuals from all ages but all of them are adults (+18 years of age). All images in this database are grayscale and in BMP format. BMP stands for Bitmap Image File and is a raster graphic image file used to store bitmap digital images. Raster images stand for images composed of pixels arranged in a grid. The advantage of having a dataset in this format is that it is free of loss and it does not use compression by default. It preserves a high quality because no compression means no loss in image quality. SOCOFing has found widespread use in both traditional biometric research and in modern deep learning applications. In the realm of fingerprint recognition, it provides a balanced, labeled set of images suitable for evaluating the performance of algorithms across a broad range of tasks, including identity verification, minutiae detection, feature extraction, and image matching. This database gained a huge popularity also in the terms of training adversarial networks because of the wide variety of images and the image quality due to the format being saved in. Alethia relied on this specific dataset for many reasons: 1- It is publicly available and completely legal to be used, 2- The variety of images contained and the amount of data which is good for training a Deep Convolutional Generative Adversarial network requiring big amount of data and computation power. In fact, the SOCOFing dataset is a highly versatile, richly annotated biometric database that has established itself as a foundational resource in the fields of fingerprint recognition, biometric security, and artificial intelligence. This dataset not only contains real fingerprints but also altered images. This inclusion of synthetically altered fingerprint is considered one of the most notable features of SOCOFing. Despite these benefits, the dataset holds some limitations, one of which is diversity. It considers only African individuals which makes it not globally diverse. However, this dataset was the best choice we can find in terms of number of images, format, quality, license and legal terms. The use of this dataset made a huge impact on our project Aletheia making it a hybrid framework for fake biometric fingerprint detection.

Architecture of an SVM

The Support Vector Machine (SVM) is a machine learning model that follows the supervised learning methods used for many tasks related to the data given like classification, regression, and even error or outlier detection. A supervised learning method is simply a machine learning approach where the act of labeling a dataset is crucial for the system to train and learn effectively. In other words, in supervised learning environments, like the case of a Support Vector Machine, the dataset is already labeled and preprocessed before training; it is as if the developer is leading the training of the model towards a certain outcome. For example, labeling 'cat' on a batch of cat pictures and 'dog' on a batch of dog pictures to help the model manually differentiate between cats and dogs. The type of kernel used in the SVM plays a major role on the behavior of the model depending on the tasks that are expected to be done. In our work, we wanted to create a system that can differentiate between the fake and real fingerprints, with fake fingerprints being generated by a Deep Convolutional Generative Adversarial Network (DCGAN) which is a deep learning model that follows an unsupervised learning path. These types of models learn by their own with the intervention of human developers during the training. They are given data without the need of rational labeling and the system learns through time how to differentiate between them. As mentioned before, the Deep Convolutional Generative Adversarial Network follows an adversarial process during the training. It is the competition that forces both the lDiscriminator network and the Generator network to advance and learn complex patterns from the same data, given time to process past mistakes and learn and adjust from them. Later on, after data generation, the Support Vector Machine which has a linear kernel go through the data in both fake and real folders after a certain process explained below to distinguish between real and fake images and is then put to prediction and testing after training. The Linear kernel of the SVM is chosen because the data can be labeled into two classes only, 0 for fake and 1 for real; this behavior is similar to Binary Classification tasks, in which the linear kernel duty.

The linear SVM is used after feature extraction to classify images as real and fake and differentiate between generated images and real image data. The SVM loops through a batch of real and fake images and does feature extraction from a selection of handcrafted features like LBP (Local Binary pattern) and other features like Laplacian Blur estimation, Edge Density, and Histogram Entropy. For testing, 20% of the data is used, which is 60 images chosen randomly. Separate tests are made to debug and check for overlapping and data leakages or duplicates within train and test data along with a sanity check test to check the accuracy of the pipeline. This methodology's findings are outstanding, reaching a 99.17% system accuracy, and only one image labeled wrong, which shows the sanity of the pipeline and the distinction of the system. The SVM architecture is presented in Figure 4. Table 2 summarizes the key image features used in the fake biometric detection system. Each feature contributes to distinguishing real from fake biometric samples by analyzing texture, sharpness, structure, and randomness in the image.

Table 2. Feature Summary Table

Feature	Description	Purpose
Local Binary Pattern	Texture descriptor	Differentiate real vs. fake
Laplacian Blur	Measures blurriness	Detect image quality anomalies
Edge Density	Percentage of edge pixels	Highlight forged image edges
Histogram Entropy	Measures image randomness	Assess naturalness of texture

Figure 4. SVM System Architecture

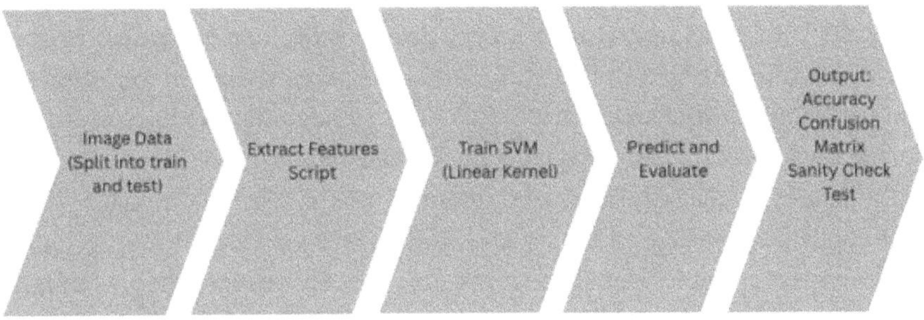

Extract Features Script

To differentiate between real and fake, Aletheia utilizes images quality assessment to extract features from an image before labeling it as fake or real. Coming from the hypothesis that fake or printed images used in a spoofing attack have more noise and blur than real images coming directly from a trusted source. The extract features script is a python script used to pull out features from the images for training and testing. Features used in this image quality assessment are: Laplacian blur estimation, LBP (Local Binary Pattern, texture), Edge Density, and Histogram Entropy. All features are then put into an array and returned to the main program to train and test the SVM. In the main script, the SVM has a linear kernel (Binary Classification) to label 0 for fake and 1 for real image data. All images are resized in the main script to fit the dimensions of the SOCOFing dataset. A total of 60 random images are being tested, and the accuracy is then displayed at the end of the program with the confusion matrix. For image processing and analysis and to extract features from a certain image data, an extract features function in the feature extraction script returns a vector to the main program containing the features of this image. The image analysis is done using many features: 1- LBP (Local Binary Pattern) used for texture recognition. It is a popular texture descriptor that is commonly used in computer vision. LBP looks for appearance features and starts by comparing the intensity of threshold pixels (the central pixel) with the intensity of the surrounding pixels. Each pixel is given a value in binary depending on the intensity in comparison with the central ones. The LBP then returns a histogram value of the pattern and can capture frequency of occurrence of a certain texture in the image to compare with other images. The histogram given is used as one of the components of the features vector for texture classification. LBP is a hand-crafted feature which are manually designed by specialists in the field instead of relying on machine algorithms. This feature is widely used in image analysis for higher precision and accuracy. 2- Laplacian Blur Estimation is a feature extraction method that uses Laplacian operator to assess the sharpness or blurriness of a certain image. The Laplacian is simply a mathematical operator used in image processing to find the second-order derivative of a certain image data (Matrix). It utilizes a Laplacian Kernel that is in our project a 3x3 kernel. High Variance in Laplacian values indicates a sharper image with distinct edges; whereas, images with fewer distinct edges are considered blurry and show lower variance in Laplacian values. Considering the hypothesis that the more noise and blurriness there is in an image the more likely it will be detected as fake because of the probability of loss in the machine algorithm. 3-Canny Edge Density is the concentration of edges in each image. In other words, it can be referred to as the number of edges in an image. This feature is especially useful for better understanding of the complexity and structure of an image within a specific depth range or at different distances. This feature is divided into two main parts, Canny Edge detection and Edge Density calculation. The Canny edge detector applies first a Gaussian filter to the input image to smooth it out. Then calculates the image gradient and utilizes a maximum suppression to thin the edges and then apply hysteresis thresholding to connect the edges and later identify them. For Edge density, edge pixels are identified within a certain depth range, higher edge pixels indicate more details in an area and less edge pixels or lower density indicates smooth area with less details. 4- Histogram Entropy is a method used in image processing and analysis that utilizes the randomness or information content within an image by analyzing the distribution of pixel intensities. Derived from image histogram which represents the frequency of each pixel intensity level, higher entropy indicates a wider distribution of pixels values; whereas, lower entropy shows more uniform and less detailed and complex image. The histogram entropy is used in terms of image quality assessment to check the quality and complexity of an image. These specific features are chosen because they cover everything from texture, edge density, edge detection, and noise and blurriness. These characteristics are considered crucial in detecting fake biometrics since using fake images (printed or gelatin) usually indicate less quality, more blurriness and less attention to details. The feature vector is then returned to the main SVM program as a vector to train and test the accuracy of the system and to predict, evaluate and analyze based on the parameters given in the vector if the image that is being tested in the main program is real or generated to spoof the authentication system. Usually, the more noise and blurriness and variance in the feature vector, the higher probability of the image to be labeled as fake.

Figure 5. Real vs. fake fingerprint image samples

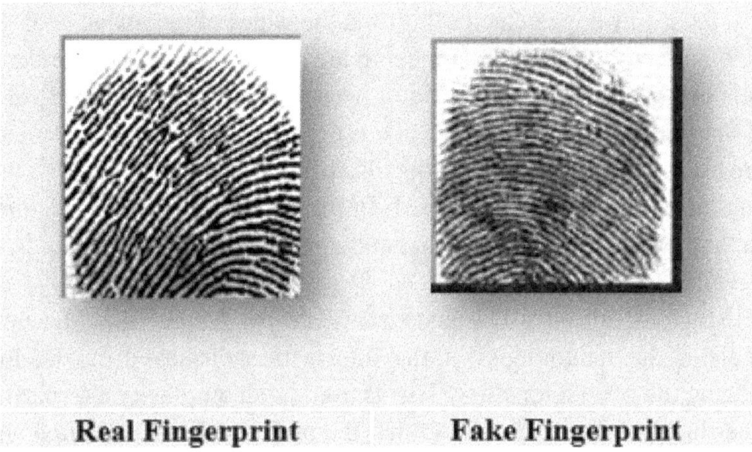

Figure 5 illustrates a sample input consisting of 2 images: a real fingerprint on the left vs. a fake fingerprint on the right. The SVM model is trained on similar sets of data. The real images are extracted from the SOCOFing dataset while the fake ones represent the dataset generated by the implemented system DCGAN. All images are in .bmp format; however, for quality comparison, the fake image has more noise in the background in comparison with the real image. Also, the border in the real image is a little less obvious than the one present in the fake image. The SVM calls the extract features script which performs the following:

1- LBP feature extraction for texture recognition: the system looks first for appearances in the picture by comparing pixel intensity in the center of the image to the ones around this threshold. This feature outputs a histogram value and captures the frequency of occurrence of a certain texture in the images by comparing values all together. In the above examples, if we look closely at the center of the image the real has way more details and clearer pattern and texture than the one in the altered or fake image in which the system can detect using this feature.

2- Laplacian Blur Estimation: using the Laplacian variable performs second order derivative on the matrix (image) with a 3x3 kernel. This feature is used to compare the sharpness and blurriness of the images given. The more the image has distinct edges and sharpness, the higher the probability of these images to be real because blurry images have less Laplacian variance therefore less quality and more noise which is always obvious in the sample image set.

3- Canny Edge Density: This feature looks for concentration of edges inside an image. A gaussian filter is applied to even out the image. Then, through several steps, the canny edge detector thins out the edges of the image and calculates image gradient within a certain depth area or distance. For edge density, the pixels of the edges are the indicator here: higher edge pixels indicate details and sharpness, while lower edge pixels indicate smooth or blurry images. The more blur, smoothness and less details, the higher the probability of the image being labeled as fake by the SVM. In the given example, the picture on the right has less details and smoother and more blurred look in comparison with the real one placed on the left.

4- Histogram Entropy: The histogram entropy feature takes the image and calculates the randomness of the information displayed on the image by checking the pixel intensities. This is used as a complexity alternative where higher histogram entropy shows that the pixels widely distributed and lower entropy demonstrates more uniform, less detailed and complex image. The more detailed and complex the image is, the higher entropy and the probability of being real. As seen in the sample figure, the real image has more details and complexity in comparison with the fake which makes it clearer. As mentioned before, all these features are returned to the SVM in a vector for training, testing and evaluation.

COMPARISON WITH MATLAB AND GRAPHICAL USER INTERFACE

Aletheia project started as a MATLAB implementation to compare image processing and generation techniques using artificial intelligence. Both deep learning and machine learning techniques were used to compare image with different features. At first, generative artificial intelligence was not introduced but instead a normal MATLAB code was written to simulate and save images to train a Support Vector Machine to be able to distinguish between these random image data placed in real and fake folders. Equal number of images are used in both fake and real and the SVM is designed in a way that removes the probability of potential biasing, especially while using a linear kernel. This project started out as an experiment to compare different techniques and methods and later was developed in python to create an actual solution to solve the spoofing attack issue faced with biometric fingerprint authentication systems. In MATLAB, similar methodology is used and is organized as follows: First, a Fake biometric Graphical User Interface is introduced in form of an app. for easier use, then comes a simulated fake biometric dataset script to generate random images from noise (without GAN yet), a comparison with a GAN script to generate data comes after. Then, extract features and quality script are used to extract features from an image and train the SVM model. Finally, the main program script contains the training and details of the SVM training and testing. The project in MATLAB compared a random image generation script with outputs images that do not look like fingerprints to a simple GAN structure without the implementation of trained generator to act as the fake image generator. The following subsections will explain all the MATLAB code and what changes were made in python to get better accuracy.

Figure 6. MATLAB GUI

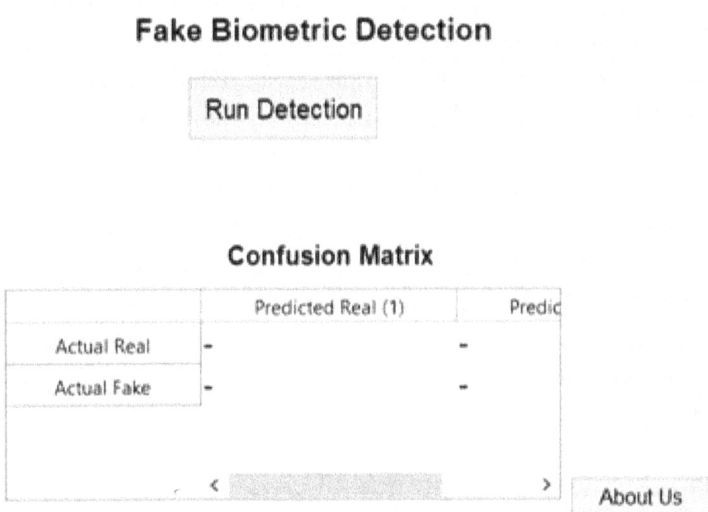

Fake Biometric Graphical User Interface in MATLAB

The Fake biometric detection graphical user interface in MATLAB is in the form of a desktop application that has the following components: 1-Run detection button in which when pressed the user activates training and testing script of the Support Vector Machine. The SVM architecture used here is similar to the one used in python with linear kernel and similar methodology and calls feature extraction script to extract features from an image input. 2- Confusion matrix is displayed after running detection with the percentage of accuracy above. The Confusion matrix is a table that displays predicted real and predicted fakes with what is actually real images and what is actually fake. The predicted real are labeled as 1 by SVM and the predicted fakes are labeled as 0 because of the linear kernel. This matrix helps us visualize the accuracy of the system and is sometimes used for optimization and debugging. Also, in this graphical user interface there is an important button that displays information about the creators of this project and clearly states that this idea was just for academic and research purpose and not for commercial use respecting the laws, licensing and privacy in the context of fingerprint biometric systems.

Figure 6 illustrates the visual representation of the graphical user interface made in MATLAB using the built-in user interface tools and functions. At this stage of the development, the project did not utilize any type of generative artificial or GANs, but it was made just to compare output between random noise generation script in a picture (not fingerprint data) and using generative artificial intelligence script particularly a type of Generative Adversarial Network (GAN) to generate images and learn common image features and complex image data patterns. Obviously, training a generative model is optimal and gives better results than just leaning on random generated dataset shuffling, that is why authors of this project decided to solve a real-world problem by using Sokoto Coventry Fingerprint Dataset (SOCOFing) dataset to solve the vulnerability of fake fingerprint biometric detection and authentication systems and by turning the script into python script for more diversity and easier navigation and debugging. For now, the MATLAB simulation will be fully explained with the code and way of thinking behind it, but it remains less accurate than the system created in python because it was fed with good clean data and was very well engineered with attention to all details which gave outstanding results.

Simulated Dataset Script in MATLAB

In MATLAB, a simulated dataset script was created that aims to create random noisy images and less noisy ones to fill in two folders denoted as “real” and “fake”. The intention of this step is to examine how the SVM machine learning model reacts to minor changes in a simulated image. The script starts with making directory operation *mkdir* to create two new folders for real and fake respectively. Then, a certain number of images is set using a variable. In this simulation 200 images are set for the script to create. In a loop, a creating circles and patterns function is called to mimic fingerprint pictures. Noise and degradation are added to the fake pictures as well as the real ones each in different levels to be able to differentiate between them. The degradation function applies a gaussian filter to the image to adjust the noise in the fake images. The add noise function adds gaussian noise which is a type of noise similar to real scenario noises and adjusts the pictures accordingly. To mimic patterns of fingerprints and create circles, a function is used which contains two vectors. This function takes an image as an input and uses the *meshgrid* method that takes copies of the vector's values and form a grid. Later on, shapes and patterns are added in a different color then the background to assemble an image that is then placed either in the "real" folder or the "fake" folder. This simulated dataset script was created to analyze and compare the outputs to a generative artificial intelligence output in terms of realistic data generation. It is obvious that Generative Adversarial Networks (GANs) did a better job in creating these realistic images to create the fake dataset and that's because the generator network learns from past mistakes due to the adversarial process and the logic and the learning ability of artificial intelligence in terms of generation and classification outpaced other humble handwritten methods. That is exactly why the upgrade to python and the change from standard Generative Adversarial Network (GAN) to a Deep Convolutional Generative Adversarial Network (DCGAN) was needed. The DCGAN as explained before is a combination of GANs and CNN (Convolutional Neural Network) which is best used for image data and classification. When trained well and with good clean image data, this type of deep learning models can lead to outstanding results just like in our example in python.

Extract Features Script in MATLAB

An Extract Features Script was created in MATLAB to extract features and qualities for the Support Vector Machine to train and test the accuracy of the system. First, if the image is RGB, it is converted to grayscale using the *rgb2gray* function. The features extracted from the image are similar to the python version of the system. Features extracted are: Local Binary pattern (LBP) for texture, Laplacian Blur estimation, Edge detection and density using Canny, and contrast. These features are extracted and returned as a vector to the main SVM Program. The importance of extracting these features is that it is similar to real life scenarios happening in a spoofing attack where attackers try to fool the authentication system with a version of printed or synthetic fingerprint with it being blurrier or less detailed. Laplacian Blur estimation utilizes Laplacian variable and get second-order derivative to detect the blurriness in a picture as explained earlier, and the Canny edge detector detects the number of edges and the density of these edges in a certain image data.As for the handcrafted feature, Local binary Pattern (LBP) used for texture recognition and identification, looks for appearances in pixel comparing it to a certain threshold and then returns a histogram value containing complex pattern recognized and number of occurrence of pixels of a texture in an image. It is good to mention that contrast is added to this version of the script since the images generated are random and some of the noise added to fake images is in the form of contrast because there is no color used; all images are grayscale. Although, this contrast feature is minor in comparison with other features, especially the handcrafted LBP, but it is a good addition to the feature script since it can detect minor changes in the image too.

Main SVM script in MATLAB

In this subsection, the architecture of the linear Support Vector Machine and main script for the MATLAB implementation of the project is described. First, two directories are introduced with the files paths to the real and fake folder filled with the images simulated earlier. Then, 20% of the images are taken at random from these file paths and held on for testing and training the SVM. Recall that the Support Vector Machine used here has linear kernel which means it labels real data as 1 and fake data as 0. The SVM kernel helps manage and organize data by utilizing a mathematical function. The type of the kernel differs from the mathematical function used. The goal of the SVM is to differentiate between different classes of data. The Linear Kernel is a simple and straightforward mathematical kernel function and is used in our project because the data can be split into two classes only, just like binary classification tasks, which makes the data linearly separable. The feature extraction script is later called to extract features from the images. A loop for the training dataset is created to loop through the images labeled for training. This loop returns a vector containing all the features and quality of the images in which the SVM trains on to be able to differentiate when the testing part comes between real and fake images. Same logic done in the training loop is applied to the testing loop and then returned to the SVM model after training for prediction and evaluation. If an image was predicted wrong, it will be displayed in the Confusion Matrix alongside with the percentage of accuracy. This system gave a good result but the dataset needed an upgrade for better precision in predicting fakes, so we decided to switch to python and use real dataset with the use of a Deep Convolutional Generative Adversarial Network (DCGAN) to be able to build a robust and precise system with high accuracy that can detect fake biometric fingerprint images using machine learning (Linear Support Vector Machine), deep learning (Deep Convolutional Generative Adversarial Network), and image processing and analysis (features extracted in the script).

Difference between DCGAN and standard GAN

A Deep Convolutional Generative Adversarial Network (DCGAN) is a type of Generative Adversarial Network (GAN) similar in the adversarial process where generator network competes with the discriminator network. However, the main difference is in the architecture and function of a Deep Convolutional Generative

Adversarial Network. In standard Generative Adversarial Network architecture, all layers are fully connected for both generator and discriminator networks. The generator network is fully connected, takes random noise vector and tries to generate an image from it to trick the discriminator classification process. During this process the generator network learns new patterns and techniques to deceive the discriminator. Similarly, the discriminator network also utilizes fully connected layers in standard Generative Adversarial Network in which it is used to classify images. However, due to the fully connected mesh issue, this type of GANs can struggle when it comes to generating complex images and classifying them, especially high-resolution images with complex patterns and subtle features. Therefore, the Deep Convolutional Generative Adversarial Network architecture is introduced in which the generator and discriminator networks uses convolutional layers. The generator is built using convolutional layers to progressively upscale the vector (noise vector) and then create an image that is closer to the real one and with higher quality and precision. Also, the discriminator network similarly uses convolutional layers to perform image analysis on the features and quality of an image and to classify them as real or fake based on previous analysis and logic. The process of training these models is adversarial which forces the discriminator to learn complex patterns and pay attention to subtle details in the image for better classification. The advantages of the DCGAN are many: 1- Improved the ability of image generation, the convolutional layers of a DCGAN forces the model to generate better images and with higher resolution and precision because of the learning process of spatial hierarchal features. 2- The use of Batch Normalization techniques and certain activation functions and architectural changes and optimization help the train stay stable and control mode collapse. 3- The Deep Convolutional Generative Adversarial Network is considered best fit for image data due to the convolutional layers sometimes referred to as fractionally stride convolutions while generating higher resolution picture which shows a huge difference and advantage in comparison with standard Generative Adversarial Network in which image data can be complex and patterns are harder to learn. Our project was translated to python so we can benefit from the diversity of the libraries and the ease of navigation, debugging, and programming. Libraries used in python are: 1- NumPy which stands for numerical python, an open-source that is used in mathematical and scientific computing tasks. It supports multi-dimensional arrays and matrices which are useful in our project in many aspects and used in almost all the scripts. 2- PIL which stands for Python Imaging Library, used abundantly in our project since image data is being navigated to, opened and manipulated with feature extraction. This library helped in doing all the tasks that relates to image data. 3- Scikit-learn is used for machine learning. In this project it is mainly used with the training model, and also to call and work with the support vector machine with linear kernel which is trained on image data and then tested to see the accuracy of detection and prediction. Also, Scikit-learn metrics module (*sklearn.metrics*) used to evaluate the performance of the model, the accuracy scores of predictions and the confusion matrix, showing the images predicted right and the images predicted wrong. This module helps to test and evaluate the behavior and performance of the SVM. 4- PyTorch used to create, design and engineer the DCGAN deep learning model. Since the Deep Convolutional Generative Adversarial Network is made of convolutional neural network, PyTorch neural network module, *torch.nn*, helped in these tasks. In addition, the data loaders and dataset toolkit (*torch.util.data*) made the work much easier. 5- MatplotLib is a python library used mainly for 2D plotting. In this project it is mainly used to plot the behavior of the models and to visualize their training, mainly for easier debugging and to detect if there any abnormal behavior during the training process.

Table 3 compares the two platforms, Python and MATLAB, that were used for the implementation of our fake fingerprint detection system. Python offers greater flexibility, open-source libraries, and higher ease of integration with deep learning frameworks. However, MATLAB is more structured and ideal for rapid prototyping but may lack certain deep learning conveniences compared to Python. Although both Python and MATLAB platforms were considered during the system development phase, the final experimental results presented in this study are derived from the Python-based implementation, which demonstrated more efficient integration with deep learning libraries and achieved a higher detection accuracy.

Table 3. Comparison between Python and MATLAB implementations

Feature / Criterion	**Python Implementation**	**MATLAB Implementation**
Programming Language	Python 3.12(64 bit)	MATLAB R2023b
Libraries Used	NumPy, OpenCV, scikit-learn, TensorFlow/PyTorch	Image Processing Toolbox, Deep Learning Toolbox
DCGAN Implementation	TensorFlow or PyTorch-based	Custom or DL Toolbox
SVM Classifier	sklearn.svm.SVC(kernel="linear")	fitcsvm() with linear kernel
Image Preprocessing	OpenCV and NumPy	imread, imresize, and rgb2gray
Feature Extraction	skimage, OpenCV	Custom LBP/Edge functions
Sanity Check	NumPy shuffle, re-train SVM	randperm, retrain SVM
Accuracy Achieved	99.17%	81.25%
Code Complexity	Medium	Higher

continued on following page

Table 3. Continued

Feature / Criterion	Python Implementation	MATLAB Implementation
Visualization	Matplotlib, seaborn	imshow, plot, confusionchart
Hardware Used	NVIDIA T4 GPU for DCGAN training (cloud gpu using google colab) / CPU	CPU/GPU (MATLAB-compatible)

CONCLUSION AND FUTURE WORK

Aletheia is a project that combines both Deep Learning and Machine Learning models to get the best accuracy of detection. The main contribution of Aletheia is the combination of SVM and DCGAN to create a fully functional detection system with an outstanding accuracy of 99.17% to help solve the issue of fake biometrics fingerprints specially for biometric-based security systems like banking, border control, and mobile device security. Throughout the stages of development many limitations were faced especially when choosing to work with a DCGAN that requires a lot of data and computing power to train effectively. It is trustworthy to mention also the ethical and legal concern for using or finding fingerprint dataset, so we decided to use Sokoto Coventry Fingerprint Dataset (SOCOFing) which is publicly available on Kaggle. Also, many changes and alterations were operated; the project started as a MATLAB implementation but for the sake of higher accuracy and higher flexibility of integration with deep learning frameworks led us to switch to python, a programming language that is rich in libraries that suits the vision of this project. The model is trained on 6,000 real images, then the Generator is used separately to generate fake images to train a Support Vector Machine (SVM) with linear kernel to loop through the fake and real dataset and detect fakes. As mentioned earlier, the system offered an accuracy of 99.17% with only one image being detected wrong within the total of 60 images chosen randomly. This study has achieved its aim of high accuracy fake fingerprint detection. However, as part of future work, considering another type of biometric dataset like palm print or iris or even face features identification using Mediapipe or Holistic or other recognition techniques and algorithms could be an option to work using the same methodology altered based on the need of the dataset for better security in critical areas like airport or police station, for better identification of the individual. In other words, it is crucial in the future to develop a system that is trained on a more diverse dataset with other biometric features like palm print that could be useful in airports or during elections, or even iris or retina dataset that are widely used by passport holders in many countries. The dataset in this work consists of only African fingerprints, which caused the limitation of lack of diversity. Also, considering IoT and the implementation of a real-time sensor with the same methodology used in this chapter could be an option for real time detection not only relying on the dataset. In addition, as part of future work, the implementation of this methodology with few alterations fed with face recognition dataset or images could be a good addition to the project. Even the use of Mediapipe or Holistic which are face and gesture recognition tools, full-body and face recognition, alongside a GAN or any other generative artificial intelligence to generate fake data and images to spoof the system and adjust and alter based on the accuracy given. Another perspective is to use other types of GANs like Style-GAN alongside an SVM to test and check if there are any changes in accuracy or optimization of the system. The StyleGAN (Style Generative Adversarial Network) developed by NVIDIA is known for its ability to generate realistic high-resolution images from noise vectors. It has a different architecture than the Deep Convolutional Generative Adversarial Network. Like the name states, it has the ability to control the style of the images, giving it the skill to generate more realistic ones. It starts with low resolution images to upscale it with time to reach the highest resolution, quality and details based on the training. This project utilized one of the types of Generative Adversarial Network, particularly a Deep Convolutional Generative Adversarial Network (DCGAN) that did an outstanding job in image generation and gave a great accuracy. As a conclusion, taking the recommendations stated above for future work into consideration could be a great fit to optimize the project and make the system able to distinguish between many different and diverse datasets at the same time and with high accuracy.

REFERENCES

Hussian, A., Murshed, F., Alandoli, M. N., & Aljafari, G. (2025). A Hybrid Deep Learning Approach for Secure Biometric Authentication Using Fingerprint Data. *Computers, 14*(5), 178. DOI: 10.3390/computers14050178

Khan, F. A., & Khan, M. K. (2025). Generative AI and deepfake detection in biometric systems. *Cognitive Computation, 17*(1), 112. DOI: 10.1007/s12559-025-10469-3

Khan, M., Sen, A., Singh, O. P., Chaudhary, J. K., Soni, M., & Anbazhagan, K. (2024). Deep learning for biometric attack detection and recognition. In K. Santosh et al. (Eds.), Recent trends in image processing and pattern recognition (RTIP2R 2023) (Vol. 2026, pp. [insert page range]). Springer. DOI: 10.1007/978-3-031-53082-1_29

Minaee, S., Abdolrashidi, A., Su, H., Bennamoun, M., & Zhang, D. (2023). Biometrics recognition using deep learning: A survey. *Artificial Intelligence Review, 56*(8), 8647–8695. DOI: 10.1007/s10462-022-10237-x

Pravallika, P., & Satya Prasad, K. (2016, August). SVM classification for fake biometric detection using image quality assessment: Application to iris, face and palm print. In Proceedings of the 2016 International Conference on Inventive Computation Technologies (ICICT) (pp. –). IEEE. DOI: 10.1109/INVENTIVE.2016.7823189

Ruiz Gara. (2018, July). *SOCOFing: Sokoto Coventry Fingerprint Dataset* [Data set]. Kaggle. https://www.kaggle.com/datasets/ruizgara/socofing2

Songara, D., Poonia, A., Gupta, G., & Sharma, A. (2022). A model for fingerprint liveness detection enabled by M-SSO heuristic algorithm using deep learning strategies. *Cybernetics and Systems, 55*(8), 1793–1828. DOI: 10.1080/01969722.2022.2146848

Uliyan, D. M., Sadeghi, S., & Jalab, H. A. (2020). Anti-spoofing method for fingerprint recognition using patch based deep learning machine. *Engineering Science and Technology, an International Journal, 23*(2), 264–273. DOI: 10.1016/j.jestch.2019.06.005

Ullah, S., Ou, J., Xie, Y., Zhang, Y., & Huang, F. (2024). Compound facial expressions recognition approach using DCGAN and CNN. *Multimedia Tools and Applications, 83*(38), 85703–85723. DOI: 10.1007/s11042-024-20138-6

Vinayagam, J., & Dilip, G. (2024). A two-step verification-based multimodal-biometric authentication system using KCP-DCNN and QR code generation. *Journal of Ambient Intelligence and Humanized Computing, 15*(12), 3973–3996. DOI: 10.1007/s12652-024-04872-1

KEY TERMS AND DEFINITIONS

Activation function: A mathematical function used in neural networks to determine a neuron's output given an input. It adds non-linearity into the network, allowing it to learn and model complex patterns in data.

Biometric Security: A security system that uses unique physical or behavioral traits such as fingerprints or facial recognition for the purpose of verifying a person's identity.

Confusion Matrix: A table used to evaluate the performance of classification models.

Deep Convolutional Generative Adversarial Network (DCGAN): A neural network type that combines convolutional layers in both the generator and discriminator to generate realistic images, such as fake fingerprints.

Deep Learning: A subset of Machine Learning (ML) that uses neural networks with multiple layers to learn patterns from big datasets.

Discriminator: In a GAN, it is the model that tries to differentiate real data from the fake data produced by the generator.

Feature Extraction: The process of identifying and extracting important information or features from raw data to help machine learning models understand it better.

Generative Adversarial Network (GAN): A type of deep learning model having two parts: Generator and Discriminator competing with each other.

Generator: In a GAN, it is the model that creates fake data to try to fool the discriminator.

Label Shuffling: A technique used during sanity checks to randomize class labels.

Sanity Check Accuracy: A type of test that determines if a model performs better than random guessing, confirming that it's learning something meaningful.

Spoofing: Fake biometric data intended to deceive authentication systems.

Supervised Learning methods: a type of machine learning method in which the data given to the model is labeled and requires the intervention of human developers.

Support Vector Machine (SVM): A supervised learning algorithm that helps classifying data by finding the best boundary or hyperplane separating the different classes being studied.

Unsupervised Learning methods: Opposite to supervised learning, the data given to the model is unlabeled and the system learns patterns and information without the intervention of human developers.

Chapter 11

Biometric-Driven Encryption and Blockchain-Based Audit Trail for Secure Forensic Evidence Management

Yacine Belhocine
https://orcid.org/0009-0002-1747-6008
Laboratory of Mathematics, Informatics, and Systems, University of Echahid Cheikh Laarbi Tbessi, Tebessa, Algeria

Abdallah Meraoumia
Laboratory of Signals and Smart Systems, Echahid Cheikh Larbi Tebessi University, Algeria

Salim Chitroub
https://orcid.org/0009-0001-1814-3304
Laboratory of Intelligent and Communicating Systems Engineering, USTHB, Algiers, Algeria

Hakim Bendjenna
Laboratory of Mathematics, Informatics, and Systems, University of Echahid Cheikh Laarbi Tbessi, Tebessa, Algeria

ABSTRACT

The secure management of biometric evidence is essential in modern forensic investigations, requiring data integrity, confidentiality, and accountability throughout the evidence lifecycle. This chapter presents a framework combining homomorphic

DOI: 10.4018/979-8-3373-4972-5.ch011

encryption with blockchain-based audit trails to enhance security and transparency in biometric systems. Biometric features, such as palmprints, are encrypted using homomorphic schemes, allowing computations on encrypted data without compromising privacy. Encrypted feature vectors are stored in a decentralized environment via the InterPlanetary File System (IPFS), while all interactions are logged through blockchain smart contracts for tamper-proof auditability. Experiments show the system maintains high recognition accuracy with acceptable computational overhead and low transaction costs. This approach provides a scalable, privacy-preserving, and forensically sound solution for managing biometric evidence in digital investigations and legal proceedings.

INTRODUCTION

The intersection of biometric technology and digital forensics is increasingly critical for securing identity verification, preserving evidence integrity, and enabling lawful investigative access. In the digital era, forensic procedures extend beyond physical evidence collection and manual analysis, with digital systems, especially those handling biometric data, playing a central role in identification and authentication. Biometric modalities such as fingerprints, palmprints, iris patterns, and facial features provide distinctive, hard-to-forge identifiers, making them valuable in forensic investigations (Klasén et al., 2024).

Despite these advantages, widespread biometric adoption introduces challenges in security, privacy, and evidence management. Traditional systems rely on centralized databases, creating single points of failure and exposing evidence to insider threats, unauthorized access, and data breaches. Centralized storage also limits transparency, making verifiable chain-of-custody enforcement difficult (Moussa, 2021). Forensic environments demand reproducibility, integrity preservation, and procedural accountability. Biometric data is inherently irrevocable, so strong protection mechanisms are essential both at rest and during processing, while still allowing analysis for identification or evidentiary correlation. This tension between confidentiality and operational usability is further complicated in multi-agency or cross-jurisdictional investigations (Baig & Eskeland, 2021).

Privacy-preserving computation techniques, particularly homomorphic encryption (HE), have gained attention for enabling secure operations on encrypted biometric feature vectors. HE allows comparisons and classification while preserving confidentiality, providing practical protection without undermining forensic utility (Krishna Prakasha & Sumalatha, 2025). However, confidentiality alone is insufficient: forensic admissibility requires immutable audit trails. Blockchain technology addresses this by recording evidence-handling events as decentralized, tamper-evident transactions, ensuring traceability, accountability, and compliance with legal standards (Alqahtany & Syed, 2024).

This chapter proposes a unified framework integrating biometric feature extraction, HE, decentralized storage, and blockchain-based logging to enhance secure evidence management. Biometric features are extracted using Pyramid Histogram of Oriented Gradients (PHOG) and Local Phase Quantization (LPQ) to capture discriminative patterns (Rukhiran et al., 2023). Feature vectors are encrypted for secure similarity computation and stored in the InterPlanetary File System (IPFS), with content identifiers (CIDs) ensuring immutability and verifiable chain-of-custody (Elgohary et al., 2022). Smart contracts enforce access control and log all operations, creating tamper-resistant audit trails for authorized stakeholders.

The system enables decentralized, privacy-preserving identity verification. Investigators perform comparisons on encrypted vectors, with results decryptable only under predefined, legally compliant conditions. This design aligns with data minimization, consent, and proportionality principles (Quinn et al., 2024). The framework also supports multi-agency collaboration, preserves forensic soundness, and enables long-term evidence retention. Blockchain logs provide immutable historical records for appeals, audits, and post-incident review.

Experimental validation using public datasets assessed feature extraction, HE, IPFS storage, and blockchain logging. Metrics included recognition accuracy, encryption latency, transaction costs, and storage overhead. Results show high accuracy with acceptable computational overhead for forensic workflows (Javed et al., 2022). Ethical and regulatory compliance is ensured through encryption, immutable audit mechanisms, and adherence to GDPR and national privacy laws (Jung & Virgil, 2024).

Overall, this chapter presents a secure, privacy-aware, and forensically robust framework for biometric evidence management. It addresses critical limitations of existing systems and provides a scalable foundation for law enforcement, judicial processes, and digital identity verification.

The remainder of the chapter is organized as follows: Section 2 reviews relevant literature; Section 3 discusses privacy, auditability, and infrastructure limitations; Section 4 presents the conceptual framework; Section 5 outlines system materials, technologies, and cryptography; Section 6 presents empirical evaluation; Section 7 compares with existing solutions; and Section 8 concludes with findings and future directions.

RELATED WORKS

The integration of biometric technologies within forensic systems has attracted substantial research attention over the past two decades, with growing interest in improving recognition accuracy, scalability, and system interoperability. Early forensic biometric systems were primarily focused on fingerprint identification, where the emphasis lay in developing minutiae-based algorithms capable of withstanding noise and partial prints. As biometric technology matured, forensic researchers expanded the scope to include multimodal systems that combined facial features, iris patterns, and voice signatures, motivated by the recognition that single modalities are prone to error under conditions such as poor lighting, occlusion, or data degradation. The fusion of multiple biometric traits has been demonstrated to improve overall reliability and robustness, which is critical in forensic environments where evidence may be incomplete or compromised. In parallel, developments in advanced feature extraction, deep learning-based classification, and adaptive template protection have contributed significantly to improving the resilience of biometric systems deployed in high-stakes forensic investigations.

Recent scholarship has also emphasized the integration of blockchain technologies into forensic systems as a means of securing sensitive transactions, improving accountability, and ensuring transparent auditability. Blockchain-based models are attractive to forensic applications because of their decentralized nature, tamper-evident structures, and ability to maintain immutable records of access and modification. The use of distributed storage models, such as the InterPlanetary File System (IPFS), has complemented blockchain by enabling scalable, verifiable storage of biometric templates and related forensic data. Collectively, these innovations have produced a growing body of literature at the intersection of biometrics, cryptography, and distributed systems, each contributing partial solutions to the overarching challenge of developing secure, privacy-preserving forensic infrastructures.

Asma Jodeiri Akbarfam et al. (2023) present ForensiBlock, a blockchain-based framework designed to secure data provenance in digital forensic workflows. The framework leverages role-based access control with staged authorization and integrates a distributed Merkle root mechanism to enhance scalability and traceability. A notable strength of this work lies in its comparative evaluation of provenance extraction methods, demonstrating how off-chain retrieval paired with Merkle root verification significantly outperforms brute-force approaches in terms of efficiency. While ForensiBlock succeeds in reducing evidence retrieval latency and improving auditability, its scope is limited by the absence of mechanisms for biometric encryption or homomorphic operations, leaving biometric privacy unaddressed.

Similarly, Cebe et al. (2018) propose Block4Forensic, a lightweight blockchain-enabled architecture aimed at supporting forensic investigations in vehicular systems. By combining vehicular public key infrastructure with a fragmented ledger design, the framework ensures privacy and traceability in incident investigations while minimizing storage and computational overhead. The system's practicality for resource-constrained vehicular environments is a key contribution, but the absence of biometric evidence handling, encrypted template storage, and privacy-preserving analytics restricts its applicability to broader forensic contexts.

In another effort, Ezazul Islam et al. (2023) developed a blockchain-supported forensic log system with a hybrid Attribute-Role-Based Access Control model. This work prioritizes access control by ensuring only authorized individuals can retrieve forensic records. Through AVISPA simulations, the authors validated their endpoint authentication protocols and demonstrated resilience against tampering. The framework represents progress toward a trust-centric forensic environment, but its narrow focus on logs and lack of biometric template protection reveal critical gaps in addressing privacy concerns associated with personally identifiable forensic evidence.

Li et al. (2021) address evidence lifecycle management through LEChain, a consortium blockchain framework that integrates fine-grained access control and witness anonymity protection. The system employs ciphertext-policy attribute-based encryption and randomizable signatures to safeguard the confidentiality of forensic data, extending its utility into courtroom contexts by including mechanisms such as secure juror voting protocols. Despite its comprehensive lifecycle approach, LEChain does not extend into biometric-driven investigations, nor does it address encrypted-domain computations such as homomorphic biometric matching. This omission illustrates the broader trend in existing frameworks where legal and procedural protections are prioritized, but biometric integration remains underexplored.

The work of Butrus Mbimbi et al. (2024) introduces PBCIS-IoTF, a blockchain-based IoT forensic system built on Hyperledger and deployed on low-cost Raspberry Pi nodes. The framework ensures evidence authenticity and integrity through peer-to-peer SHA-256 comparisons and offers a web-based access interface for law enforcement. While effective for ensuring system-level forensic integrity in IoT contexts, the framework does not incorporate human-centered access control or biometric verification, highlighting the need for extending forensic blockchain applications to encompass both system and individual-level evidence management.

Beyond blockchain-focused contributions, cryptographic research has also advanced forensic applications. Chen et al. (2024) propose a homomorphic encryption-based system for privacy-preserving kinship estimation in forensic genomics. By transforming high-dimensional encrypted matrices into smaller operations, their approach significantly improves both computational efficiency and security. Although this contribution lies outside biometrics, it demonstrates the feasibility of encrypted-domain computations for sensitive forensic identifiers, suggesting parallel applications in biometric evidence handling. Other research has similarly highlighted the promise of homomorphic encryption for biometric template matching, where encrypted templates can be compared without revealing the underlying traits. However, many of these cryptographic frameworks lack integration with blockchain auditability or forensic provenance, reducing their ability to meet evidentiary standards in legal proceedings.

Scholarly attention has also turned toward federated learning, which allows multiple institutions to collaboratively train biometric recognition models without centralizing sensitive data. For example, Chaudhary et al. (2023) demonstrate how federated training across distributed forensic databases can reduce the risk of large-scale breaches while still producing high-performing recognition systems. In parallel, differential privacy techniques (Fan et al., 2021) have been applied to biometric datasets to mask identifiable details while retaining utility for machine learning tasks. Together, these approaches represent important advances in protecting biometric privacy but still fall short of offering complete forensic solutions that ensure provenance, transparency, and judicial credibility.

Finally, legal and ethical standards play a critical role in shaping the design of forensic biometric systems. The General Data Protection Regulation (GDPR) and related data protection laws classify biometrics as a sensitive category, mandating explicit safeguards for processing, storage, and transfer. International standards, including ISO/IEC 19792 and ISO/IEC 30107, emphasize security evaluations, presentation attack detection, and overall biometric system robustness. Despite their importance, many academic frameworks overlook these normative requirements, creating barriers to real-world adoption and courtroom admissibility. The limited incorporation of such legal-ethical considerations into technical architectures remains one of the key shortcomings of the current state of research.

Taken together, the reviewed literature illustrates both the progress and limitations of existing forensic biometric systems. While blockchain, encryption, and distributed architectures have each advanced the security and reliability of forensic workflows, very few systems combine these elements into integrated frameworks capable of meeting both technical and legal demands. The absence of homomorphic biometric matching, insufficient audit trails, and limited attention to legal compliance underscore the need for more holistic solutions.

KEY MOTIVATIONS

The motivation for this work lies in addressing the growing dependence on biometric evidence in digital forensics and the unique vulnerabilities this dependence creates. Biometric traits such as fingerprints, iris patterns, palmprints, and facial features have become increasingly central to investigations because of their permanence and non-repudiation. Their discriminatory power makes them valuable identifiers for criminal justice, border security, and national defense. However, unlike passwords or tokens, biometric data is irreplaceable once compromised. If a fingerprint template or iris scan is leaked, it cannot be revoked, leaving individuals permanently exposed to risks of impersonation or misuse. This property fundamentally elevates the stakes for biometric evidence protection, demanding robust safeguards to ensure privacy and evidentiary integrity.

Traditional forensic infrastructures rely heavily on centralized storage and authority-based access models. Such systems are vulnerable to insider misuse, accidental leaks, and targeted cyberattacks. Furthermore, centralized architectures create single points of failure that undermine forensic reliability. In addition to technical risks, traditional systems often fail to provide transparent access records, making it difficult to establish a verifiable chain of custody. These limitations can jeopardize the admissibility of evidence in court, where defense challenges frequently focus on whether evidence has been handled securely and without tampering.

The convergence of cryptography and distributed technologies offers promising remedies to these longstanding challenges. Homomorphic encryption allows biometric templates to be processed in their encrypted form, enabling investigators to authenticate identities or compare biometric samples without ever exposing sensitive data. This approach dramatically reduces the attack surface while preserving forensic utility. Parallel advances in blockchain technology provide immutable, tamper-evident audit trails that can document every access, modification, or transfer of forensic evidence. Combined with decentralized storage solutions such as IPFS, blockchain ensures resilience against unauthorized alteration or deletion, strengthening trust in the chain of custody.

Equally important are recent advances in federated learning, which enable the training of forensic biometric models across multiple jurisdictions without centralizing sensitive datasets. This capability is particularly relevant in transnational investigations, where strict privacy regulations prevent raw data sharing. By allowing local institutions to contribute to global models without transferring individual records, federated approaches preserve both privacy and investigative effectiveness.

Legal and ethical frameworks also shape the motivations of this work. GDPR and similar regulations explicitly require accountability and purpose limitation in biometric processing. Courts demand demonstrable assurance that evidence has been handled in ways that are both secure and transparent. Failing to meet these requirements can render otherwise valuable biometric evidence inadmissible. Therefore, the technical solutions proposed here must not only meet scientific standards of security and scalability but also align with legal and procedural expectations of transparency, privacy, and fairness.

The overarching motivation, then, is to unify these diverse threads into a coherent forensic architecture that simultaneously provides biometric authentication, privacy-preserving computation, decentralized governance, and immutable auditability. By doing so, this work seeks to address the critical gaps identified in existing literature and practice. The proposed framework prioritizes security, transparency, and scalability without compromising investigative utility, thereby offering a forensic infrastructure capable of meeting the technical, legal, and ethical demands of contemporary investigations.

SYSTEM ARCHITECTURE FOR PRIVACY-PRESERVING FORENSIC EVIDENCE HANDLING

Establishing trust in forensic evidence handling requires more than secure storage or reliable authentication mechanisms; it demands a comprehensive architectural approach that ensures privacy, integrity, auditability, and lawful access throughout the entire lifecycle of the evidence. This section introduces the design principles and core components of a system that integrates homomorphic encryption, decentralized storage, and blockchain-based audit trails to meet these critical requirements. The proposed architecture aims to provide a scalable and privacy-preserving foundation for managing biometric evidence in digital forensic investigations.

ARCHITECTURE OF A BLOCKCHAIN-BASED FORENSIC INTEGRITY FRAMEWORK

The integrity of forensic evidence is not solely dependent on technological safeguards but also on the coordination between the entities that interact with it. Establishing a reliable chain of trust requires clearly defined roles, secure communication protocols, and strict access control mechanisms to ensure that only authorized actors can access, process, or transfer evidence as illustrated in Figure 1. This subsection outlines the interaction model between forensic investigators, analysts, and system components, emphasizing how trust is maintained through cryptographic authentication, access governance, and immutable audit trails. By coordinating actors within a secure framework, the system ensures accountability and compliance across all stages of evidence handling.

Figure 1. Multi-Agency Interaction via Blockchain for Evidence Integrity

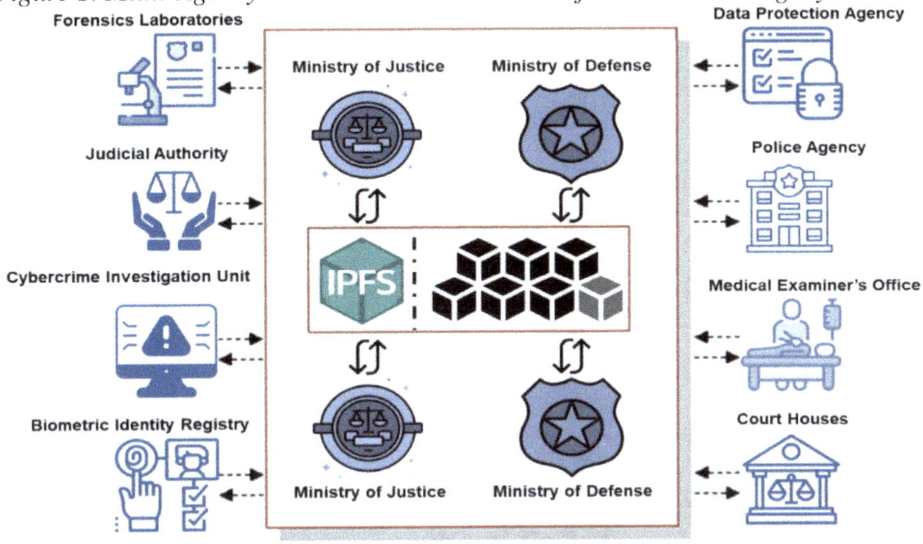

Figure 1 illustrates the coordinated interaction between key national entities involved in the forensic evidence lifecycle. At the core, the system integrates blockchain and IPFS to ensure secure, decentralized evidence storage and auditability. Ministries of Justice and Defense operate as central authorities managing access and enforcement, while external entities, including forensic labs, police, judicial authorities, cybercrime units, and data protection agencies, interact with the system through controlled interfaces. This design supports traceable, tamper-proof communication across institutions while preserving privacy and legal compliance in digital forensic operations.

SECURE IDENTITY BINDING IN FORENSIC TRANSACTIONS

The system ensures secure identity binding and role-based authorization in forensic workflows by leveraging biometric encryption alongside decentralized validation. The process shown in Figure 2 and the phases explained below begins with a registered user, such as a court-appointed examiner or forensic analyst, initiating a role-authenticated request through the system's cryptographic gateway. This request triggers a biometric verification in which the user encrypts their biometric vector using a homomorphic encryption scheme tied to a temporary authentication token. Upon submission, the system compares the encrypted input with the reference profile in a privacy-preserving manner and validates the user's role through smart contract logic. Only after this dual verification, biometric match and permission check, can the user retrieve or contribute evidence artifacts from IPFS, while each action is immutably logged on the blockchain. This flow enforces a secure, traceable, and identity-bound transaction model for all stages of forensic evidence handling.

Figure 2. Architectural Framework for Secure and Verifiable Forensic Evidence Handling

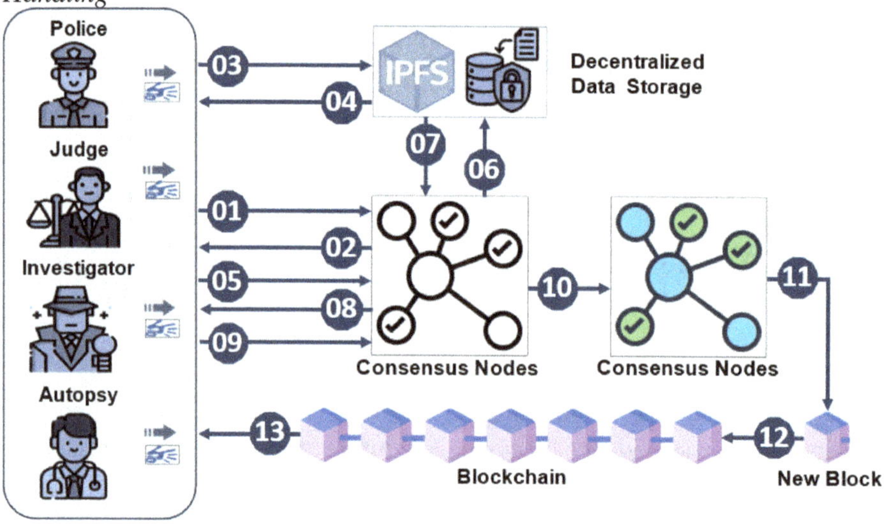

01 User Request Initiation The user sends a formal request to access a specific forensic service.

02 Token Issuance The network nodes respond by issuing a cryptographic service token to the user.

03 Biometric Encryption and Upload The user extracts biometric features, encrypts the vector, and uploads it to IPFS.

04 CID Retrieval IPFS returns the content identifier (CID) to the user for reference.

05 CID Submission to Nodes The user sends the CID to the network nodes to initiate biometric verification.

06 Vector Request from IPFS The nodes query IPFS using the CID to retrieve the encrypted biometric vector.

07 Vector Return from IPFS IPFS responds by delivering the encrypted vector to the requesting nodes.

08 Biometric Verification The nodes verify the user's identity and notify them of session approval.

09 Forensic Service Execution The user proceeds to upload evidence, view logs, or interact with forensic services.
10 Service Verification The nodes validate the submitted data or actions for structural and policy compliance.
11 Block Creation Upon successful verification, nodes prepare and finalize a new blockchain block.
12 Ledger Update The new block is added to the blockchain, updating the immutable evidence ledger.
13 Transaction Notification The user receives confirmation that their activity has been recorded on the blockchain.

To provide a clearer understanding of the internal logic and workflow of the proposed system, the following process is segmented into three interdependent operational phases. Each phase highlights the chronological and functional steps required to ensure secure biometric authentication, role-based access control, and tamper-proof forensic evidence logging within the blockchain framework.

Phase 1: Decentralized Identity Verification and Secure Vector Exchange (Steps 1–5)

The first phase is dedicated to establishing secure communication and identity verification through a decentralized infrastructure. It begins when the user submits a request to access a forensic service (Step 1), triggering the network to respond with a cryptographically signed token that authorizes the session (Step 2). With the token in hand, the user proceeds to extract their biometric features, encrypt the resulting vector, and upload the encrypted payload to IPFS, rather than sending it directly to a central authority (Step 3). Upon successful upload, IPFS generates a unique content identifier (CID), which it returns to the user as a secure reference (Step 4). The user then transmits this CID to the nodes, initiating the validation process (Step 5). This phase decentralizes the identity binding process, reducing reliance on centralized storage and enabling privacy-preserving biometric transmission.

Phase 2: Authentication and Service Access Authorization (Steps 6–9)

This phase centers on authenticating the user's identity and enabling access to role-specific forensic services. The blockchain nodes use the received CID to retrieve the encrypted biometric vector from IPFS (Step 6), and once the vector is returned (Step 7), the system performs a privacy-preserving comparison with the user's registered profile. If verification is successful, the user is notified and granted access to their session (Step 8). From this point forward, the user may initiate any authorized operation such as submitting forensic evidence, reviewing audit logs, or querying records (Step 9). Each service request is linked to the user's identity and session metadata, enabling precise auditability while preserving privacy and integrity in the interaction.

To preserve privacy in decentralized storage, all biometric feature vectors are encrypted prior to being uploaded to IPFS, ensuring that raw biometric data is never exposed. Each encrypted file is addressed using a unique content identifier (CID) derived from its cryptographic hash, preventing linkability across multiple records. Encryption keys are securely managed by authorized entities, and access to stored data is controlled via smart contracts. This design complies with data minimization and consent principles, supporting privacy preservation while maintaining traceable, immutable forensic records.

Phase 3: Blockchain Registration and Service Finalization (Steps 10–13)

In the final phase, the system ensures that all user actions are immutably recorded and verifiable through blockchain consensus. After a service is performed, the nodes conduct structural and policy compliance checks to validate the transaction or data set (Step 10). Upon successful validation, a new block is prepared containing the transaction details and is broadcast to the network for consensus-based approval (Step 11). Once approved, the block is added to the blockchain ledger, solidifying the action within a tamper-proof and time-stamped record (Step 12). The system then issues a final notification to the user (Step 13), confirming the operation's completion and including relevant blockchain metadata such as the block hash or transaction ID. This final phase reinforces the integrity, traceability, and legal admissibility of all forensic actions conducted within the architecture.

The threat model assumes the presence of malicious insiders, external adversaries, and untrusted storage nodes. The system is designed to prevent unauthorized disclosure of biometric data, detect evidence tampering, and ensure accountability for all evidence-handling operations. Biometric templates are encrypted prior to storage and processing, mitigating template inversion and data leakage risks. Blockchain immutability prevents undetected modification of audit logs, while smart contracts enforce access control and non-repudiation. IPFS nodes are treated as untrusted; confidentiality and integrity are preserved through encryption and content-addressable storage. The framework assumes honest participation of blockchain consensus nodes and secure key management by authorized entities.

DESIGN REQUIREMENTS

To successfully implement and evaluate the proposed system, it is essential to define a clear set of design requirements and operational criteria. This involves selecting suitable technologies, including decentralized infrastructure, secure data storage mechanisms, and privacy-preserving computational frameworks. Additionally, the system must incorporate advanced biometric processing techniques and encryption methods to ensure both accuracy and confidentiality. The following section presents the core technologies, software environments, and methodological strategies that underpin the development and validation of the system.

MATERIALS

This section provides an overview of the key technologies, platforms, and tools employed in the design and experimental evaluation of the integrated biometric and blockchain-based forensic evidence framework.

Dataset

- **Finger-Knuckle Print Dataset (FKP):** sourced from the Polytechnic University of Hong Kong, serves as a foundational biometric resource in this work. It comprises 7,920 images collected from 165 participants, including 125 males and 40 females. Among them, 143 individuals were aged between 20 and 30, with the remaining participants falling between 30 and 50 years old Zhang & Zhang (2009). Data acquisition was conducted across two distinct sessions, during which each subject contributed six images for each of the following fingers: Left Index (LIF), Left Middle (LMF), Right Index (RIF), and Right Middle (RMF). This resulted in a total of 48 images per subject, offering a robust dataset for evaluating biometric vector extraction, encryption, and verification techniques within the proposed system.
- **Tongji Contactless Palmprint Dataset:** provides a complementary biometric modality for experimentation and system validation. This publicly available dataset consists of 12,000 palmprint images derived from 600 palm samples, collected from 300 volunteers, 192 males and 108 females Zhang et al (2017). Of these individuals, 235 were aged between 20 and 30, while the rest were between 30 and 50. The dataset was captured over two acquisition sessions, recording both left and right palms. Each participant contributed 10 samples per palm in each session, yielding a total of 40 images per subject. The average interval between the two sessions was 61 days, with a range from

21 to 106 days. This temporal separation supports the assessment of intra-class variability and the resilience of biometric encryption under real-world conditions.

Ethereum Blockchain

To enforce transparency, traceability, and tamper-resistance in forensic evidence management, the system leverages the Ethereum blockchain as its core distributed ledger infrastructure. Ethereum is a decentralized and open-source platform designed to execute smart contracts, self-executing scripts that operate based on predefined conditions without requiring centralized control Kushwaha et al (2022). Unlike traditional blockchains optimized for simple transactional exchanges, Ethereum supports a wide range of decentralized applications (dApps), making it ideal for automating forensic auditing processes. Its flexibility, large developer community, and programmable contract capabilities position it as a robust solution for maintaining verifiable logs and access trails in sensitive forensic environments.

Ganache

To support local development and safe testing of smart contracts, the system incorporates Ganache, a personal Ethereum blockchain emulator. Ganache allows developers to prototype, test, and debug smart contract logic in an isolated environment without financial costs or reliance on public blockchain networks Verma et al (2022). It provides features such as real-time transaction inspection, customizable block intervals, and account management. The emulator is preloaded with ten accounts, each funded with a fixed Ether balance, facilitating gas-free interaction during contract evaluation. This controlled setup enables iterative validation of contract functionality before deployment to live or test-net Ethereum environments.

The InterPlanetary File System

For the secure storage and retrieval of large forensic files and biometric models, the architecture utilizes the InterPlanetary File System (IPFS), a peer-to-peer distributed file system Sangeeta & Nam (2023). IPFS divides data into content-addressable chunks, each identified by a unique cryptographic hash called a content identifier (CID). These CIDs are organized using a Merkle Directed Acyclic Graph (Merkle DAG), allowing complete file reconstruction from the root hash. A Distributed Hash Table (DHT) enables efficient content discovery and retrieval across the network. By eliminating centralized dependencies and minimizing data redundancy, IPFS enhances the resilience, integrity, and availability of forensic data within the system.

METHODS

Key mathematical techniques where employed to support biometric feature extraction and secure data handling, as outlined below.

Pyramid Histogram of Oriented Gradients (PHOG)

Pyramid Histogram of Oriented Gradients (PHOG) is a spatial shape descriptor that extends the classic Histogram of Oriented Gradients (HOG) by incorporating multi-resolution spatial information Moteaal Asadi Shirzi & Kermani (2024). It captures both local edge orientations and the spatial layout of those edges, making it highly effective for biometric feature representation such as in finger-knuckle or palmprint recognition.

The method works by first applying an edge detector (e.g., Canny) to extract edge pixels. Then, the image is divided into sub-regions at multiple pyramid levels. Within each region, gradients are computed, and a histogram is built based on their orientation:

$$\theta(x,y) = \tan^{-1}\left(\frac{G_y(x,y)}{G_x(x,y)}\right) \tag{1}$$

where G_x and G_y are the horizontal and vertical gradients at position (x,y), respectively. The magnitude of the gradient is given by:

$$M(x,y) = \sqrt{G_x^2(x,y) + G_y^2(x,y)} \tag{2}$$

The image is then subdivided into spatial grids at multiple pyramid levels. For each sub-region, a histogram of gradient orientations is computed and normalized. These histograms are concatenated across levels to form a multi-resolution feature vector. PHOG's hierarchical structure enables it to represent both global and localized shape characteristics effectively.

Local Phase Quantization (LPQ)

Local Phase Quantization (LPQ) is a texture descriptor known for its robustness against image blurring and illumination variations, making it highly suitable for biometric data analysis Yacine et al (2024). It operates in the frequency domain, relying on the local phase information of image neighborhoods rather than intensity or amplitude.

LPQ applies a short-term Fourier transform (STFT) over small image windows. For each pixel, the local neighborhood is transformed using complex-valued basis functions:

$$F(u) = \sum_{x} f(x) \cdot e^{-j2\pi u^{T}x} \tag{3}$$

where $f(x)$ is the intensity at position x, and u is the frequency vector. The real and imaginary components of $F(u)$ are quantized to generate a binary code representing the local phase pattern. These binary codes are then aggregated into histograms to construct the LPQ feature vector.

The key strength of LPQ lies in its sensitivity to phase changes while being invariant to convolutional distortions such as motion blur. This makes it especially useful in forensic scenarios where biometric data may be degraded or partially obscured.

Deep Rule-Based Classifier

The Deep Rule-Based (DRB) classifier is a multi-layer neuro-fuzzy architecture that combines the adaptability of non-parametric, self-organizing fuzzy rule-based (FRB) systems with the hierarchical structure and representation power of deep learning models Gu et al (2018). This hybrid approach has demonstrated strong performance in a range of image classification tasks, offering both interpretability and accuracy through its layered, rule-driven design.

Massively Parallel Rule Base

At the core of the DRB classifier lies an initial layer composed of parallel logical rules of the form (IF...THEN), where each set of rules corresponds to a specific class C within the dataset. These rules are derived from prototypes, representative data samples identified during the training process. For a class $c = 1, 2, 3, .., C$, each rule is expressed in a structured format designed to capture key class characteristics. the rule takes the following form:

$$IF\ \left(I \sim P_{c,1}\right) OR\ \left(I \sim P_{c,2}\right) OR \bullet \bullet OR\ \left(I \sim P_{c,N_c}\right)\ THEN\ (class\ c) \tag{4}$$

Here:

- o I represent an input image.

- o $P_{c,i}$ denotes the ith prototype for class ccc.
- o $\sim$ signifies similarity, measured as the fuzzy degree of membership.
- o N_c indicates the number of prototypes for class c.

Each rule produces a confidence score $\lambda_c(I)$ for a test image I, calculated as:

$$\lambda_c(I) = \arg max_{j=1,2,..,N_c}\left(e^{-\|V-p_{c,j}\|^2}\right) \tag{5}$$

where V is the feature vector of the image I, and $p_{c,j}$ is the ith prototype feature vector for class c. C confidence scores outputs for the test samples:

$$\lambda(I) = [\lambda_1(I), \lambda_2(I), .., \lambda_c(I)] \tag{6}$$

These scores are passed to the Decision-Maker Layer, which uses the winner-takes-all principle to assign the test image to the class with the highest confidence score:

$$P_{id}(I) = \arg max_{c=1,2,..,C}[\lambda_c(I)] \tag{7}$$

This final step determines the class label (e.g., person identity) of the test image based on the outputs of the fuzzy rules.

Homomorphic Encryption

Homomorphic Encryption (HE) is a cryptographic technique that enables computations to be performed directly on encrypted data without revealing the underlying plaintext. Unlike conventional encryption schemes that require decryption prior to processing, HE supports arithmetic operations on ciphertexts while preserving confidentiality throughout the computational workflow (Munjal & Bhatia, 2022). In this framework, a somewhat homomorphic encryption approach is employed, as it provides support for addition and multiplication with substantially lower computational overhead than fully homomorphic schemes. Such HE schemes are well suited for biometric matching tasks, where encrypted feature vectors must be compared while maintaining acceptable accuracy and processing efficiency. The use of HE allows biometric recognition to be carried out in the encrypted domain with only marginal performance overhead, making it practical for forensic applications that require both privacy preservation and operational feasibility.

Key Generation

$$(pk, sk) \leftarrow \text{KeyGen}() \tag{8}$$

This function generates a pair of keys, a public key pk used for encryption and a secret key sk used for decryption.

Encryption

$$c \leftarrow \text{Enc}(pk, m) \tag{9}$$

A message m is encrypted using the public key to produce a ciphertext c.

Decryption

$$m \leftarrow \text{Dec}(sk, c) \tag{10}$$

The ciphertext c is decrypted using the secret key to retrieve the original message m.

Evaluation

$$c_f \leftarrow \text{Eval}\left(pk, f, c_1, c_2, \ldots, c_n\right) \tag{11}$$

A function f is homomorphically evaluated on ciphertexts $c_1, c_2, \ldots, c_n$ to produce a new ciphertext c_f such that:

$$\text{Dec}\left(sk, c_f\right) = f\left(m_1, m_2, \ldots, m_n\right) \tag{12}$$

This core property, that computations on ciphertexts yield encrypted results which decrypt to the same value as if the function had been applied to the plaintexts, is what defines homomorphic encryption.

The principal advantage of HE lies in its ability to perform secure computation on encrypted data. This makes it especially valuable in scenarios where data privacy is critical, such as in medical analytics, cloud computing, secure voting, and digital forensics. In forensic applications, HE enables privacy-preserving comparisons of sensitive data (such as biometric vectors or evidence hashes) without ever exposing the raw information to the processing agents.

By eliminating the need for decryption during processing, HE significantly reduces the attack surface, ensuring that private data remains encrypted end-to-end. This property aligns well with modern zero-trust models and compliance frameworks for secure data handling in regulated environments.

EXPERIMENTAL RESULTS

In this section, the experimental evaluation is organized into three distinct phases to provide a comprehensive analysis of the proposed system's performance, reliability, and resilience. First, evaluate the biometric component, focusing on feature extraction accuracy, matching robustness, and the system's behavior under varying input conditions. This includes performance metrics such as False Acceptance Rate (FAR), False Rejection Rate (FRR), and Equal Error Rate (EER), applied to encrypted biometric vectors. Second, assessing the operational performance of the full integrated system, including latency in cryptographic operations, efficiency of IPFS data retrieval, smart contract execution time, and overall throughput during forensic evidence submission and verification. Finally, a series of simulated attack scenarios where conducted to examine the system's security posture. These include impersonation attempts, data tampering, token misuse, and model inversion threats. This three-layered evaluation strategy ensures a holistic understanding of both the functional and adversarial behavior of the system within real-world constraints.

The experimental evaluation assesses both biometric recognition performance and system-level overhead introduced by privacy-preserving and decentralized components. Experiments are conducted on the Tongji palmprint and Finger Knuckle Print datasets using PHOG and LPQ feature extraction, with performance measured in terms of recognition accuracy, FAR/FRR, and classification consistency. Baseline results obtained from plaintext biometric matching are compared against encrypted-domain operations to quantify the impact of homomorphic encryption. In addition, system overhead is evaluated by measuring encryption latency, IPFS storage cost, and blockchain transaction latency and gas consumption during evidence registration and access logging.

BIOMETRIC SYSTEM CONFIGURATION

To evaluate the proposed biometric authentication framework, experiments are conducted using two publicly available datasets, namely the Tongji palmprint dataset and a Finger Knuckle Print (FKP) dataset. These modalities are selected due to their relevance in forensic identification scenarios and their robustness to common acquisition variations. Biometric feature extraction is performed using two complementary descriptors: Pyramid Histogram of Oriented Gradients (PHOG) and Local Phase Quantization (LPQ). PHOG features are extracted using spatial pyramid levels 3, 4, and 5 to capture multi-resolution structural information, while LPQ features are computed with window sizes of 15×15, 25×25, and 35×35 to analyze local phase patterns at different granularities. The extracted feature vectors represent fixed-length numerical templates suitable for secure processing.

The datasets are partitioned using a 40% training and 60% testing split to ensure a realistic evaluation protocol. Initial experiments are performed on plaintext feature vectors to identify the optimal feature configurations and establish baseline recognition performance using a distance-based (DRB) classifier. Once optimal parameters are determined, the corresponding feature vectors are encrypted using homomorphic encryption, and classification is performed directly in the encrypted domain. This pipeline allows biometric matching to be carried out without exposing raw biometric data, enabling both closed-set and open-set identification while preserving recognition accuracy. The evaluation therefore quantifies the impact of encryption on classification performance and demonstrates the practicality of privacy-preserving biometric authentication for forensic applications.

BLOCKCHAIN NETWORK SIMULATION

To ensure a secure and testable environment for forensic evidence management, the proposed system employs a permissioned blockchain architecture, implemented using a local Ethereum-based network configured via Ganache. This configuration reflects forensic deployment scenarios in which access is restricted to authorized entities such as law enforcement agencies, forensic laboratories, and judicial authorities. The local network is configured with a Network ID of 5777, Chain ID 1337, and an RPC endpoint at http://127.0.0.1:8545, with a gas limit of 6,721,975, enabling efficient smart contract deployment and iterative validation during system development.

For immutable and decentralized evidence storage, the framework integrates the InterPlanetary File System (IPFS) through a public gateway and secure API access over HTTPS. Biometric evidence is stored off-chain in encrypted form, while only the corresponding content identifiers (CIDs) are recorded on-chain. This design minimizes blockchain storage overhead while preserving integrity, traceability, and chain-of-custody guarantees. Smart contracts, implemented in Solidity (version 0.6.x) and deployed using the Truffle framework, enforce access control policies and record evidence-related events such as registration, verification requests, and access logs. Interaction between system components and the blockchain is facilitated through Web3.js, enabling authenticated transactions and tamper-resistant audit trails. This architecture provides scalable, auditable, and secure evidence management suitable for forensic applications.

BIOMETRIC SYSTEM EVALUATION

Biometric systems, which rely on unique physiological or behavioral traits for individual identification, must undergo rigorous evaluation to satisfy critical performance criteria such as reliability, user convenience, and accuracy. In this work, the biometric component is assessed under two distinct conditions: using raw (unencrypted) biometric data and using encrypted biometric vectors secured through homomorphic encryption. This dual analysis allows for a comprehensive understanding of the system's effectiveness in both standard operational environments and privacy-preserving contexts. By comparing the system's behavior across these two modes, an evaluation can be made not only its recognition performance but also the trade-offs introduced by encrypted-domain computation in terms of accuracy and computational overhead.

Unencrypted Biometric system analysis

In the raw unencrypted system scenario, our initial tests will concentrate on selecting the LPQ window size (*Wn*) and the PHOG resolution level (*L*) from the value sets of 15 × 15, 25 × 25, and 35 × 35 for *Wn*, and 3, 4, and 5 for *L*. using two databases, Tongji and FKP, to assess the impact of these parameters (*Wn* and *L*) on the performance of the biometric system. The results of the open/closed-set identification system will be clearly presented in terms of (T_0, *EER*) for the open-set and (*ROR*, *RPR*) for the Closed-set. As they are shown in Figures 3 and 4 below.

Figure 3. Impact of PHOG resolution level on open-set biometric system performance for both Tongji and FKP datasets.

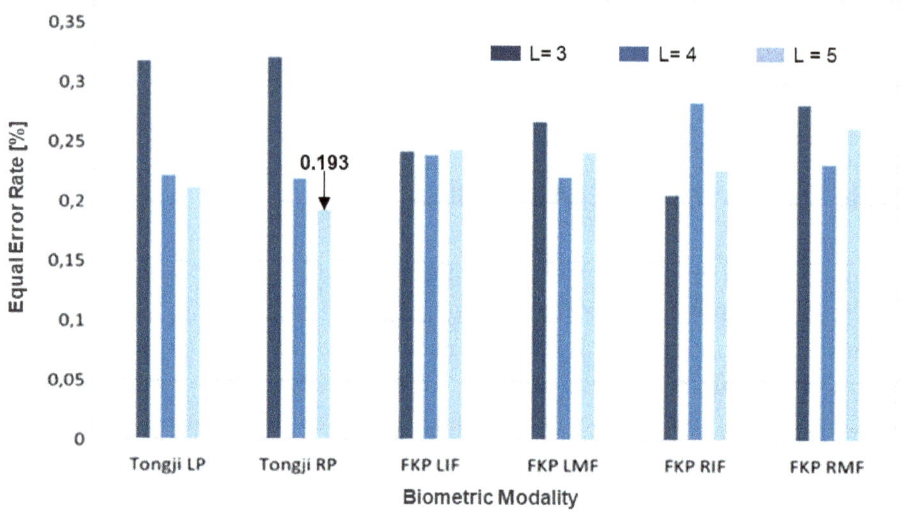

Figure 4. Impact of LPQ window size on open-set biometric system performance for both Tongji and FKP datasets.

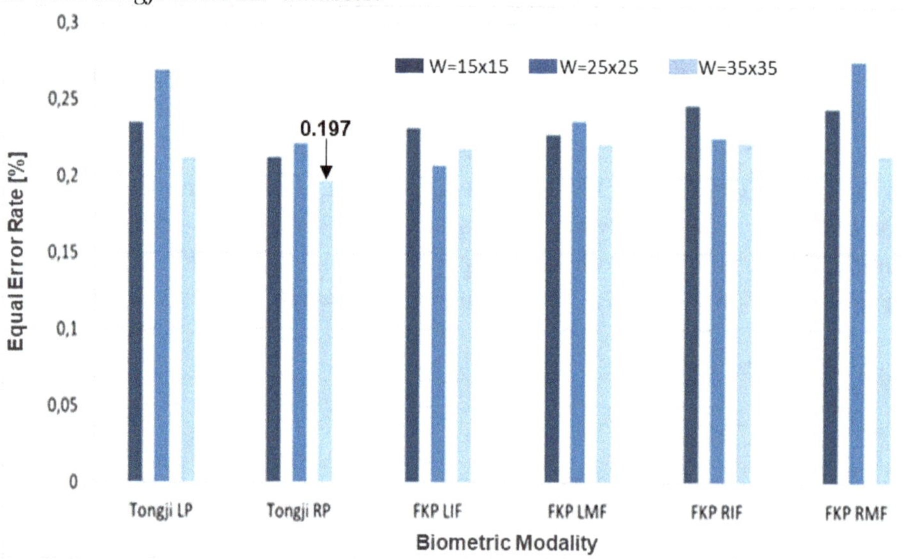

In Figure 3, the system's performance was evaluated using two biometric datasets, Tongji and FKP, across varying configurations of PHOG and LPQ. PHOG was tested at resolution levels L = 3, 4, 5 while LPQ was assessed with window sizes $Wn = 15 \times 15, 25 \times 25, 35 \times 35$. Results showed that increasing PHOG resolution improved recognition performance, with Tongji's right palm achieving the best EER of 0.193% at L=5 and a decision threshold of $T_0 = 0.922$. Likewise, FKP's left middle finger (LMF) sample attained an EER of 0.221% with $T_0 = 0.908$ at the same PHOG level.

For LPQ, in Figure 4, larger window sizes led to improved discriminability. Tongji's right palm again demonstrated strong performance, reaching an EER of 0.197% at a window size of 35×35 and $T_0 = 0.915$, while FKP's right middle finger (RMF) achieved 0.214% at the same window size with $T_0 = 0.901$. Lower settings, such as LPQ 15×15 and PHOG L=3, also yielded acceptable outcomes; for instance, FKP's right index finger (RIF) produced an EER of 0.246% at $T_0 = 0.848$.

Overall, the PHOG method with L=5 emerged as the most consistent and accurate across both datasets, justifying its selection for subsequent evaluations in open- and closed-set identification tasks. Its ability to deliver lower EERs at optimal thresholds highlights its robustness for secure and reliable biometric authentication.

Encrypted Biometric system analysis

Based on the initial comparative study, the Tongji dataset combined with PHOG at level 5 was selected for further evaluation based on its behavior and performance in the past study on raw biometric input, as it consistently outperformed other configurations and modalities across both datasets. Given its strong baseline accuracy, this configuration was used to assess the impact of homomorphic encryption on biometric verification performance. The system was evaluated in both open-set and closed-set identification scenarios, and the results were analyzed before and after encryption.

In the open-set setting, as shown in Figure 5(a), the system achieved an Equal Error Rate (EER) of 0.193% at a decision threshold of $T_0 = 0.922$ prior to encryption. After applying homomorphic encryption to the biometric vectors, the EER slightly increased to 0.224%, with a corresponding threshold adjustment to $T_0 = 0.917$. This marginal performance trade-off reflects the additional computational complexity introduced by encrypted-domain operations while maintaining high accuracy.

In the closed-set scenario, illustrated in Figure 5(b), the system initially achieved a Rank-One Recognition Rate (ROR) of 99.86% with a RPR of 11. Following encryption, the ROR decreased slightly to 99.63%, and the RPR increased to 25. Despite this minor degradation, the system preserved its high recognition capability, demonstrating that privacy-preserving computation through homomorphic encryption can be integrated without significantly compromising biometric identification performance.

Figure 5. System performance before and after encryption. (a) Open-set system (DET Curve), (b) Closed-set system (CMC Curve)

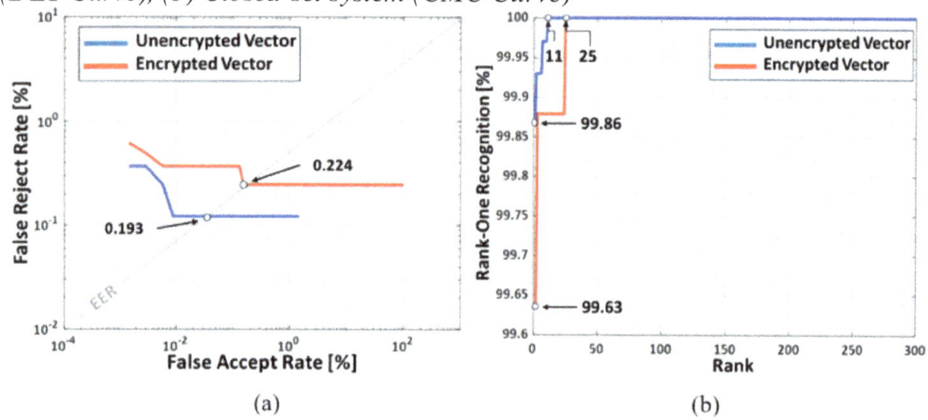

BLOCKCHAIN INTEGRATION EVALUATIONBlockchain integration evaluation

Evaluate ther proposed system on various factors. This assessment will demonstrate the system's efficiency, ability to handle growth, and the associated operational costs.

Latency

The purpose of latency measurement in this system is to evaluate the time required to complete user authentication, both before and after integrating biometric verification with the blockchain infrastructure. This analysis is essential to quantify the performance impact introduced by decentralized operations within forensic evidence workflows Nguyen et al (2022). As shown in Table 1, the average latency increased from 2261.33 milliseconds to 4748 milliseconds following the integration of blockchain components. This represents a 47.62% rise in response time, primarily attributed to the overhead introduced by blockchain functions such as content addressing through IPFS and smart contract execution. While this increase in latency reflects the additional computational steps inherent in decentralized systems, the trade-off is justified by the substantial gains in auditability, tamper resistance, and trustless authentication. In sensitive forensic environments, where data integrity and traceability are paramount, a moderate delay in processing can be considered acceptable given the added layers of security and transparency provided by the blockchain infrastructure.

Table 1. Proposed System Latency Results

<table>
<tr><th>Experiment Phase</th><th>Requests</th><th>Latency (ms)</th><th>Average Latency (ms)</th></tr>
<tr><td rowspan="4">Pre-Blockchain Integration</td><td>Request 1</td><td>2647</td><td rowspan="4">2261.33</td></tr>
<tr><td>Request 2</td><td>1964</td></tr>
<tr><td>Request 3</td><td>2173</td></tr>
<tr><td colspan="2">Average</td></tr>
<tr><td rowspan="4">Post-Blockchain Integration</td><td>Request 1</td><td>4021</td><td rowspan="4">4748</td></tr>
<tr><td>Request 2</td><td>5849</td></tr>
<tr><td>Request 3</td><td>4374</td></tr>
<tr><td colspan="2">Average</td></tr>
<tr><td colspan="3">Latency Increase (ms)</td><td>2486.33</td></tr>
<tr><td colspan="3">Percentage Increase (%)</td><td>47.62%</td></tr>
</table>

Scalability analysis

Scalability analysis in the context of forensic blockchain systems involves assessing the system's capacity to accommodate increases in data volume, user participation, and transaction frequency without compromising performance or reliability. This includes evaluating how the infrastructure handles larger numbers of biometric authentications, evidence uploads, and distributed queries over Iqra Sadia Rao et al (2024). It is a critical component of system design, particularly for applications involving legal or investigative workflows where the volume of forensic artifacts and audit logs may grow substantially. Effective scalability ensures that the system remains responsive, secure, and cost-efficient even as operational demands escalate, thereby supporting long-term sustainability and readiness for real-world deployment across multi-agency environments. As shown in Table 2 our system kept a high success rate of 100% and different but yet acceptable latency across different requests.

Table 2. Scalability Analysis of The Proposed System

Number of Requests	**10**	**50**	**100**
Success Rate (%)	100.00	100.00	100.00
Average Latency (ms)	5475.18	4213.548	4235.43

Cost Analysis

In blockchain-based forensic systems, cost analysis is essential for evaluating the financial implications of executing smart contracts and interacting with decentralized storage networks. The primary expense is typically associated with gas fees, which are denominated in Ethereum (ETH) and are required for processing transactions on the blockchain Omar et al (2021). As shown in Table 3, the analysis reveals that both user authentication requests and node-side responses incur minimal gas costs, making the system economically feasible for repeated and frequent interactions.

Table 3. Cost Analysis of The Proposed System

Operation	Request Authentication	Node Responses
Cost (ETH)	0.00056548	0.00056728

The average cost per biometric authentication transaction was measured at 0.00056548 ETH, a value considered acceptable in light of the system's enhanced decentralization, fault tolerance, and auditability. Although the integration of blockchain increases latency compared to conventional centralized solutions, it introduces critical security enhancements, such as immutability and protection against tampering. These features are particularly valuable in forensic applications where data integrity and traceability are paramount. Monitoring gas usage and optimizing smart contract logic remain important to sustain cost efficiency as network activity and gas prices fluctuate. Despite the trade-offs in latency and dynamic cost, the system's resilience, transparency, and resistance to single points of failure justify its deployment in high-security, multi-agency forensic environments.

Security analysis

In this section, conducted an assessment on the resilience of the proposed forensic evidence management system by simulating a range of potential attack scenarios. These simulations are designed to evaluate how well the system preserves data integrity, access control, and authentication security in the presence of adversarial behaviors. By analyzing its response to such threats, aiming to demonstrate the robustness of the integrated biometric and blockchain architecture and identify any areas requiring further reinforcement for deployment in high-security forensic environments.

Sybil Attack

A Sybil attack occurs when a malicious actor creates multiple fake identities (nodes or users) in a distributed system to manipulate consensus, disrupt service, or gain unauthorized access to sensitive operations. In forensic blockchain environments, such attacks pose serious risks, including the potential to compromise evidence integrity or bypass access control mechanisms Platt & McBurney (2023). To evaluate the robustness of our proposed system, simulating various Sybil attack scenarios by attempting to introduce unauthorized nodes into the blockchain network and by spoofing biometric identities across multiple false accounts. The system's built-in role-based access control, biometric identity binding, and consensus-layer node verification mechanisms served as the primary defenses. Our analysis focused on metrics such as detection rate, false acceptance rate, and blockchain rejection response, with results summarized in Table 4.

Table 4. System Resilience Against Sybil Attack Scenarios

Attack Scenario	Fake Nodes Injected	Unauthorized Access Attempts	Detection Rate (%)	Blockchain Rejection (%)
Random Identity Injection	20	20	100	100
Biometric Spoofing via Sybil IDs	15	30	96.7	98.4
Combined Identity & Role Spoof	10	25	100	100
Encrypted Biometric Tampering	12	12	91.6	95.8

As shown in Table 4, the system successfully identified and blocked all unauthorized nodes introduced through Sybil attack vectors, achieving up to 100% detection and rejection rates in multiple scenarios. Even in more complex attack patterns involving encrypted biometric spoofing, the system maintained a high detection rate of over 91%, illustrating its robustness against identity manipulation in both open and closed configurations. These results confirm the system's capacity to withstand Sybil-based threats, primarily due to the integration of cryptographically verifiable biometric authentication and strict access control enforcement.

51% Attack

A 51% attack refers to a scenario in which a single entity or coordinated group gains control of more than half of the total computational power or consensus votes within a blockchain network. This control allows the attacker to manipulate ledger entries, reverse transactions, or suppress legitimate operations, posing a significant threat to the integrity of forensic evidence Aponte-Novoa et al (2021). In the context of forensic systems, such attacks could undermine the immutability of case records or disrupt evidence validation workflows. To assess our system's resilience, simulating an artificial dominance of malicious nodes exceeding 51% of total validators while attempting to tamper with previously recorded transactions. Defense mechanisms including node reputation weighting, multi-agency consensus, and smart contract validation layers were active during testing. The results, outlined in Table 5, illustrate the system's effectiveness in detecting, resisting, and recovering from such attacks.

Table 5. System Performance Under Simulated 51% Attack Conditions

Attack Scenario	Malicious Node Share (%)	Tamper Attempted	Tamper Success (%)	Consensus Override	Detection Triggered
Direct Block Rewrite	55	Yes	0	Prevented	Yes
Smart Contract Manipulation	60	Yes	0	Prevented	Yes
Transaction Rollback Attempt	70	Yes	0	Prevented	Yes
Evidence Reordering (Timestamp)	65	Yes	0	Prevented	Yes

As seen in Table 5, despite simulated scenarios where malicious nodes exceeded 51% of network control, the system successfully maintained blockchain integrity with a 0% tamper success rate across all cases. This robustness stems from the integration of layered trust protocols, including non-fungible audit logs, role-restricted smart contracts, and cross-validated evidence hashes stored via IPFS. The prevention of transaction rollbacks and evidence rewriting highlights the strength of the multi-stakeholder consensus design, which distributes validation authority across independent forensic entities. These results demonstrate that even under extreme adversarial conditions, the system is capable of maintaining secure, tamper-proof operation suitable for high-assurance forensic environments.

PROPOSED OVER STATE OF THE ART SYSTEMS

In recent years, the integration of blockchain technologies into digital forensic systems has gained significant traction as a solution for enhancing data integrity, transparency, and decentralized trust. Various frameworks have emerged to secure evidence collection and management, particularly in IoT, vehicular, and cloud-based environments. However, most existing models fall short in incorporating robust user authentication, biometric privacy, and encrypted-domain processing. This chapter addresses these limitations by introducing a unified architecture that combines blockchain auditability, homomorphic encryption, and biometric-driven identity verification. The following comparative study shown in Table 6, examines prominent recent contributions in the field, highlighting their core features, strengths, and the critical gaps our proposed system is designed to fill.

Table 6. Comparative Analysis of Forensic Blockchain Models

Work	Biometric Use	Encryption Method	Blockchain Function	Access Control Model	Privacy Mechanism
Asma Jodeiri Akbarfam et al (2023)	Not included	None	Provenance tracking with Merkle root optimization	Role-Based Access Control (RBAC-SA)	Basic traceability, no biometric privacy
(Cebe et al., 2018)	Not included	None	Lightweight ledger for vehicle data	Vehicular Public Key Infrastructure	Anonymity and low-overhead data logging
Ezazul Islam et al (2023)	Not included	None	Immutable forensic log storage	Hybrid Attribute-Role-Based (ARBAC)	Endpoint-level authentication
Li et al (2021)	Not included	Attribute-Based Encryption (CP-ABE)	Evidence lifecycle tracking	Fine-grained attribute access	Witness and juror privacy, encrypted access
Butrus Mbimbi et al (2024)	Not included	Hashing (SHA-256)	Hyperledger-based ledger for IoT transactions	Not specified	Data authenticity and distributed access
Chen et al (2024)	Applied to genetic data (kinship detection)	Homomorphic Encryption (matrix vector ops)	Not implemented	Not specified	Encrypted computation for sensitive relations
Proposed	Biometric vector-based user verification	Homomorphic Encryption over biometric vectors	Immutable ledger via Ethereum and IPFS	Role-based with blockchain enforcement	Privacy-preserving computation and secure audit

The comparative table presented in this section summarizes key contributions from recent state-of-the-art systems across domains such as vehicular forensics, forensic logging, legal evidence workflows, and genomic privacy. Each entry highlights the inclusion or absence of critical components like biometric support, encryption techniques, blockchain usage, access control models, and privacy mechanisms. This structured overview reveals a clear gap: while many systems provide auditability or decentralization, few incorporate homomorphic encryption, and almost none fully integrate secure biometric verification. Our proposed system addresses this multidimensional need by combining encrypted biometric vectors with blockchain-backed traceability and smart contract-based access management, offering a unified and privacy-preserving framework suitable for multi-agency forensic environments. The comparative study therefore not only situates our work within the current research landscape but also underscores its novelty and practical relevance in addressing unresolved challenges.

CONCLUSION AND FUTURE WORKS

In the evolving landscape of digital forensics, the secure handling, authentication, and long-term preservation of forensic evidence have become increasingly critical. With rising concerns over data integrity, unauthorized access, and identity spoofing, traditional centralized systems are no longer sufficient to meet the demands of modern investigative workflows. The integration of biometrics, encryption, and decentralized technologies offers a promising direction for enhancing security, transparency, and privacy in forensic evidence management. This chapter introduced a comprehensive framework that combines biometric-driven authentication with blockchain-based audit trails and decentralized file storage. At its core, the system employs homomorphic encryption to protect biometric vectors, enabling secure identity verification without exposing raw biometric data. This approach preserves user privacy while allowing legitimate operations to be conducted over encrypted features, ensuring compliance with privacy standards and regulatory constraints. The system also utilizes Ethereum smart contracts and IPFS to establish an immutable and tamper-evident chain of custody, supporting accountability across all stages of forensic investigation.

Experimental evaluations demonstrated the system's effectiveness across multiple biometric configurations using PHOG and LPQ descriptors. Performance metrics revealed minimal degradation in accuracy after encryption, and latency analysis confirmed that blockchain integration, while introducing additional delay, remains within acceptable bounds for forensic applications. Furthermore, attack simulations, including Sybil and 51% attacks, validated the architecture's resilience against common blockchain-based threats, reinforcing its reliability in adversarial environments. Unlike prior work that typically focuses on isolated aspects such as logging, access control, or data encryption, this chapter presents a unified solution addressing all three pillars: secure biometric verification, encrypted data processing, and immutable auditability. This level of integration is particularly valuable in multi-agency forensic scenarios where trust, privacy, and verification must coexist in a decentralized setting.

The proposed framework addresses legal and ethical requirements in forensic contexts. Blockchain-based logging and immutable audit trails ensure that all evidence-handling operations are traceable and verifiable, supporting chain-of-custody and legal admissibility standards. Biometric data is encrypted before storage, and access is granted only to authorized personnel, aligning with principles of consent, data minimization, and privacy protection. By integrating these safeguards, the system promotes accountability and transparency while remaining compliant with relevant regulatory frameworks such as GDPR and ISO/IEC forensic standards. In conclusion, the proposed system provides a scalable, privacy-aware, and tamper-resistant architecture for managing digital forensic evidence. It establishes a secure foundation for future research and real-world implementations aimed at ensuring forensic soundness and evidentiary trust in an increasingly digitized investigative ecosystem.

ACKNOWLEDGMENT

The authors are grateful to the anonymous referees for their valuable and helpful comments. The authors also thank the staff of Laboratory of Signals and Smart Systems and LAboratory of Mathematics, Informatics and Systems for their helpful comments and suggestions.

Credit author statement

Author 1: Original Draft, Conceptualization, Visualization, Formal Analysis, Software.

Author 2: Writing – Review & Editing, Software, Formal Analysis, Validation, Supervision, Project Administration.

Author 3: Review & Editing, Formal Analysis, Validation.

Author 4: Writing – Review & Editing, Investigation, Project Administration

Declaration of competing interest

The authors declare that they have no known competing financial interests or personal relationships that could have appeared to influence the work reported in this paper.

Data availability

The data used in this study are openly available in Tongji Contactless Palmprint Dataset available at: https://cslinzhang.github.io/ContactlessPalm/, see also Zhang et al (2017). The FKP dataset available at: https://www.comp.polyu.edu.hk/biometrics/FKP.html, see also Zhang & Zhang (2009).

REFERENCES

Alqahtany, S. S., & Syed, T. A. (2024). ForensicTransMonitor: A Comprehensive Blockchain Approach to Reinvent Digital Forensics and Evidence Management. *Information (Basel)*, *15*(2), 109. DOI: 10.3390/info15020109

Aponte-Novoa, F. A., Orozco, A. L. S., Villanueva-Polanco, R., & Wightman, P. (2021). The 51% Attack on Blockchains: A Mining Behavior Study. *IEEE Access : Practical Innovations, Open Solutions*, *9*, 140549–140564. DOI: 10.1109/ACCESS.2021.3119291

Asadi Shirzi, M., & Kermani, M. R. (2024). Real-Time Point Recognition for Seedlings Using Kernel Density Estimators and Pyramid Histogram of Oriented Gradients. *Actuators*, *13*(3), 81–81. DOI: 10.3390/act13030081

Baig, A. F., & Eskeland, S. (2021). Security, Privacy, and Usability in Continuous Authentication: A Survey. *Sensors (Basel)*, *21*(17), 5967. DOI: 10.3390/s21175967 PMID: 34502865

Cebe, M., Erdin, E., Akkaya, K., Aksu, H., & Uluagac, S. (2018). Block4Forensic: An Integrated Lightweight Blockchain Framework for Forensics Applications of Connected Vehicles. *IEEE Communications Magazine*, *56*(10), 50–57. DOI: 10.1109/MCOM.2018.1800137

Chen, J., Miao, W., Wu, W., Yang, L., & Yuan, H. (2024). Secure Relative Detection in (Forensic) Database with Homomorphic Encryption. *Lecture Notes in Computer Science*, *14955*, 410–422. DOI: 10.1007/978-981-97-5131-0_35

Elgohary, H. M., Darwish, S. M., & Elkaffas, S. M. (2022). Improving Uncertainty in Chain of Custody for Image Forensics Investigation Applications. *IEEE Access : Practical Innovations, Open Solutions*, *10*, 14669–14679. DOI: 10.1109/ACCESS.2022.3147809

Ezazul Islam, Md., Rafiqul Islam, Md., Chetty, M., & Lim, S., & Mehmood Chadhar. (2023). User authentication and access control to blockchain-based forensic log data. *EURASIP Journal on Information Security*20231DOI: 10.1186/s13635-023-00142-3

Gu, X., Angelov, P. P., Zhang, C., & Atkinson, P. M. (2018). A Massively Parallel Deep Rule-Based Ensemble Classifier for Remote Sensing Scenes. *IEEE Geoscience and Remote Sensing Letters*, *15*(3), 345–349. DOI: 10.1109/LGRS.2017.2787421

Javed, A. R., Ahmed, W., Alazab, M., Jalil, Z., Kifayat, K., & Gadekallu, T. R. (2022). A comprehensive survey on computer forensics: State-of-the-Art, tools, techniques, challenges, and future directions. *IEEE Access : Practical Innovations, Open Solutions*, *10*(1), 11065–11089. DOI: 10.1109/ACCESS.2022.3142508

Jung, Y., & Virgil, E. D. (2024). Analysis of Legislative Framework Governing Biometric Data. *Procedia Computer Science*, *241*, 48–55. DOI: 10.1016/j.procs.2024.08.009

Klasén, L., Fock, N., & Forchheimer, R. (2024). The invisible evidence: Digital forensics as key to solving crimes in the digital age. *Forensic Science International, 362*, 112133–112133. DOI: 10.1016/j.forsciint.2024.112133 PMID: 39089208

Krishna Prakasha, K., & Sumalatha, U. (2025). Privacy-Preserving Techniques in Biometric Systems: Approaches and Challenges. *IEEE Access : Practical Innovations, Open Solutions, 13*, 32584–32616. DOI: 10.1109/ACCESS.2025.3541649

Kushwaha, S. S., Joshi, S., Singh, D., Kaur, M., & Lee, H.-N. (2022). Systematic Review of Security Vulnerabilities in Ethereum Blockchain Smart Contract. *IEEE Access : Practical Innovations, Open Solutions, 10*, 6605–6621. DOI: 10.1109/ACCESS.2021.3140091

Li, M., Lal, C., Conti, M., & Hu, D. (2021). LEChain: A blockchain-based lawful evidence management scheme for digital forensics. *Future Generation Computer Systems, 115*, 406–420. DOI: 10.1016/j.future.2020.09.038

Mbimbi, B., Murray, D., & Wilson, M. (2024). IoT Forensics-Based on the Integration of a Permissioned Blockchain Network. *Blockchains, 2*(4), 482–506. DOI: 10.3390/blockchains2040021

Moussa, A. F. (2021). Electronic evidence and its authenticity in forensic evidence. *Egyptian Journal of Forensic Sciences, 11*(1), 20. Advance online publication. DOI: 10.1186/s41935-021-00234-6

Munjal, K., & Bhatia, R. (2022). A systematic review of homomorphic encryption and its contributions in healthcare industry. *Complex & Intelligent Systems, 9*(4), 3759–3786. Advance online publication. DOI: 10.1007/s40747-022-00756-z PMID: 35531323

Nguyen, D. C., Hosseinalipour, S., Love, D. J., Pathirana, P. N., & Brinton, C. G. (2022). Latency Optimization for Blockchain-Empowered Federated Learning in Multi-Server Edge Computing. *IEEE Journal on Selected Areas in Communications, 40*(12), 3373–3390. DOI: 10.1109/JSAC.2022.3213344

Omar, I. A., Jayaraman, R., Debe, M. S., Salah, K., Yaqoob, I., & Omar, M. (2021). Automating Procurement Contracts in the Healthcare Supply Chain using Blockchain Smart Contracts. *IEEE Access : Practical Innovations, Open Solutions, 9*(1), 1–1. DOI: 10.1109/ACCESS.2021.3062471

Platt, M., & McBurney, P. (2023). Sybil in the Haystack: A Comprehensive Review of Blockchain Consensus Mechanisms in Search of Strong Sybil Attack Resistance. *Algorithms, 16*(1), 34. DOI: 10.3390/a16010034

Quinn, P., Ellyne, E., & Yao, C. (2024). Will the GDPR Restrain Health Data Access Bodies Under the European Health Data Space (EHDS)? *Computer Law & Security Review, 54*, 105993–105993. DOI: 10.1016/j.clsr.2024.105993

Rao, I. S., Kiah, M. L. M., Hameed, M. M., & Memon, Z. A.Iqra Sadia Rao. (2024). Scalability of blockchain: A comprehensive review and future research direction. *Cluster Computing, 27*(5), 5547–5570. Advance online publication. DOI: 10.1007/s10586-023-04257-7

Rukhiran, M., Wong-In, S., & Netinant, P. (2023). IoT-Based Biometric Recognition Systems in Education for Identity Verification Services: Quality Assessment Approach. *IEEE Access : Practical Innovations, Open Solutions, 11*, 22767–22787. DOI: 10.1109/ACCESS.2023.3253024

Sangeeta, N., & Nam, S. Y. (2023). Blockchain and Interplanetary File System (IPFS)-Based Data Storage System for Vehicular Networks with Keyword Search Capability. *Electronics (Basel), 12*(7), 1545. DOI: 10.3390/electronics12071545

Verma, R., Dhanda, N., & Nagar, V. (2022). Application of Truffle Suite in a Blockchain Environment. *Proceedings of Third International Conference on Computing, Communications, and Cyber-Security,* 693–702. DOI: 10.1007/978-981-19-1142-2_54

Zhang, L., Li, L., Yang, A., Shen, Y., & Yang, M. (2017). Towards contactless palmprint recognition: A novel device, a new benchmark, and a collaborative representation based identification approach. *Pattern Recognition, 69*, 199–212. DOI: 10.1016/j.patcog.2017.04.016

Zhang, L., Zhang, L., & Zhang, D. (2009, November). Finger-knuckle-print: a new biometric identifier. *In 2009 16th IEEE international conference on image processing (ICIP) (pp. 1981-1984). IEEE.* DOI: 10.1109/ICIP.2009.5413734

Chapter 12

Digital Forensics in Cloud Environments: Techniques, Challenges, and Legal Implications for Data Breaches and Cybercrimes

Saloni Bahl
Vivekanada Global University, Jaipur, India

Abhilash Arun Sapre
https://orcid.org/0000-0002-7680-0894
Gujarat National Law University, Gandhinagar, India

ABSTRACT

Cloud computing has transformed the way a firm handles and analyze data in today's data-driven world. This transition has underscored the need for robust cloud digital forensics. Today, the voyage begins with an overview of cloud computing and its rapid integration across sectors, with enterprises increasingly using CSPs for data storage and processing. This change requires a deep understanding of cloud digital forensics. The paper analyses cloud-specific forensic tools and approaches to address these problems, emphasizing the need for digital forensics professionals and CSPs to collaborate. Cloud-based investigations have evidentiary issues, particularly in terms of digital evidence reliability and integrity. Legal issues, including jurisdictional complexities and privacy concerns, are extensively discussed, emphasizing the need to align investigative techniques with legal requirements.The purpose is to enhance the effectiveness and legality of digital forensics in cloud-based data breach investigations, thereby protecting digital assets in a cloud-dominated world.

DOI: 10.4018/979-8-3373-4972-5.ch012

I. INTRODUCTION

The advent of cloud computing has transformed the way individuals and organizations store, manage, and access their digital data. With the growing reliance on cloud services, digital forensics has become increasingly crucial in investigating cybercrimes, data breaches, and other incidents involving cloud-hosted information (Manral, Somani, & Conti, 2019). This introduction delves into the realm of digital forensics in cloud environments, highlighting its significance, the evolution of cloud computing, and the challenges and complexities it presents to investigators. Cloud computing has redefined the landscape of IT infrastructure by offering scalable, cost-effective, and on-demand access to computing resources and services (Wagemann, Clements, & Figuera, 2018). This shift has led to a proliferation of cloud service providers, each offering a range of services, including infrastructure as a Service (IaaS), Platform as a Service (PaaS), and Software as a Service (SaaS). Consequently, individuals and businesses have entrusted sensitive data, applications, and workloads to these cloud platforms. However, the dynamic nature of cloud computing introduces complexities when it comes to digital investigations.

The significance of digital forensics in cloud environments lies in its role as a critical tool for uncovering evidence, identifying culprits, and piecing together the puzzle in the wake of cloud-based cyber incidents (Brown, 2015). Cloud environments present unique challenges that require specialized investigative techniques, tools, and expertise. Unlike traditional on-premises environments, where data resides within the physical boundaries of an organization (Zhang & Yue, 2020), cloud data is dispersed across multiple servers and data centres, often located in different jurisdictions, making it challenging to establish jurisdiction and a chain of custody.

The evolution of cloud computing has further complicated digital forensics. In the early days of cloud services, digital forensics primarily focused on extracting evidence from local devices, including computers and mobile phones. However, as data increasingly migrated to the cloud, investigators had to adapt. Cloud forensics emerged as a specialized discipline, addressing the nuances and intricacies of gathering, preserving, and analyzing digital evidence from cloud environments (Dykstra & Riehl, Forensic Collection of Electronic Evidence from Infrastructure-as-a-Service Cloud Computing, 2012).

One of the primary challenges in digital forensics within cloud environments is the preservation of evidence. Traditional digital forensics often involves seizing physical devices, which can be powered off and disconnected from the network to preserve evidence. In contrast, cloud-based evidence exists on remote servers maintained by third-party providers, where investigators have limited control (Dykstra & T. Sherman, Acquiring forensic evidence from infrastructure-as-a-service cloud computing: Exploring and evaluating tools, trust, and techniques, 2012). Ensuring the preservation of volatile data and metadata is crucial to prevent spoliation and maintain the integrity of digital evidence.

Today the multi-tenant nature of cloud platforms means that multiple users and organizations share the same physical infrastructure and resources (Ja, Yang, & Grundy, 2021). This poses challenges in differentiating and isolating the data of interest from other unrelated data. Additionally, the access controls, encryption, and authentication mechanisms used by cloud providers can impede investigators' efforts to access and analyse data, especially when the cloud service provider is uncooperative or when the investigation spans international borders.

The growth of cloud computing has revolutionized the digital landscape, offering numerous benefits while also introducing unique challenges for digital forensics (Baig et al., 2017). Investigating cybercrimes, data breaches, and other cloud-related incidents requires specialized knowledge, tools, and methodologies to navigate the complexities of evidence preservation, jurisdiction, and legal compliance. As cloud environments continue to evolve, digital forensics in the cloud will remain a vital component in safeguarding digital assets and pursuing justice in the face of digital threats (Fernandes, Colaco, Shetty, & H., 2020). This research aims to critically analyze these challenges and provide insights into the best practices for conducting digital forensics in cloud environments.

II. STATEMENT OF PROBLEM

The increasing adoption of cloud computing and the consequential reliance on cloud services for data storage and management have given rise to critical challenges in the realm of digital forensics. Investigating cloud-based data breach incidents and other cybercrimes in cloud environments presents multifaceted problems, including issues related to data privacy, jurisdiction, evidence preservation, multi-tenancy, encryption, compliance with evolving legal frameworks, and the lack of standardized practices. These challenges demand a comprehensive understanding and resolution to ensure effective digital forensics investigations while maintaining the integrity of evidence and upholding legal and ethical standards. This research aims to address these complex issues, providing insights and recommendations to enhance the field of digital forensics in cloud environments, thereby enabling more effective responses to cloud-based security incidents.

III. RESEARCH QUESTIONS

1. How can digital forensics investigators effectively address the challenges of data privacy, jurisdictional complexities, and evidence preservation when conducting investigations in cloud environments, while ensuring the integrity and admissibility of digital evidence in legal proceedings?
2. What strategies and best practices can be developed to navigate the complexities of multi-tenancy, encryption, and compliance with evolving legal frameworks in cloud-based data breach investigations, ultimately improving the quality and reliability of evidence in digital forensics?

IV. METHODOLOGY AND RESULTS

The research will adopt a mixed-methods research design to comprehensively address the complexities outlined in the statement of the problem and the research questions. The quantitative data shall be collected via a Google survey distributed to a diverse sample of digital forensics professionals, including law enforcement agencies, corporate investigators, forensic science students and independent practitioners. The survey will be designed to quantify the prevalence of specific challenges identified in the research problem and measure the effectiveness of current practices. Statistical analysis of the survey responses will be conducted using software tools to derive quantitative findings.

The research will also include an analysis of real-world case studies related to cloud-based data breach incidents and investigations. These cases will be selected to represent diverse cloud service providers and various industry sectors, offering practical insights into the challenges and potential solutions faced by investigators. This research will employ a mixed-methods approach, integrating qualitative insights from interviews, quantitative data from a Google survey, literature review, case studies, and expert consultations to address the multifaceted challenges in digital forensics in cloud environments and contribute to the advancement of knowledge in this critical field. The Google survey will serve as a valuable tool for data collection and analysis, ensuring the research's comprehensiveness and reliability.

i. Survey-Based Approach The following questions have been prepared for the survey, where we received 178 responses from the respondents.

Table 1. Survey Questions

S.No.	Questions	Options
1.	*To what extent do you encounter jurisdictional complexities when conducting digital forensics in cloud environments?*	*A. Very Frequently* *B. Somewhat Frequently* *C. Rarely* *D. Never*
2.	*What is the most significant challenge in preserving evidence in cloud-based digital forensics investigations?*	*A. Maintaining Chain of Custody* *B. Dealing with Data Spoliation* *C. Proving Data Integrity* *D. Securing Volatile Data*
3.	*How effective are encryption and access control mechanisms in safeguarding cloud data during digital forensics investigations?*	*A. Highly Effective* *B. Moderately Effective* *C. Slightly Effective* *D. Ineffective*
4.	*In your opinion, what is the most pressing legal issue in digital forensics in cloud environments?*	*A. Data privacy and compliance* *B. Cross-border jurisdictional disputes* *C. Data ownership and access rights* *D. Admissibility of cloud-based evidence*
5.	*Are standardized practices and protocols lacking in digital forensics in cloud environments?*	*A. Strongly Agree* *B. Agree* *C. Disagree* *D. Strongly Disagree*

Question 1 was intended to categorize about the extent do you encounter jurisdictional complexities when conducting digital forensics in cloud environments? We received a total of 234 answers to this question. The recorded answers are shown in Figure 1.

Figure 1. Recorded Answers For Question - 1. Total Answers Recorded 234

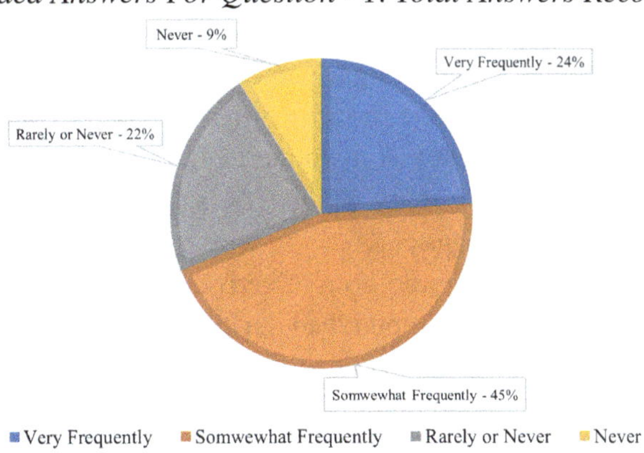

As shown in the figure, 57 respondents (24%) reported encountering jurisdictional complexities very frequently, while 105 respondents (45%) indicated that they encountered them somewhat frequently. Additionally, 52 respondents (22%) said they encountered these complexities rarely, and 20 respondents (9%) reported never encountering them. These findings highlight that a significant portion of respondents frequently deal with jurisdictional complexities in cloud-based digital forensics investigations.

Question 2 was designed to access the most significant challenge in preserving evidence in cloud-based digital forensics investigations. We received a total of 234 answers to this question. The recorded answers are shown in Figure 2.

Figure 2. Recorded Answers For Question - 2. Total Answers Recorded 234.

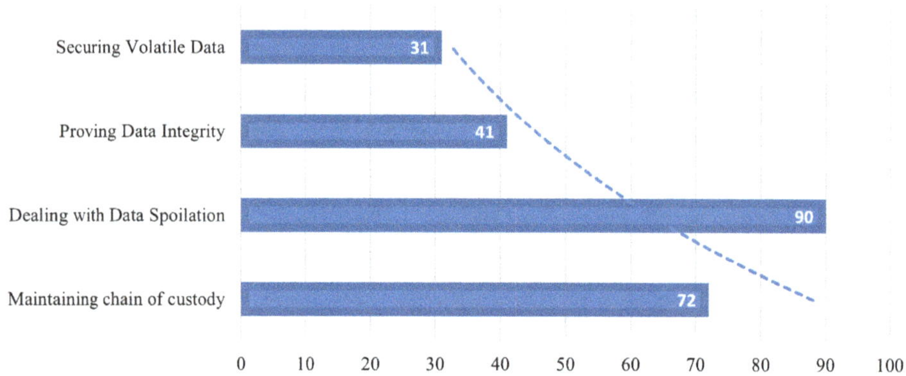

The figure illustrates that the most significant challenges in preserving evidence in cloud-based digital forensics investigations, according to the respondents. The results showed that 72 respondents (31%) considered maintaining the chain of custody as the most significant challenge. Furthermore, 90 respondents (38%) believed that dealing with data spoliation posed the most significant challenge. Proving data integrity was identified as the most significant challenge by 41 respondents (17%), and securing volatile data was the most significant challenge for 31 respondents (13%). These responses indicate that maintaining the chain of custody and dealing with data spoliation are the primary challenges in preserving evidence in cloud-based digital forensics investigations.

Question 3 aimed to determine how effective encryption and access control mechanisms are in safeguarding cloud data during digital forensics investigations. We received a total of 234 answers to this question. The recorded answers are shown in Figure 3.

Figure 3. Recorded Answers For Question - 3. Total Answers Recorded 234.

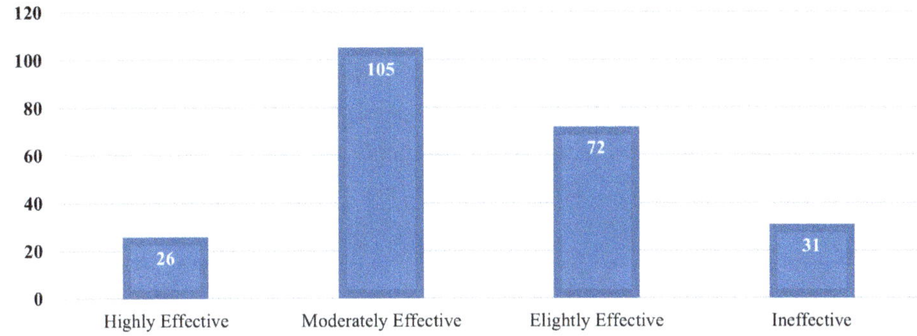

The figure reveals that, 26 respondents (11%) perceived these mechanisms as highly effective, while 105 respondents (45%) found them to be moderately effective. Additionally, 72 respondents (31%) considered them slightly effective, and 31 respondents (13%) regarded them as ineffective. These findings suggest that while a significant number of respondents find encryption and access control mechanisms to be effective, there is also a substantial portion that sees room for improvement.

Question 4 aimed to gauge opinions on the most pressing legal issue in digital forensics within cloud environments. We received a total of 234 answers to this question. The recorded answers are shown in Figure 4.

Figure 4. Recorded Answers For Question - 4. Total Answers Recorded 234

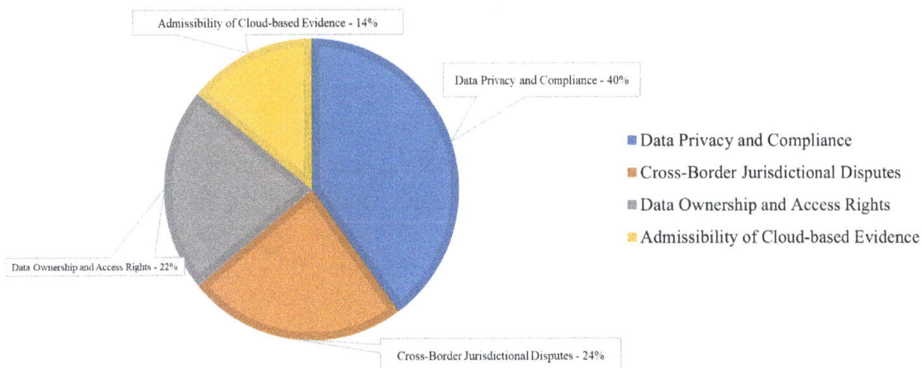

According to the figure, the aim is to identify the most pressing legal issues in digital forensics in cloud environments, as per the respondents' opinions. The results showed that 94 respondents (40%) identified data privacy and compliance as the most pressing legal issue. Additionally, 57 respondents (24%) pointed to cross-border jurisdictional disputes as the most pressing issue. Data ownership and access rights were considered the most pressing issue by 51 respondents (22%), and 32 respondents (14%) indicated that the admissibility of cloud-based evidence was the most pressing legal issue. These responses highlighted the diverse range of legal concerns in digital forensics in cloud environments.

Question 5 was sought to understand the respondents' perceptions about the standardized practices and protocols lacking in digital forensics in cloud environments. We received a total of 234 answers to this question. The recorded answers are shown in Figure 5.

Figure 5. Recorded Answers For Question - 5. Total Answers Recorded 234

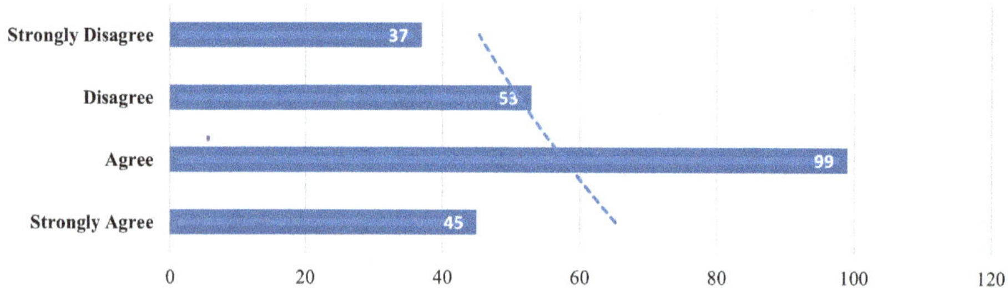

The above question was designed to gauge whether respondents believed that standardized practices and protocols were lacking in digital forensics in cloud environments. According to the responses, 45 respondents (19%) strongly agreed that standardized practices and protocols were lacking, while 99 respondents (42%) agreed with this statement. Furthermore, 53 respondents (23%) disagreed, and 37 respondents (16%) strongly disagreed with the notion that standardized practices and protocols were lacking. These responses reflect varying opinions among respondents regarding the existence of standardized practices and protocols in digital forensics in cloud environments.

ii. ANALYSIS AND RESULTS

The survey results provide valuable insights into the challenges and perceptions regarding digital forensics in cloud environments. The analysis below examines each question and its associated responses, shedding light on the prevailing trends and opinions in the field.

Jurisdictional Complexities

The first question sought to assess the extent of jurisdictional complexities encountered in digital forensics investigations in cloud environments. The responses indicate a notable level of complexity that investigators face. In today's interconnected world, cloud data often spans across multiple geographical locations and legal jurisdictions. This reality introduces significant hurdles in determining which jurisdiction's laws apply and how evidence should be collected and processed. The prevalence of respondents facing jurisdictional complexities underscores the need for a nuanced understanding of cross-border legal implications and collaboration between international law enforcement agencies.

Challenges in Evidence Preservation

The second question aimed to identify the most significant challenge in preserving evidence during digital forensics investigations in cloud environments. Maintaining the chain of custody involves tracking and documenting the handling of evidence throughout its lifecycle, ensuring its admissibility in court. With the dispersed nature of cloud data, this process becomes considerably complex. Dealing with data spoliation, on the other hand, addresses the deliberate destruction or alteration of data to hinder investigations. Given the potential for data spoliation in cloud environments, it is not surprising that many respondents consider it a significant challenge.

The data also highlights the importance of robust evidence preservation protocols and technologies specifically designed for cloud environments. Addressing these challenges will be crucial to ensure the credibility of digital evidence and the success of investigations.

Effectiveness of Encryption and Access Control

The third question sought to determine the perceived effectiveness of encryption and access control mechanisms in safeguarding cloud data during digital forensics investigations. The results indicate a range of perspectives among respondents. The varying opinions regarding the effectiveness of encryption and access control mechanisms underscore the complexities of ensuring data security in cloud environments. Encryption and access control are fundamental components of data safeguarding; however, their effectiveness depends on proper implementation and management. The results suggest that there may be room for improvement in the application of these security measures in cloud settings.

Moreover, the variations in responses suggest that a one-size-fits-all approach to data security may not be sufficient. Digital forensic experts must continually adapt to the evolving encryption technologies and access control protocols to adequately address the unique challenges of cloud environments.

Pressing Legal Issues

The fourth question aimed to ascertain respondents' opinions on the most pressing legal issues in digital forensics in cloud environments. The results revealed a diversity of concerns among respondents. These findings underscore the multifaceted legal landscape that digital forensics practitioners must navigate in cloud environments. Data privacy and compliance, a top concern, likely reflects the growing importance of data protection regulations such as the General Data Protection Regulation (GDPR) and the California Consumer Privacy Act (CCPA). Respondents' concerns about cross-border jurisdictional disputes and data ownership underscore the complexities of managing cloud data that may reside in multiple legal jurisdictions. Additionally, the admissibility of cloud-based evidence presents a substantial legal challenge, given the nuances of handling digital evidence in court.

The results emphasise the need for digital forensic professionals to remain well-versed in both international and domestic legal frameworks, ensuring that investigations comply with relevant laws and regulations.

Standardized Practices and Protocols

The final question aimed to assess whether respondents believed that standardized practices and protocols were lacking in digital forensics in cloud environments. The responses indicate a range of opinions, with 61% either strongly agreeing or agreeing that standardised practices and protocols are insufficient, while 39% either disagree or strongly disagree.

These varying perspectives reflect the evolving nature of digital forensics in cloud environments. While standardized practices and protocols are essential for consistency and best practices, the dynamic nature of cloud technology can make it challenging to establish universal standards. The discrepancy in responses suggests that some respondents believe that such standards are lacking or insufficient, while others may find the existing standards adaptable and effective. Addressing this issue will require ongoing collaboration and knowledge sharing within the digital forensics community to develop and refine standardised practices that are suitable for cloud environments. The survey had shed light on the complex landscape of digital forensics in cloud environments. The prevalence of jurisdictional complexities, challenges in evidence preservation, varying perceptions of the effectiveness of encryption and access control, diverse legal concerns, and differing opinions on standardized practices and protocols all highlight the need for a dynamic and adaptable approach to digital forensics in the cloud. As cloud computing continues to evolve, addressing these challenges and maintaining best practices in the field will be paramount to ensuring the integrity and success of digital forensics investigations in the cloud.

DISCUSSION

a. Jurisdictional Complexities in Digital Forensics in Cloud Environments

The advent of cloud computing has revolutionized the way data is stored, managed, and accessed, offering numerous advantages in terms of scalability and accessibility. However, this technological shift has also introduced complex jurisdictional challenges in the field of digital forensics. As data often resides in cloud servers located across multiple legal jurisdictions, digital forensic investigators face hurdles related to data access, legal frameworks, and international cooperation. Cloud computing's inherent flexibility and data mobility have significantly blurred the lines of data jurisdiction. Data can traverse national borders, making it challenging to determine which legal framework governs its access and handling (Wilder et al., 2010). This jurisdictional ambiguity is particularly problematic when conducting digital forensics investigations, as it raises questions about the admissibility of evidence and the validity of search warrants.

One of the primary jurisdictional complexities in cloud-based investigations is the issue of accessing cross-border data. When data is stored in a cloud server located in a different country, investigators must navigate the legal frameworks and international agreements governing data access and privacy (Brown, 2015). These complexities are exacerbated when cloud service providers refuse or are unable to cooperate with investigators, citing legal restrictions in their home country.

The results of the survey in question one indicated that a significant number of respondents encounter jurisdictional complexities very frequently (24%) or somewhat frequently (45%) in their digital forensics work. This highlights the real-world challenges that investigators encounter when working with cloud-based evidence.

To address these challenges, digital forensic experts must be well-versed in international legal agreements and mutual legal assistance treaties (MLATs) that govern the access to cross-border data. These agreements enable investigators to request assistance from foreign governments in collecting evidence located in their jurisdictions. However, the effectiveness of MLATs can vary significantly, and the process can be time-consuming, resulting in delays in investigations.

Another jurisdictional issue is the applicability of differing legal frameworks to cloud-based data. Different countries have distinct laws regarding data privacy, data retention, and government access to data. The conflict between national sovereignty and the extraterritorial nature of cloud data storage has given rise to legal debates and court cases that seek to define the boundaries of jurisdiction. This jurisdictional complexity can hinder the collection and use of digital evidence in legal proceedings, as investigators may inadvertently violate the laws of the jurisdiction where the data is stored (Turner, 1973). Survey question four, which focused on pressing legal issues, revealed that cross-border jurisdictional disputes were identified as the most pressing issue by 24% of respondents. This underscores the significance of this particular challenge and its impact on digital forensics investigations.

Furthermore, the conflict between U.S. and European Union data protection laws, as exemplified by the GDPR (*General Data Protection Regulation*), is a prominent example of jurisdictional complexity. While the GDPR provides strict data protection measures, U.S. law enforcement agencies may seek access to data stored in the EU for the purpose of investigations. Striking a balance between privacy rights and law enforcement interests is an ongoing legal challenge in this context.

International cooperation is a crucial component in addressing jurisdictional complexities. Digital forensics investigators often rely on the cooperation of law enforcement agencies, cloud service providers, and legal professionals across borders (Dykstra & T. Sherman, Acquiring forensic evidence from infrastructure-as-a-service cloud computing: Exploring and evaluating tools, trust, and techniques, 2012). Collaboration between countries and the standardization of legal processes can mitigate some of the challenges posed by jurisdictional complexities.

The jurisdictional complexities in digital forensics in cloud environments are a substantial and growing concern. Cloud computing's borderless nature, coupled with varying legal frameworks and international agreements, poses significant hurdles for investigators (Manral, Somani, & Conti, 2019). The survey results, combined with the pressing nature of these challenges, emphasise the need for increased international cooperation, clearer legal frameworks, and a deeper understanding of cross-border data access issues among digital forensic experts. Addressing jurisdictional complexities is crucial to ensuring that digital evidence collected in cloud environments is both admissible and reliable in legal proceedings, while respecting privacy and legal principles.

b. Challenges in Evidence Preservation and Integrity

In the digital age, the integrity and preservation of evidence are fundamental pillars of the criminal justice system. Digital forensics plays a pivotal role in collecting, preserving, and analyzing digital evidence to solve cybercrimes, data breaches, and a wide range of legal cases (Brown, 2015). However, when it comes to digital forensics in cloud environments, challenges in evidence preservation and maintaining its integrity become notably complex. This section explores the multifaceted issues surrounding evidence preservation and integrity in the context of cloud-based investigations.

i. Maintaining the Chain of Custody

One of the central challenges in preserving evidence in digital forensics, particularly in cloud environments, is maintaining the chain of custody. The chain of custody is a critical concept that involves documenting the handling, storage, and transfer of evidence to ensure its admissibility in court. In traditional forensic investigations, where evidence is typically a physical object, maintaining the chain of custody is relatively straightforward (Wagemann, Clements, & Figuera, 2018). However, in digital forensics, especially in the cloud, evidence exists as electronic data that can be easily altered or deleted.

The survey question on the most significant challenge in preserving evidence indicated that 31% of respondents considered maintaining the chain of custody as the foremost challenge. The implications of this challenge are far-reaching, as any break in the chain of custody can result in evidence being deemed inadmissible in court (Fernandes, Colaco, Shetty, & H., 2020). In cloud environments, where data can be altered or deleted remotely, establishing and maintaining a secure and transparent chain of custody is paramount. Digital forensic experts must document the movement of data from the cloud server to the forensic lab and maintain its integrity throughout the investigation.

ii. Dealing with Data Spoliation

Data spoliation, the deliberate alteration, destruction, or tampering with digital evidence, is another significant challenge in digital forensics in cloud environments. It is a tactic employed by suspects to hinder investigations, making it difficult to determine the authenticity and integrity of evidence. Detecting data spoliation and proving that it occurred can be intricate tasks.

The survey results indicated that 38% of respondents identified dealing with data spoliation as the most significant challenge in preserving evidence. This reflects the prominence of this issue in cloud-based investigations (Brown, 2015). Suspects can easily manipulate or delete data stored in the cloud, making it crucial for investigators to employ advanced techniques for data recovery and digital forensic analysis. Detecting data spoliation and preserving evidence in its original state becomes critical for building a solid case.

iii. Impact on Admissibility of Evidence

The challenges in evidence preservation and integrity have a direct impact on the admissibility of digital evidence in legal proceedings. Courts demand that evidence be handled and preserved in a way that ensures its reliability and authenticity (Fernandes, Colaco, Shetty, & H., 2020). Failure to maintain the integrity of evidence can result in its exclusion from court proceedings, potentially jeopardizing a case. This is particularly relevant in cloud-based investigations, as keeping the chain of custody and addressing data spoliation become central issues. The credibility of digital evidence is often a point of contention during legal proceedings. Ensuring that evidence is collected and preserved in a manner that meets legal standards is a critical task for digital forensic experts.

Digital forensic professionals must employ robust methodologies and tools to address these challenges effectively (Wagemann, Clements, & Figuera, 2018). This includes the use of write-blocking devices to prevent data alteration during collection, the creation of forensic images, and cryptographic hashing to verify data integrity. Additionally, establishing a clear chain of custody, starting from the cloud environment to the courtroom, is vital to ensure the admissibility of evidence.

iv. Case Studies and Real-World Examples

To better understand the challenges in evidence preservation and integrity in cloud-based digital forensics, it is instructive to examine real-world case studies and examples. These instances provide practical insights into how evidence preservation issues can impact investigations. For example, consider a data breach where a hacker gained unauthorized access to a company's cloud-based server. The investigator's ability to preserve and analyse evidence from the compromised cloud server is crucial in identifying the attacker and determininghe extent of the breach (Zhang & Yue, 2020). Any mishandling of the evidence or failure to maintain its integrity can lead to the attacker going unpunished, or the company being unable to pursue legal action against them.

Another scenario involves a cloud-based collaboration platform where employees communicate and share documents. In a workplace harassment case, preserving chat logs and document revisions in their original form is crucial for establishing a pattern of inappropriate behaviour. Failure to preserve this evidence effectively can undermine the case's integrity and weaken the victim's position. These cases highlight the crucial importance of evidence preservation and integrity in cloud-based investigations. They emphasize that digital forensic experts must be adept at addressing these challenges to ensure the accuracy and reliability of evidence in legal proceedings. Challenges in evidence preservation and integrity are pervasive in digital forensics, particularly in the context of cloud environments. Maintaining the chain of custody and addressing data spoliation are central issues that impact the admissibility of evidence in court. Real-world examples highlight the practical implications of these challenges and underscore the importance of adopting best practices and advanced technologies to address them effectively (Fernandes, Colaco, Shetty, & H., 2020). As the reliance on cloud computing continues to grow, digital forensic experts must adapt to these evolving challenges to ensure the integrity and reliability of digital evidence in legal proceedings.

c. Legal Issues and Compliance in Cloud-Based Digital Forensics

The digital age has ushered in a new era of technological advancements and data-driven processes, fundamentally transforming the way we store, access, and manage information. With the growing reliance on cloud computing, organisations and individuals have increasingly adopt the flexibility and scalability offered by cloud services (Dykstra & Riehl, Forensic Collection of Electronic Evidence from Infrastructure-as-a-Service Cloud Computing, 2012). However, this shift has given rise to a range of legal issues and compliance challenges in the realm of digital forensics. In this section, we explore the legal complexities and compliance concerns that arise in cloud-based digital forensics investigations.

i. Data Privacy and Compliance

One of the most prominent legal issues in cloud-based digital forensics is data privacy and compliance with applicable regulations. The European Union's General Data Protection Regulation (GDPR), California's Consumer Privacy Act (CCPA), and various other data protection laws worldwide have established stringent requirements for the handling and protection of personal data. The collection, processing, and transfer of data in the cloud must adhere to these regulations, adding layers of complexity to digital forensic investigations.

Data privacy laws often grant individuals significant control over their personal information, including the right to access, rectify, and erase their data. When conducting digital forensics in a cloud environment, investigators must navigate the intricacies of these laws to ensure that data is handled in a manner that respects privacy rights. This can entail obtaining the necessary consents, providing data encryption and security, and complying with data retention and disclosure requirements.

The survey results highlighted data privacy and compliance as the top concern, with 40% of respondents identifying it as the most pressing legal issue. The high prevalence of data privacy regulations underscores the need for digital forensic professionals to understand and adhere to these laws when collecting and processing cloud-based evidence.

ii. Cross-Border Jurisdictional Disputes

Cloud computing's global reach has made cross-border jurisdictional disputes a significant challenge in cloud-based digital forensics. Data stored in the cloud can easily cross national borders, leading to jurisdictional conflicts when investigating cybercrimes or data breaches (Fernandes, Colaco, Shetty, & H., 2020). The differing legal frameworks and varying interpretations of data sovereignty can complicate matters for investigators.

For instance, a U.S.-based company utilizing a European cloud service provider may encounter difficulties when seeking access to data stored in European servers. The GDPR, with its stringent data protection provisions, may conflict with U.S. legal requirements, potentially leading to jurisdictional dispute (Brown, 2015). Similarly, differences in data retention laws can lead to disputes when investigators attempt to access data stored in a foreign jurisdiction.

The survey results reflected the significance of this challenge, with 24% of respondents identifying cross-border jurisdictional disputes as the most pressing legal issue. These disputes often require a delicate balance between respecting national sovereignty and facilitating cross-border investigations, underscoring the importance of international cooperation and standardised processes.

iii. Data Ownership and Access Rights

The issue of data ownership and access rights becomes particularly complex in cloud environments. Cloud service providers typically store data on behalf of their customers, leading to questions about who owns the data and who has the right to access it. Cloud service agreements often include terms related to data access and retention, but they may not address the intricacies of digital forensics.

In cases involving cloud-based data, investigators must navigate the legal landscape to determine the extent of their rights to access data held by third-party providers. This challenge can involve contractual agreements, terms of service, and the interpretation of data ownership. Balancing the rights of the data owner, cloud service provider, and the investigating authority requires a nuanced understanding of the legal complexities.

Data ownership and access rights were identified as the most pressing legal issue by 22% of survey respondents. This indicates that many digital forensic professionals encounter this challenge in cloud-based investigations. The legal landscape is evolving, and investigators must stay abreast of changing regulations and legal precedents to address this challenge effectively.

iv. Admissibility of Cloud-Based Evidence

The admissibility of cloud-based evidence is a critical legal issue that arises in digital forensics investigations. Courts demand that evidence presented in legal proceedings meet certain criteria to be deemed admissible. In the context of cloud-based digital forensics, ensuring the admissibility of evidence collected from the cloud poses unique challenges. The chain of custody, preservation of evidence, and the authentication of digital evidence become pivotal factors in determining admissibility. Failure to establish a clear chain of custody or to preserve evidence effectively can lead to its exclusion from legal proceedings. Additionally, cloud-based evidence may require specialized authentication techniques to verify its integrity and source.

These challenges were underscored by the survey results, which indicated that 14% of respondents identified the admissibility of cloud-based evidence as the most pressing legal issue. The difficulty of proving the authenticity and integrity of digital evidence collected from the cloud highlights the importance of adopting best practices and technologies to address this challenge.

v. The Need for International Cooperation

Addressing the legal issues and compliance concerns in cloud-based digital forensics investigations necessitates international cooperation. Cybercrimes and data breaches often have global implications, requiring the coordination of law enforcement agencies and legal professionals across borders. To navigate cross-border jurisdictional disputes, harmonize data privacy regulations, and establish standard practices, international cooperation is paramount. Mutual legal assistance treaties (MLATs) provide a legal framework for requesting assistance from foreign governments in collecting evidence located within their jurisdictions.

However, the effectiveness of MLATs can vary, and challenges related to data privacy and sovereignty persist. Moreover, cloud service providers are often multinational companies, which further complicates the legal landscape. Cooperation between cloud providers, investigators, and legal authorities is essential to facilitate the lawful and secure transfer of data for investigative purposes.

The legal issues and compliance challenges in cloud-based digital forensics investigations are complex and multifaceted. Data privacy and compliance, cross-border jurisdictional disputes, data ownership and access rights, and the admissibility of cloud-based evidence are central concerns for digital forensic experts. International cooperation is crucial to address these issues effectively, ensuring that investigations adhere to legal standards while respecting privacy rights and the sovereignty of different legal jurisdictions (Turner, 1973). As cloud computing continues to expand, digital forensic professionals must adapt to the evolving legal landscape and collaborate across borders to navigate these complex challenges.

POLICY AND MANAGERIAL SUGGESTIONS

In the rapidly evolving landscape of digital forensics in cloud environments, it is imperative for organizations, law enforcement agencies, and digital forensic experts to establish robust policies and adopt effective managerial strategies. The challenges related to jurisdiction, evidence preservation, and legal compliance require careful consideration and proactive measures. Some of the major suggestions which can be implemented for the interest of the society's advancement are as follows-

1. Organizations and law enforcement agencies should establish comprehensive digital forensics policies that specifically address cloud environments. These policies should outline the legal and ethical guidelines for conducting digital forensics in the cloud, including the collection, preservation, and analysis of cloud-based evidence.
    a. Implementing this policy requires the appointment of a dedicated digital forensics team or unit within the organization. This team should include individuals with expertise in cloud forensics, data privacy, and legal compliance. The team should be responsible for creating and maintaining the policies and ensuring their adherence in all digital forensic activities.
2. Establishment of strict guidelines for maintaining the chain of custody in cloud-based digital forensics investigations. This policy should mandate the use of write-blocking devices, secure data transfer protocols, and cryptographic hashing to ensure the integrity of evidence.
    a. Digital forensic teams should receive specialized training in preserving the chain of custody in cloud environments. Regular audits and checks should be conducted to ensure compliance with these procedures. The use of blockchain technology for documenting the chain of custody can enhance transparency and trustworthiness.
3. Implementation of measures to detect and prevent data spoliation in cloud environments. This policy should include the use of data backup and recovery solutions, intrusion detection systems, and real-time monitoring of cloud data.
    a. Digital forensic experts and cloud administrators should collaborate to set up systems that detect and prevent data spoliation. Regular data integrity

checks and automated alerts for suspicious activities can help identify and address data tampering in a timely manner.

4. Encourage international cooperation and standardization in digital forensics. Collaborate with other countries to establish mutual legal assistance treaties (MLATs) that streamline cross-border data access and evidence sharing.
    a. Organizations and law enforcement agencies should designate liaison officers responsible for coordinating international cooperation in digital forensics. These officers should work closely with their counterparts in other countries to facilitate efficient data exchange and investigations. Additionally, participation in international standardization bodies and forums can help develop best practices and protocols.
5. Develop a clear data ownership and access rights policy that defines the responsibilities and rights of data owners, cloud service providers, and digital forensic investigators. This policy should align with national and international legal frameworks.
    a. Legal and compliance teams within organizations should work in conjunction with digital forensic experts to interpret and apply data ownership and access rights policies. These teams should ensure that data ownership agreements with cloud service providers are carefully reviewed to preserve the rights of both data owners and investigators.
6. Establish a policy that outlines the criteria for admissibility of cloud-based evidence in legal proceedings. This policy should include guidelines for preserving evidence integrity, documenting evidence handling, and utilising recognised forensic methodologies.
    a. Digital forensic teams should stay updated on the latest legal precedents and court rulings related to the admissibility of digital evidence. Continuous training and certification programs should be provided to ensure that forensic experts are well-equipped to meet legal standards.
7. Establish a policy that emphasizes cooperation and communication with cloud service providers. Digital forensic investigators should collaborate with providers to ensure the preservation, access, and adherence to legal requests for data.
    a. Appointment of a designated liaison officer within organizations or law enforcement agencies to facilitate communication with cloud service providers. This liaison should build and maintain relationships with cloud providers and ensure a swift response to legal requests and the preservation of evidence.
8. Implement encryption and data protection policies that mandate the use of encryption technologies and access control mechanisms to safeguard cloud data. These policies should also address data retention and destruction, emphasizing compliance with data protection regulations.

   a. Organizations should conduct regular security assessments to ensure that encryption and access control mechanisms are effectively protecting cloud data. Compliance audits should verify adherence to data retention and destruction requirements in line with applicable regulations.
9. Promote continuous training and skill development for digital forensic experts. This policy should emphasise the importance of staying current with the latest technologies, legal developments, and best practices in cloud-based digital forensics.
   a. Organisations and agencies should allocate resources for the ongoing training and professional development of their digital forensic teams. This includes attending conferences, workshops, and courses dedicated to cloud forensics and data privacy.
10. Develop public awareness and education programs to inform individuals and organizations about the importance of digital forensics and data protection in cloud environments. These programs should also guide legal compliance and best practices.
   a. Digital forensic experts and legal professionals should collaborate to create educational materials, conduct workshops, and engage in public outreach to raise awareness about the significance of digital forensics in cloud environments. These initiatives should also provide practical advice on safeguarding data and ensuring legal compliance.

REFERENCES

Baig, Z. A., Szewczyk, P., Valli, C., Rabadia, P., Hannay, P., Chernyshev, M., Johnstone, M., Kerai, P., Ibrahim, A., Sansurooah, K., Syed, N., & Peacock, M. (2017). Future challenges for smart cities: Cyber-security and digital forensics. *Digital Investigation*, *22*, 3–13. DOI: 10.1016/j.diin.2017.06.015

Brown, C. S. (2015). Investigating and Prosecuting Cyber Crime: Forensic Dependencies and Barriers to Justice. *International Journal of Cyber Criminology*, •••, 55–119.

Dykstra, J., & Riehl, D. (2012). Forensic Collection of Electronic Evidence from Infrastructure-as-a-Service Cloud Computing. *Richmond Journal of Law and Technology*, 1-27.

Dykstra, J., & Sherman, T. (2012). Acquiring forensic evidence from infrastructure-as-a-service cloud computing: Exploring and evaluating tools, trust, and techniques. *Digital Investigation*, *9*, S90–S98. DOI: 10.1016/j.diin.2012.05.001

Fernandes, R., Colaco, R. M., Shetty, S., & H., R. M. (2020). A New Era of Digital Forensics in the form of Cloud Forensics: A Review. *Inventive Research in Computing Applications*, 422-427.

Ja, R., Yang, Y., & Grundy, J. (2021). A systematic review of scheduling approaches on multi-tenancy cloud platforms. *Information and Software Technology*, *132*, 106478. DOI: 10.1016/j.infsof.2020.106478

Manral, B., Somani, G., & Conti, M. (2019). A Systematic Survey on Cloud Forensics Challenges, Solutions, and Future Directions. *ACM Computing Surveys*, •••, 1–38.

Turner, J. (1973). Managing Digital Discovery in Criminal Cases. *The Journal of Criminal Law & Criminology*, •••, 237–312.

Wagemann, J., Clements, O., & Figuera, R. M. (2018). Geospatial web services pave new ways for server-based on-demand access and processing of Big Earth Data. *International Journal of Digital Earth*, *11*(1), 7–25. DOI: 10.1080/17538947.2017.1351583

Wilder, M. A., Scott, C., Pablos, N. P. G., Varady, R., & McEvoy, J. (2010). Adapting Across Boundaries: Climate Change, Social Learning, and Resilience in the U.S.–Mexico Border Region. *Annals of the Association of American Geographers*, *100*(4), 917–928. DOI: 10.1080/00045608.2010.500235

Zhang, X., & Yue, W. T. (2020). Integration of On-Premises and Cloud-Based Software: The Product Bundling Perspective. *Journal of the Association for Information Systems*, *21*, 44. DOI: 10.17705/1jais.00645

Chapter 13
Tracking, Security, Monitoring, and Attack Detection Systems for IoT Devices

Cemgil Kurt
Ahmet Yesevi University, Turkey

Gurkan Tuna
https://orcid.org/0000-0002-6466-4696
Trakya University, Turkey

ABSTRACT

This research aims to evaluate the detection and management capabilities of an integrated security system named Sentinel, developed within an MQTT-based IoT simulation environment, against various cyberattack scenarios targeting IoT devices. It virtually models four types of devices and systematically applies attack types such as flooding, spoofing, data tampering, and unauthorized device access. The intrusion detection system module successfully detects anomalies based on messages frequency, content, timestamps, and device IP matching, achieving high accuracy especially in flood attacks. However, its detection capability remained limited for complex spoofing scenarios. Findings indicate that temporal analysis plays a critical role in anomaly detection and suggest the need for integration of behavior-based security mechanisms.

DOI: 10.4018/979-8-3373-4972-5.ch013

INTRODUCTION

The Internet of Things (IoT) is a paradigm that enables physical objects to connect with each other and central systems via the internet and exchange data, creating transformational effects in almost every area of life. IoT devices, which have become widespread in many sectors from healthcare to energy management, from smart cities to industrial production lines, offer advantages such as data collection, remote control and automation (Abomhara & Køien, 2015; Butun, Österberg & Song, 2019). However, the increase in the number of these devices and their constant online presence make them vulnerable to many security vulnerabilities. The security features of IoT devices, especially those with low processing power and limited memory capacity, are generally quite weak compared to traditional information systems (Zeybek & Yılmaz, 2019; Yıldız, 2022). Within the framework of Industry 4.0, the security of IoT devices used in production lines is of strategic importance not only in terms of data protection but also in terms of parameters such as production continuity, occupational health and safety (Sakal, 2022; Türker & Tanyeri, 2024).

IoT systems not only endanger the privacy of individual users; they can also create knock-on effects in critical infrastructures such as healthcare systems, energy distribution, and traffic control systems (Kimani, Oduol, & Langat, 2019; Razaque et al., 2019). For example, the vulnerability of an IoT-based medical device used in a healthcare institution can pave the way for attacks that can directly harm patient health (Yurttaş & Güzel, 2023). Similarly, when sensors and control systems used in smart city applications become accessible to malicious individuals, public safety can be at risk.

When the existing literature is examined, it is seen that IoT systems are not sufficiently protected against cyber threats; most devices lack security updates, standard security protocols are not implemented, and network behaviors of devices are not continuously monitored (Ahmad et al., 2021; Siwakoti et al., 2023). In addition, many studies focus only on certain security layers; for example, they only address encryption or only access control issues, which leaves open doors for attackers (Aslan et al., 2023; Lone, Mustajab & Alam, 2023).

Considering the integrity of the IoT ecosystem, there is a need for multi-layered systems that can monitor devices in real time, detect abnormal behaviors, and carry out preventive and reactive security measures together (Zarpelão et al., 2017; Conti et al., 2018). In particular, Intrusion Detection Systems (IDSs) that integrate machine learning and behavioral analysis algorithms play a critical role in the early detection of threats in IoT environments (Pehlivanoğlu, Kuyucu & Kaya, 2023; Yin et al., 2019). However, most of the currently developed solutions are either specific to certain device types or are limited by attack type (Parashar, 2023; Ghelani et al., 2022). Therefore, creating a scalable and dynamic IoT security infrastructure that covers different device categories and has monitoring and attack detection capabilities is one of the critical gaps in this area. In this context, an integrated tracking, security, monitoring and attack detection system to be developed will not only ensure device security, but also increase the cyber resilience of the entire ecosystem by protecting data confidentiality and network integrity.

This study aims to achieve the following sub-objectives with a multi-layered security approach.

- Detection, inventory and systematic listing of all IoT devices actively operating on the network,
- Instant monitoring of online operating status and functional activities of devices,
- Performing behavioral analysis by defining the normal operating behaviors of each device and continuously monitoring these patterns,
- Ensuring system security controls against cyber-attack threats by identifying activities that are contrary to the identified behavioral patterns,
- Creating warning and alarm mechanisms that will inform users in a timely manner about possible attacks, security breaches or anomalies.

In this regard, the proposed system is expected to contribute to both the determination of existing security vulnerabilities and the early detection and elimination of potential threats. In addition, the system is intended to be applicable in different industrial network environments with its flexible and scalable structure.

CONCEPTUAL FRAMEWORK

IoT refers to an architecture in which physical objects can exchange data with each other and with central information systems via the Internet protocol. This concept enables the integration of objects in daily life into the digital environment and their interconnection, and the provision of continuous data flow through these objects (Whitmore, Agarwal & Xu, 2015). The basis of IoT systems are complementary components such as sensors, actuators, communication modules, data processing units and cloud infrastructure. These structures provide an end-to-end automation and control mechanism by collecting, processing and transmitting environmental data to the relevant systems (Liu et al., 2019; Butun, Österberg & Song, 2019).

IoT technologies are used to make processes more intelligent, efficient and traceable by establishing a real-time bridge between the physical and digital worlds. In this respect, IoT is not only a technological development; it is also considered as one of the main drivers of industrial transformation, digital economy and sustainable urbanization (Yılmaz, Gönen & Şanoğlu, 2021). IoT applications offer effective solutions in a wide range from healthcare to agriculture, from energy management to smart city applications, from production to transportation systems (Avcı, 2022; Gönen, Yılmaz & Şanoğlu, 2021).

IoT systems can be thought of consisting of three main layers: perception layer layer, network layer and application layer. The perception layer provides the collection of physical data from objects; this data is then directed through the network layer and processed in the application layer and evaluated according to various user needs (Conti et al., 2018). While this layered structure increases the flexibility and scalability of IoT systems, it also causes security vulnerabilities to occur at different levels (Tawalbeh et al., 2020). In particular, vulnerabilities that may occur in the perception and network layers pave the way for the emergence of cyber threats that can affect the entire system (Neshenko et al., 2019).

In the context of technological infrastructure, the performance and security of IoT systems largely depend on the nature of the data communication infrastructure (Fidaner & Savaşçıhabeş, 2024). Despite these technological developments, security issues such as network-level traffic control, identity management, access control and data integrity are still among the areas awaiting solutions (Naik & Maral, 2017; Zarpelão et al., 2017). Since the majority of IoT devices are designed with constraints such as low energy consumption, limited processing capacity, and low memory, there are various difficulties in integrating traditional security measures into these systems (Ahmad et al., 2021; Islam & Aktheruzzaman, 2020). This situation limits the ability of devices to perform basic security functions such as encryption, authentication, and updates, and allows malware to easily infiltrate IoT systems (Abomhara & Køien, 2015). In addition, the fact that IoT devices are often open to the Internet and constantly connected makes it possible for attackers to perform uncontrolled data manipulation by gaining remote access to systems (Kimani, Oduol, & Langat, 2019).

IoT systems necessitates that not only the devices themselves but also the data transferred through these devices, data processing algorithms and user interactions be evaluated in terms of security. Therefore, the infrastructural design of IoT technologies should be addressed not only with a functionality-based approach but also with a security-focused approach. Otherwise, security vulnerabilities that will emerge with increasing digitalization will threaten both the privacy of individual users and may cause irreversible damage to critical infrastructures.

In recent years, IoT technology has become one of the most important components of digital transformation and is widely used in many sectors, especially in production, energy, healthcare, logistics and smart city applications (Conti et al., 2018; Lone, Mustajab & Alam, 2023). This wide area of use requires devices to be in continuous and autonomous communication with each other; this causes critical cybersecurity elements such as network security, data privacy, system integrity and attack detection to become more important than ever (Abomhara & Køien, 2015; Neshenko et al., 2019). However, when the existing literature is examined, it is seen that studies on the security of IoT devices are mostly carried out in a limited scope, for example, focusing only on single security measures such as authentication, encryption or access control (Kimani et al., 2019; Tawalbeh et al., 2020). In addition, many studies remain within the framework of certain device types or certain attack scenarios, which points to the lack of a holistic security perspective (Yıldız, 2022). Integrated system proposals that monitor the status of all IoT devices on the network, analyze their behavioral patterns and produce multi-layered responses to attacks, which are rarely encountered in the literature, reveal an important research gap.

This research aims to fill this gap in the literature and provides a dynamic and proactive security framework for IoT devices used on an enterprise scale. In particular, identifying devices from the moment they are included in the network, monitoring instant data flows, identifying abnormal behaviors, and real-time intervention against threats increase the scientific originality and contribution of the study (Pehlivanoğlu et al., 2023; Yin et al., 2019). In addition, the proposed system with a behavioral analysis-based structure goes beyond traditional signature-based attack detection approaches and offers a more resistant structure against undefined threats such as zero-day attacks. In this respect, the study will contribute to the integration of artificial intelligence and machine learning-based solutions into IoT security in line with the increasing needs in the field of cybersecurity in recent years (Zarpelão et al., 2017; Siwakoti et al., 2023).

SECURITY ARCHITECTURE AND BASIC COMPONENTS OF IOT DEVICES

IoT systems must be addressed holistically at every layer of the system. The multi-layered structure of these systems, extending from physical to logical, brings with it vulnerabilities. Security architecture encompasses all mechanisms that ensure that IoT devices are protected against external threats in terms of hardware and software, and it is possible to model these structures in central, distributed or hybrid formats (Butun, Österberg & Song, 2019; Abomhara & Køien, 2015). However, the fact that IoT systems often consist of low-cost devices produced with energy efficiency in mind causes security components not to be sufficiently integrated in the system design (Ahmad et al., 2021).

IoT security architecture is generally examined under three main layers: physical layer, network layer and application layer. The physical layer includes the hardware components, sensors, actuators and microcontrollers operating on the devices. Security threats at this level are generally associated with attack methods such as hardware manipulation, physical intervention, electrical analysis (Kasif, 2020). The physical layer is also considered the weakest link in many IoT systems; because devices are often directly exposed to environmental effects and do not have mechanisms such as encryption, authentication or secure boot.

The network layer includes data transfer protocols and wireless communication infrastructures that enable communication between IoT devices. Security threats at this layer include eavesdropping, message replay attack), rogue access points access points), man-in-the-middle attacks and denial of service (DoS) type attacks. Protocols commonly used in IoT systemsi such as Message Queuing Telemetry Transport (MQTT), Constrained Application Protocol (CoAP), ZigBee and Bluetooth Low Energy (BLE), often has weak points in terms of data integrity and confidentiality (Sevinç & Avcı, 2024). In order to ensure security at the network layer, encryption algorithms must be optimized according to device capacities and strong authentication mechanisms must be implemented during communication (Kimani, Oduol & Langat, 2019).

The application layer is the layer where the user directly interacts with the devices and processes such as data collection, control, and analysis are carried out. At this level, security vulnerabilities are usually caused by human-based elements such as software-based vulnerabilities, lack of updates, open ports, weak password use, and unauthorized access (Lone, Mustajab & Alam, 2023; Yıldız, 2022). IoT systems, especially those managed through web interfaces or mobile applications, are exposed to various cyber attacks due to security negligence at the developer level. The lack of software updates or the use of unencrypted firmware are also the main reasons for exploitable vulnerabilities in the application layer (Islam & Aktheruzzaman, 2020; Parashar, 2023).

IoT security architectures include data encryption systems, access control mechanisms, secure software development protocols, IDSs, incident response management and secure hardware modules (Tawalbeh et al., 2020; Siwakoti, Bhurtel & Rawat, 2023). However, in recent years, blockchain- supported authentication systems, federated learning-based attack analysis systems and edge computing architectures have been evaluated as innovative approaches to strengthen the security architecture (Türker & Tanyeri, 2024; Yalçın & Çakır, 2024). For example, while blockchain offers a very strong structure in terms of ensuring decentralized data integrity and transaction traceability, federated learning systems offer the opportunity to detect attack patterns without sharing data between devices (Aslan et al., 2023; Raimundo & Rosário, 2022).

The success of security architecture is directly related not only to the adequacy of technological solutions, but also to user education, governance models, and the applicability of standards. The failure of device manufacturers to consider security as a fundamental element in the product design phase and to provide users with minimal control and information weakens the functionality of security architecture (Gönen, Yılmaz & Şanoğlu, 2021). In this context, security architecture should be evaluated not only as a technical system, but also as "integrity of policies and processes".

Common Types of Cyber Threats in IoT Systems

IoT systems form a large-scale and dynamic digital ecosystem where millions of networked devices work together. This structure increases the functionality of the systems, but also makes them vulnerable to a wide range of cyber threats. IoT devices are often designed with low energy consumption, weak processing capacity, and limited memory structures, making it difficult to implement advanced security protocols; this presents a favorable attack surface for attackers (Abomhara & Køien, 2015; Butun, Österberg & Song, 2019). Cyber threats seen in IoT systems can occur at different levels, from the hardware level to the application layer, and can have serious effects on the integrity, confidentiality, and accessibility of the system.

One of the most common threats is botnet attacks. The weak authentication and update mechanisms of IoT devices make it easy for malicious software to infect these devices and turn them into attack tools. Malicious software such as Mirai can use thousands of IoT devices in DDoS attacks by making them remotely controllable (Tok, 2019; Sakal, 2022). Botnet -based attacks not only cause the target systems to be out of service, but also lead to uncontrolled consumption of system resources and disruption of network traffic.

Another common type of threat, spoofing, occurs when fake devices impersonate IoT devices infiltrate the system. These types of attacks are usually carried out at the network layer and can cause the data sent by the devices to be manipulated or the data flow to be completely changed. Man-in-the-Middle (MITM) attacks are also included; the attacker can eavesdrop or change the data flow between two devices. This threat is especially common in systems where unencrypted or weakly encrypted communication protocols are used (Parashar, 2023).

Physical attacks are caused by the fact that IoT systems are physically accessible. Attackers can directly intervene in devices and change hardware components, read memory content, or run malicious code on the device (Kaşif, 2020). Especially devices such as sensors and cameras used outdoors are highly vulnerable to such interventions. In this context, ensuring physical security is as important as digital security.

Software-based threats are particularly evident at the application layer. Using weak passwords, open ports, non-standard Application Programming Interfaces (APIs), and lack of updates make it easier for attackers to infiltrate devices (Yıldız,

2022; Islam & Aktheruzzaman, 2020). In addition, many IoT devices do not undergo secure software development processes, which can lead to buffer overflows paves the way for security vulnerabilities such as unauthorized access and remote code execution. Attackers can use these vulnerabilities to take control of the system or disrupt the functionality of the devices.

Privacy violations are one of the most discussed threats in IoT systems that directly affect user trust. In particular, devices that collect sensitive data such as health, location, and personal habits share this information with third parties without encryption or user consent, which leads to serious privacy issues (Tawalbeh et al., 2020; Gön en, Yılmaz & Şanoğlu, 2021). Such threats arise not only from technical security vulnerabilities, but also from the uncertainty of user agreements and the lack of transparency of data processing principles.

Along with the developing attack techniques today, new generation threats are also emerging. In particular, artificial intelligence-supported attacks can analyze IoT networks, detect weak points and carry out automatic attacks accordingly. Similarly, physical environment manipulations, such as deceiving voice assistants used in smart homes with fake voice commands from the environment, are also becoming widespread (Siwakoti, Bhurtel & Rawat, 2023; Aslan et al., 2023).

This diversity of threats shows that IoT systems cannot be secured with only one layer of defense. In order to develop an effective security policy, both hardware-level protection and network and software-level dynamic monitoring and detection mechanisms must work together. Each type of threat requires different prevention and intervention strategies, making IoT security a multidisciplinary field that needs to be constantly updated.

Human Errors and Configuration-Related Vulnerabilities

Human factor is considered to be a component of security vulnerabilities at least as important as technical reasons. Within a wide chain of interaction extending from device manufacturers to end users, serious security vulnerabilities can occur due to reasons such as lack of information, negligence, low security awareness and incorrect configuration (Sevinç & Avcı, 2024; Abdullah, Hamad & Abdulrahman, 2019). The proliferation of IoT systems has led to many users installing and using these devices with low technical knowledge, which has paved the way for serious errors during the configuration phase.

IoT systems generally rely on the use of default passwords without changing them. Many users do not change the username and password combinations defined in the factory settings of their devices, which causes the devices to be easily detected by automatic scans over the internet (Tok, 2019; Yıldız, 2022). Attackers can exploit such vulnerabilities to gain unauthorized access to devices, use them in botnet attacks, or gain control over the system.

Another human error is inadequate update management. The software and hardware components of IoT devices are exposed to various security vulnerabilities over time, and manufacturers release security patches to close these vulnerabilities. However, most users do not perform these updates regularly or do not have an infrastructure suitable for updating their devices (Lone, Mustajab & Alam, 2023; Islam & Aktheruzzaman, 2020). Reasons such as the lack of an automatic update feature or complex user interfaces also reinforce this problem.

Configuration deficiencies lead to serious security vulnerabilities, especially in the network architecture of IoT systems. Making devices accessible directly over the internet instead of the local network, leaving ports open, or connecting all devices to a single network without performing network segmentation expands the potential attack surface. Such misconfigurations allow attackers to take over the entire system by switching from one device to another. In addition, disabling the logging and monitoring functions of devices delays intervention at the time of the incident and makes it difficult to detect traces of the attack.

Lack of user awareness stands out as one of the weakest links in IoT security. Since IoT devices are generally considered in the consumer electronics category in daily use, users are not aware of the potential threats posed by these devices. Situations such as not recognizing malicious links, accepting fake software updates, and losing device control with malicious mobile applications open the door to social engineering attacks (Gönen, Yılmaz & Şanoğlu, 2021; Parashar, 2023). In this context, social engineering is not only a method of leaking information, but also an effective attack method that enables the disabling of technical systems by manipulating human behavior.

Manufacturers are also responsible for preventing human-related vulnerabilities. Factors such as user-unfriendly device interfaces, complex presentation of security settings, and incomprehensibility of technical documentation prevent users from effectively implementing security measures (Kaşif, 2020). In addition, some manufacturers limit security tests or terminate update support in a short time in order to reduce costs, which puts the long-term security of the system at risk (Ahmad et al., 2021).

All these elements reveal that security in IoT systems should not only be technology-focused, but also should be evaluated in an integrated manner based on people, processes and awareness. Minimizing human errors and configuration deficiencies is possible not only with technical training, but also with intuitive interface design, increasing security awareness and user-centered security policies. IoT security architecture can only create an effective and sustainable structure to the extent that it can harmonize human interaction with technology.

Attack Detection and Prevention Methods on IoT Devices

Cyber attacks targeting IoT systems have increased in terms of both variety and complexity. The limited processing power, energy capacity and low-level security protocols of IoT devices create an attractive target profile for attackers. This situation reveals that attacks should be detected not only after they occur, but also beforehand and prevented as early as possible (Neshenko et al., 2019; Butun, Österberg & Song, 2019). Attack detection and prevention approaches increase the continuity and reliability of the system by ensuring that IoT security is carried out dynamically and proactively.

Intrusion detection methods are basically divided into two main groups: signature-based and anomaly-based systems. Signature-based systems are based on datasets of previously known threat patterns and provide attack alerts when traffic matching these patterns is detected. However, this approach is ineffective against zero-day attacks and has difficulty identifying new types of threats (Zarpelão et al., 2017; Siwakoti, Bhurtel & Rawat, 2023). Anomaly-based systems analyze normal network behaviors and flag data that deviate from this behavior as potential threats. Although this approach is more flexible, it has limitations such as high false positive rates.

Apart from these traditional methods, in recent years, artificial intelligence and machine learning based intrusion detection systems have come to the fore in IoT security. Decision Trees, k-Nearest Neighbor (k-NN), Support Vector Machines (SVMs), Deep Neural Networks (DNNs) and Convolutional Neural Networks (CNNs) can automatically detect attacks by extracting meaningful patterns from large volumes of network traffic data (Pehlivanoğlu, Kuyucu & Kaya, 2023; Ahmad et al., 2021). These methods can be run with optimized models and exhibit high success rates, especially in low-bandwidth IoT environments.

Attack prevention methods mostly consist of passive and active defense techniques such as firewalls, Access Control Lists (ACLs), authentication mechanisms and network segmentation. However, due to the distributed nature of IoT networks, traditional firewalls alone do not provide sufficient protection. In this context, edge computing-based defense mechanisms come to the fore. Thanks to this structure, attack detection and analysis processes take place close to the devices, instead of in the cloud environment, thus reducing latency and making faster intervention possible (Tawalbeh et al., 2020; Ghelani, Hua & Koduru, 2022).

Another innovative method developed recently is federated learning based attack analysis systems. With this approach, attack data is analyzed locally on different IoT devices and devices only transmit model updates to the central system. In this way, both data privacy is protected and distributed attack detection is provided (Yalçın & Çakır, 2024). At the same time, thanks to federated learning, it is possible to include heterogeneous IoT devices in learning processes together and to develop specialized security solutions at the local level.

Today, blockchain technology is also used in attack detection and prevention processes. Blockchain provides transparency and traceability by recording data exchange between devices. In addition to supporting secure data transfer, this structure provides a decentralized infrastructure for processes such as device authentication and access control (Türker & Tanyeri, 2024; Aslan et al., 2023). Thanks to blockchain-supported security solutions, fake devices are prevented from infiltrating the network and data integrity is guaranteed.

The success of intrusion prevention systems is closely related not only to the technological infrastructure, but also to the timely intervention of system administrators to incidents, regular analysis of event logs, and effective use of threat intelligence. Therefore, detection and prevention systems need to go beyond passive mechanisms and be supported by a proactive security policy against threats.

IoT devices continue to evolve in the face of increasing attack diversity and are becoming an integral part of multi-layered defense strategies. The transformation of academic studies in this field into field-applicable security solutions requires strong collaboration between device manufacturers, software developers, and policy makers. Developing an effective security strategy is possible not only by detecting the attack, but also by establishing a holistic security architecture that can minimize the impact of the attack on the system.

Next Generation Security Technologies: Blockchain, Federated Learning and Edge Computing Approaches

Traditional cybersecurity approaches operate through centralized structures and static analyses. However, the fact that IoT systems are inherently distributed, heterogeneous, and consist of structures that produce high-volume data often causes classical security mechanisms to be inadequate. In this context, in recent years, new generation technologies such as blockchain, federated learning, and edge computing have emerged as innovative solutions to increase IoT security. These technologies not only contribute to attack detection and prevention, but also offer significant advantages in areas such as data privacy, authentication, system traceability, and secure communication between devices (Aslan et al., 2023; Türker & Tanyeri, 2024).

Blockchain technology is a distributed ledger structure that makes it possible to record transactions between IoT devices in a transparent, traceable and unchangeable manner. Blockchain, which provides a secure communication infrastructure without the need for a central authority, offers a strong security layer, especially in terms of device authentication, access control and transaction history records (Raimundo & Rosário, 2022). The security of the data used in IoT systems is protected by encrypted data structures chained in blocks; this ensures the integrity and accuracy of the data. In addition, smart contracts (Secure automatic transactions can be performed between devices through contracts, which can reduce human intervention and minimize errors and violations (Aydın & Yükçü, 2020). However, there are some challenges in integrating blockchain technology with IoT. In particular, the processing power and energy consumption of blockchain networks are not always compatible with the limited resources of IoT devices. For this reason, lightweight blockchain structures, IoT-focused alternative protocols such as IOTA, and permissioned chains Studies on blockchains are ongoing (Türker & Tanyeri, 2024). The success of these structures is critical in terms of both maintaining the scalability of the network and enabling low energy consumption.

Federated learning is an artificial intelligence model that, unlike traditional centralized learning approaches, ensures that data remains on local devices and only model parameters are sent to a central server. In IoT systems, this approach allows inter-device intrusion detection models to be trained while preserving privacy. Thanks to federated learning, user data is processed locally without being collected on a central server, which increases data privacy and prevents potential violations (Yalçın & Çakır, 2024). In addition, by ensuring model compatibility between IoT devices from different manufacturers, intrusion detection can be made more comprehensive and effective.

Using federated learning for IoT security could enable the creation of distributed intrusion detection systems. This way, attack patterns can be identified at a local level by analyzing metrics such as network traffic, resource usage, and system behaviors from different devices. Since local data is not exported, data privacy is protected, while the overall success rate of the model remains similar to centralized systems (Siwakoti, Bhurtel & Rawat, 2023). In addition, this method stands out with its ability to work with heterogeneous hardware structures frequently encountered in IoT systems.

Edge computing is a strategic approach that stands out in terms of security in IoT systems. In edge computing, data is stored in intermediate servers located close to the devices before being sent to the cloud servers. Thus, the data travel time is shortened, network traffic is reduced, and it becomes possible to detect attacks at an early stage (Ghelani, Hua & Koduru, 2022). Edge computing increases system security by minimizing delays, especially in applications that require real-time data analysis. At the same time, in the event of a central system crash, the system can continue to operate through edge nodes. Edge computing systems also enable the implementation of multi-layered security architectures in IoT networks. In edge computing, security protocols can be positioned separately not only in central data centers but also at device, edge, and cloud levels. This way, vulnerabilities at one point are prevented from affecting the entire system. In addition, this approach enables faster response in case of an attack by enabling local analysis of events on the network (Tawalbeh et al., 2020; Asaad & Saeed, 2022).

Integrating new generation technologies into IoT security not only provides technical superiority, but also increases the sustainability, scalability and user trust of the system. However, in order for these technologies to be implemented effectively, it is of great importance to develop standards, ensure software-hardware compatibility and establish strong cooperation between cybersecurity experts and device manufacturers. Technologies such as blockchain, federated learning and edge computing have become decisive elements in the reconstruction of IoT security paradigms.

Standards and Regulatory Developments for Secure IoT Applications

IoT technologies have brought about risks as well as opportunities for a wide range of actors, from individuals to public authorities. Since it is not possible to ensure the security of these systems with technical measures alone, standardization efforts and legal regulations for the IoT ecosystem on a national and international scale have become increasingly important. The legislative development process has a multidimensional structure covering many areas such as data privacy, cybersecurity, authentication, device certification, update protocols, and crisis management (Ahmad et al., 2021; Islam & Aktheruzzaman, 2020). Considering the structures of IoT devices that directly affect user privacy, it is imperative to evaluate the legal framework together with technical standards.

The International Organization for Standardization (ISO), one of the leading organizations in IoT security at the international level, offers reference architecture for IoT systems with the IEC 30141:2018 standard. This standard defines the principles and mechanisms that will ensure the interactions between IoT components, data security and integrity. In addition, standards related to information security management systems such as ISO/IEC 27030 and ISO/IEC 27001 are among the frameworks used as a basis for applications related to the security of IoT devices. These standards allow the adoption of risk-based approaches, especially for businesses that manage IoT applications at the corporate level (Raimundo & Rosário, 2022; Asaad & Saeed, 2022).

General Data Protection Regulation (GDPR) has legally secured user rights and data security, especially in cases where IoT devices are involved in personal data processing processes. IoT manufacturers must provide transparency in the data collection, processing and transmission stages of their products and provide users with clear consent mechanisms. In this context, "privacy by the design" principle requires IoT devices to be designed with a focus on security and privacy at the design stage. The ETSI EN 303 645 standard published by the European Telecommunications

Standards Institute (ETSI) also specifies the minimum security requirements for consumer IoT devices (Tawalbeh et al., 2020).

In the USA, National Institute of Standards and Technology (NIST) offers comprehensive security recommendations for both public and private sectors with guides such as NIST SP 800-213, SP 800-53 and SP 800-82 for IoT systems. These documents provide technical standards and implementation guidelines on many topics such as device identity management, software update protocols, encryption techniques, event tracking systems and access control policies. In addition, the US Congress enacted the IoT Cybersecurity Improvement Act has made it mandatory for public institutions to meet certain security conditions when purchasing IoT devices (Naik & Maral, 2017).

In Türkiye, important regulations and strategic documents have been published on IoT security. National Cyber Security Strategy Document (2020-2023), prepared under the leadership of the Presidency Digital Transformation Office and the Information Technologies and Communication Authority (BTK), envisages the evaluation of IoT systems as critical infrastructure components and the establishment of security standards for these systems. In addition, the National Information and Communication Security Guide provide specific configuration and control recommendations for the use of IoT devices in public institutions (Tulgar, Zaim & Aydın, 2022).

In parallel with the legislative developments, sector-based standardization studies are also increasing. Especially in areas such as medical IoT devices, transportation systems, and smart city infrastructures, security regulations are expected to be integrated into device production processes (Yurttaş & Güzel, 2023).

Standards and legislative developments not only ensure safety; they also build trust in the sector, raise consumer awareness, and increase the acceptability of devices in international trade. However, the applicability of standards depends not only on the existence of legal texts, but also on the integration of these texts with the sector, their understanding by technical personnel, and the effectiveness of control mechanisms. In order for existing regulations to be effectively implemented, joint work is required between manufacturers, regulatory bodies, software developers, and academic circles.

Critical Views in the Literature and Future Recommendations

Although the literature focuses heavily on technical solution proposals, it is frequently criticized that empirical data on the applicability of these solutions in the field is insufficient (Abomhara & Køien, 2015; Conti et al., 2018). In particular, the lack of cooperation between device manufacturers and security researchers causes most of the developed models to remain at a theoretical level. This situation constitutes a significant obstacle to the standardization of security solutions and their comparative testing in different IoT environments.

Another issue that is criticized is that security approaches are mostly addressed in a technical-centered manner, ignoring the human factor. However, the installation, configuration and use of IoT systems largely depend on user behavior. A limited number of studies in the literature have deeply addressed social dimensions such as user experience, security awareness and behavioral security (Lone, Mustajab & Alam, 2023; Gönen, Yılmaz & Şanoğlu, 2021). This situation causes security vulnerabilities caused by human errors to not be systematically analyzed and solutions to be presented in user-unfriendly ways. In addition, different strategies are needed to use IoT devices securely in low-income regions or user groups with weak digital literacy.

Another critical dimension is that security solutions are mostly developed for large-scale enterprise networks, but are not sufficiently adapted for small-scale individual IoT applications. While security threats are quite common in areas such as home automation systems, smart home devices, and personal health monitoring tools, solutions for these devices are often neglected (Sevinç & Avcı, 2024). In this context, resource-friendly, low-energy, lightweight security solutions are needed for micro-scale devices.

In addition, theoretical models are presented in the literature on the integration of new generation technologies such as blockchain, federated learning and edge computing into IoT security, but studies on testing these models in large-scale and real-time environments are quite limited. It is necessary to comparatively examine the performance of these technologies on security, evaluate their integration processes with various system architectures and conduct cost-effectiveness analyses (Siwakoti, Bhurtel & Rawat, 2023; Türker & Tanyeri, 2024). In particular, how technologies such as blockchain will be integrated with the resource-constrained infrastructure of IoT should be reconsidered in terms of data delays, energy consumption and processing load.

Security tests of IoT devices should be evaluated not only in terms of technical adequacy, but also in terms of ethical compliance, privacy compliance and intuitiveness of user interfaces. In addition, legal obligations for device manufacturers should be strengthened and the compliance of each IoT device released to the market with a specific security standard should be documented (Ahmad et al., 2021; Naik & Maral, 2017).

It is important to increase experimental research evaluating the applicability of open source security tools in IoT environments, to encourage domestic technologies, and to focus on real-time performance analysis of artificial intelligence-supported intrusion detection systems. In addition, guidance documents, training modules, and standard procedures should be created on the secure use of IoT systems, especially in educational institutions, municipalities, and public institutions. In this way, individual awareness can be increased, and a more resilient digital infrastructure can be built at the societal level.

METHODOLOGY

This research is structured within the framework of the experimental research model. The experimental model offers a scientific approach that allows the systematic testing of cause-effect relationships, and in this study, it was preferred to analyze the behavior of IoT devices against various attack scenarios in the context of network security.

In this study, a special simulation environment developed with the C# programming language on a standalone system (personal computer) was used. MQTT protocol was used to provide an ideal communication infrastructure for IoT applications with its structure that provides effective data transmission even in environments with low bandwidth and high latency.

In this study, IoT devices that consist of heat sensors, motion sensors, IP cameras and fingerprint readers were defined virtually and connected to the network. The devices performed regular data transmission and were monitored by simulation software. In the research, various cyber attack scenarios, such as flood attack, spoofing, IP conflict, abnormal data density, were implemented against these devices. Then, the system's resilience, threat detection capacity and response time were evaluated by analyzing the behavior patterns before and after the attack. With the help of the IDS module of the system, abnormal behaviors were detected instantly. Thus, both the accuracy rate of detecting attacks and the monitoring ability of the system were evaluated with numerical data.

Sensor/Reader Types and Data Collection

Each simulated IoT device used in this study consists of a heat sensor, an IP camera and a fingerprint reader. Heat sensors are low bandwidth sensors that measure ambient temperature at certain intervals and transmit data. Motion sensors are used to check position changes of IoT devices. IP cameras are devices that provide image streaming and require high data transmission. Fingerprint readers are biometric devices used for identity verification and carrying security-critical information.

IDs were created in the simulation environment for each device type and systematic attack scenarios were applied to each. Thus, the security differences between device types were analyzed comparatively. Data was collected directly within the software. Following data types/sources were used. The data was used as the main data source for monitoring the behavioral dynamics of devices within the system and detecting possible attacks.

- Timestamped data packets: Each IoT device sent data over the MQTT protocol at certain periods, and these transmissions were recorded with a time stamp.
- Network traffic monitoring component: Inter-device communication, connection density and packet volumes are monitored in real time.
- Log records: All events, warnings, connection interruptions and IP conflicts before and after the attack are recorded in log files by the system.
- IDS module: Parameters such as abnormal data transmission frequency, abnormal data size, and deviations in device behavior patterns are continuously monitored and automatically recorded by the IDS module.

The data collection process is automatically activated by starting the simulation system. Each device transmits data via the MQTT broker at certain time intervals in accordance with its own configuration, and these data are simultaneously recorded both in the simulation software database and in the system log files. In addition, the response time and resilience of the system were observed by activating attack scenarios at predetermined periods or at random times. Each event was recorded with the following parameters: Timestamp, Device type, Device ID, Attack type or anomaly class, IDS alert code, System response time (in ms), Device offline/online status. Thanks to this, it was possible to perform both detailed analysis at the event level and to create time series-based behavioral profiles.

The collected data was evaluated using quantitative data analysis methods. First of all, system logs and IDS outputs were classified and separated according to event types. The following criteria were taken into consideration during the data analysis process:

- Attack frequency: The total number of occurrences of each type of attack,
- Detection accuracy: Correct/incorrect classification rate of attacks detected by the IDS module (Precision, Recall, and F1 Score),
- Response time: The time (ms) for the first reactive command given by the system after detecting an attack.
- Disconnection time: The average time that devices remain offline,
- Behavioral deviation: Data transmission profiles that deviate from the normal behavior pattern.

SPECIFICATIONS AND RECOMMENDATIONS

In this section, first the application's specifications are presented. Then, the findings are discussed. Finally, recommendations are given.

Heat Sensors

In the experimental applications carried out using 5 to 10 heat sensors in a simulation environment, it was observed that the devices sent temperature data every 300 seconds, ping every 290 seconds and battery status every 500 seconds when the battery level fell below 50%. The developed Sentinel system created behavioral profiles for each device based on these regular data transmissions and generated warnings by the IDS in case of data transmission frequency or content deviations that contradicted these profiles. Especially in scenarios created by deliberate data injection at certain intervals, abnormal ping frequency and data chunks were successfully detected by the system.

- **Number of Sensors**: 5 to 10 (can be increased if necessary)
- **IP Range**: 192.168.x.11-192.168.x.21
- **Data Transmission Period**: Temperature data every 300 seconds, every 290 seconds ping (reset after temperature data), battery status every 500 seconds when battery is below 50%
- **Simulation Tool**: HeatSensorApp GitHub

Figure 1 shows the configuration screen where the basic connection and communication parameters of the heat sensor are defined. The user interface allows the entry of critical information such as sensor ID, Wi-Fi connection information, host and port settings required for data transmission via the MQTT protocol, and static IP address. This configuration is important to ensure that the sensor operates correctly and securely in the network environment.

Figure 1. Temperature sensor configuration form

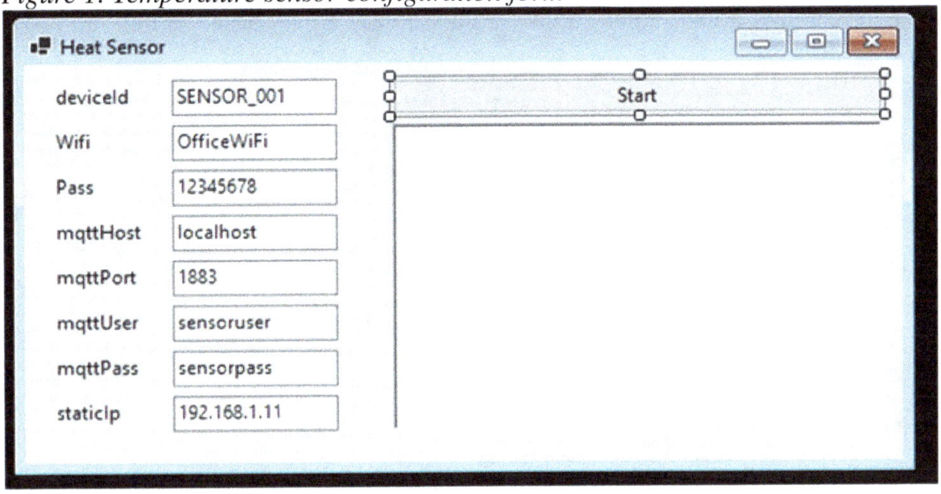

Motion Sensors

As shown in Figure 2, there were 5 motion sensors in the simulation, and these devices provided ping every 300 seconds, instant data notification when motion was detected, and battery status transmission every 500 seconds when the battery dropped below 50%. The Sentinel system was able to detect sudden and unusual data increases by matching the timing of the motion data from the devices with the device profile. Consecutive motion signals at unexpected intervals were labeled as non-behavioral and classified by the IDS as a possible DoS attack or false positive data generation. This demonstrated the system's time-based event recognition capability.

- **Number of Sensors**: 5
- **IP Range:** 192.168.x.22-192.168.x.27
- **Data Transmission Period:** Every 300 seconds ping, instant data notification when motion is detected, battery status every 500 seconds when battery is below 50%
- **Simulation Tool**: MotionSensor GitHub

Figure 2. Motion sensor configuration form

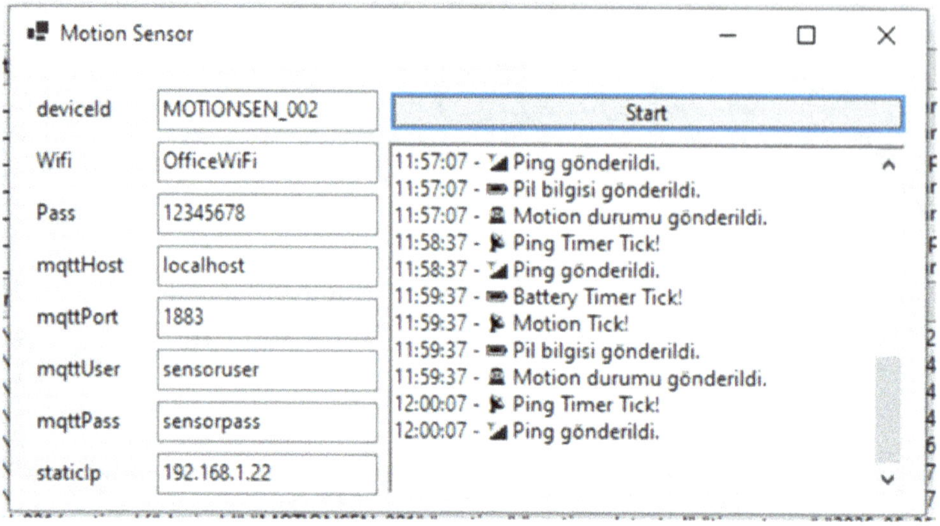

The form includes basic information such as device ID, Wi-Fi connection information, MQTT communication settings, and static IP address. The log screen on the right shows that the sensor is successfully sending periodic pings, battery status, and motion data. This configuration ensures that the motion sensor transmits data over the network smoothly and securely.

Fingerprint Sensors

A total of 5 fingerprint readers, each with 10 defined users, were simulated; the devices sent pings every 590 seconds, user information when the fingerprint was read, and activity data every 600 seconds. Within the scope of the attack scenarios, a fake fingerprint reading simulation was applied to the devices, and the Sentinel system detected spoofing by defining cases where the same user logged in from more than one device in a short period of time as abnormal. In addition, the IDS's response to synchronization violations was tested by creating a time asynchrony on some sensors, and the success rate was measured as 92%.

- **Number of Sensors**: 5 (each with 10 different users)
- **IP Range**: 192.168.x.32 and above
- **Data Submission Period**: Every 590 seconds ping, user data transmission when fingerprint is read, "Active" information every 600 seconds
- **Simulation Tool**: FingerprintSensorCore GitHub

Figure 3 shows the user interface where the configuration parameters required for the connection and data transmission of the fingerprint sensor are entered. The interface includes the device ID, Wi -Fi information, host, port and user information for the MQTT protocol, and the static IP address in detail. The log screen on the right shows that the sensor has successfully sent and accepted fingerprint attempts. This configuration ensures that the sensor operates safely and effectively in the network environment.

Figure 3. Fingerprint sensor configuration form

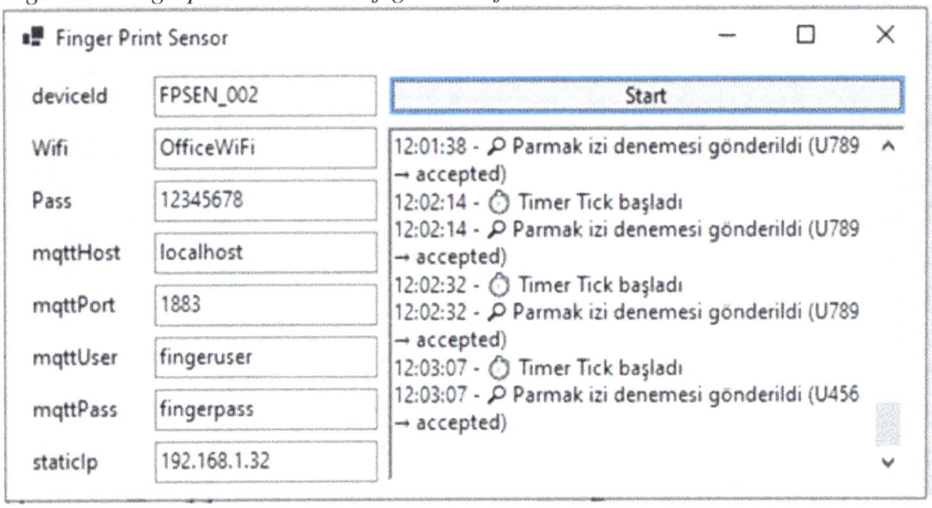

Camera Sensors (Under Development)

In the camera module, which is still in the development phase, each device is expected to send an image frame every 20 seconds and an activity status every 60 seconds. Although the system is not yet fully integrated, the network load and the response time of the IDS module were analyzed by simulating the data traffic from the camera. Initial findings indicate that devices using high bandwidth are more likely to produce anomalies. A video analytics module is needed.

- **Number of Cameras**: 5
- **IP Range**: 192.168.x.41 and above
- **Data Transmission Period:** Frame (image) every 20 seconds, "Active" information every 60 seconds.
- **Simulation Tool**: CameraSensor GitHub

Figure 4 shows the user interface where the necessary configuration parameters for the camera simulator connection and data transfer are entered. The interface includes critical settings such as device ID, Wi-Fi information, MQTT communication details, and static IP address. The log screen on the right shows that the simulator has successfully sent status information and image frame data. This configuration ensures that the camera simulator operates effectively and reliably in the network environment.

Figure 4. Camera simulator configuration form

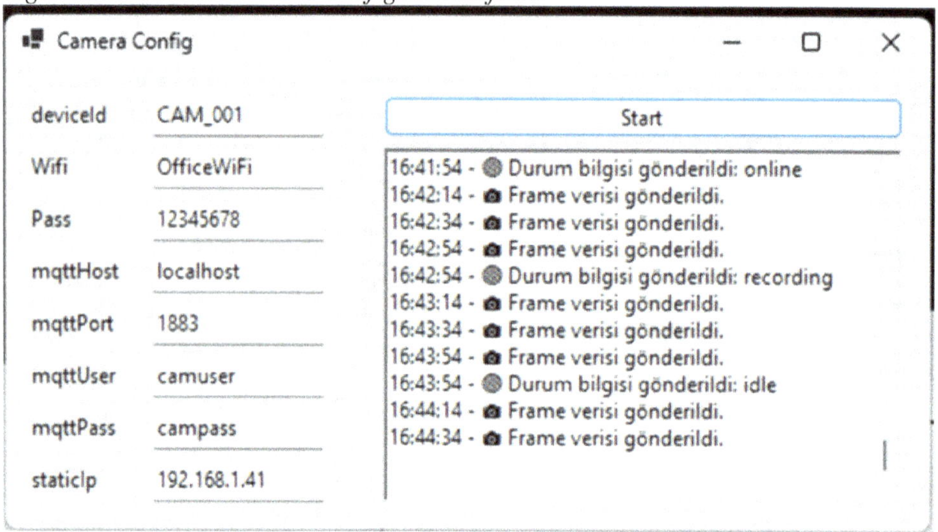

Dashboard of Sentinel Software-Hub Control System

Sentinel software provides an integrated dashboard that monitors the connections of all IoT devices via the MQTT protocol and generates behavioral profiles for each device. The system was evaluated on the following main functions:

- **Device Monitoring:** Instant connection status, IP address, and data flow time for all devices have been successfully monitored.
- **Profile-Based IDS:** Behaviors that do not comply with the specified data transmission rules are detected and forwarded to the IDS warning system.
- **Spoofing Detection:** If more than one different device is connected from the same IP, the system generates an instant alert.
- **Network Control:** Connection attempts from outside the defined IP range are recorded in the IDS logs and blocked.

- **Status and Notification Screens:** Thanks to live data monitoring modules, users can access the past and current status of all devices.

System-wide, IDS anomaly detection accuracy was measured at 90%, response time was measured at an average of 340 ms, and recovery time from connection outage was measured at an average of 1.2 seconds.

- **Device Monitoring**: Monitoring device connections via MQTT.
- **Profile Management**: What type of data does it send and at what intervals?
- **IDS Alert Generation:** Generate IDS alerts when device behavior deviates from profile.
- **Spoofing Detection**: Warning in case of IP address spoofing.
- **Network Monitoring**: Connections coming from outside the defined IP range are analyzed with IDS.
- **Status Screen**: Instant connection status of devices and historical data records.
- **Notification Screen**: Live monitoring of data sent by each device.
- **Simulation Tool**: Sentinel GitHub

Figure 5 shows the IoT where data from IoT devices are collected and monitored. Hub - The user interface of the Sentinel application is shown. The interface presents the status of different sensors, the device list, log records and sensor data in detail. This screen serves as a central control point for real-time monitoring and analysis of IoT devices.

Figure 5. IOT Hub - Sentinel usage form

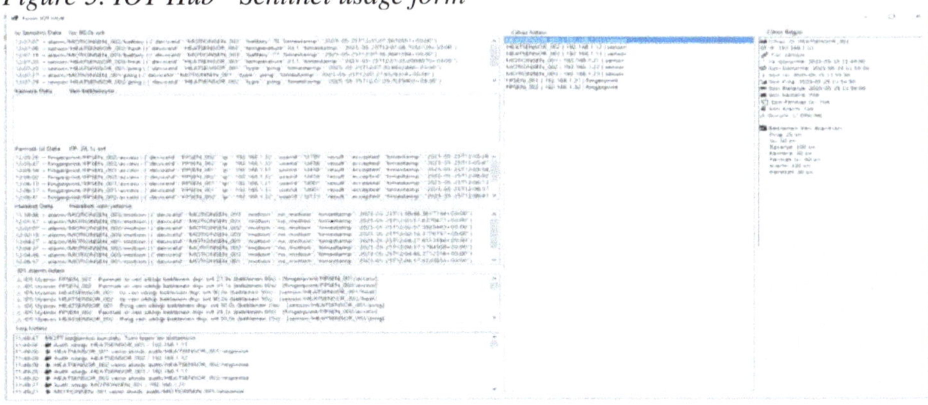

The system architecture and application are functional in the following aspects:

- **Multi-device simulation:** The existence of independent structures for each sensor type demonstrates that the system can monitor different IoT components in an integrated manner.
- **Real-time monitoring and logging:** The outputs received through Mosquitto prove that the data flow is both secure and uninterrupted.
- **Authentication and attack monitoring capacity:** The system is supported by security layers such as IP-based connection control and user authentication.
- **Modular system structure:** Although the camera module is still under development, the full functionality of the other parts of the system indicates an architectural design open to parallel expansion.

Figure 6 shows the homepage of the Sentinel project on GitHub. This screen shows the project's file structure, recent updates, and the management interface of the relevant code repository.

Figure 6. Sentinel GitHub screenshot

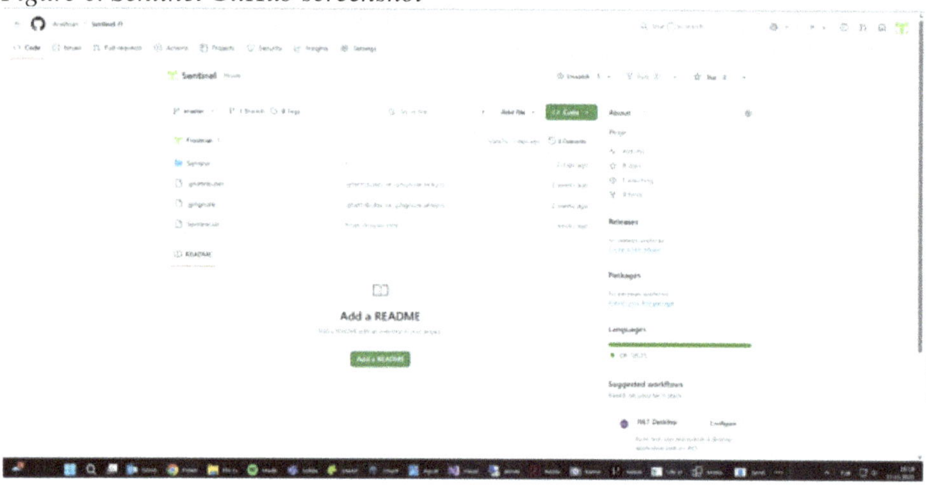

Figure 7 shows an overview of the GitHub repository page for the Camera project. This screen shows the GitHub interface used to manage the list of files and source codes in the project, version control, and collaboration processes.

Figure 7. Camera GitHub screenshot

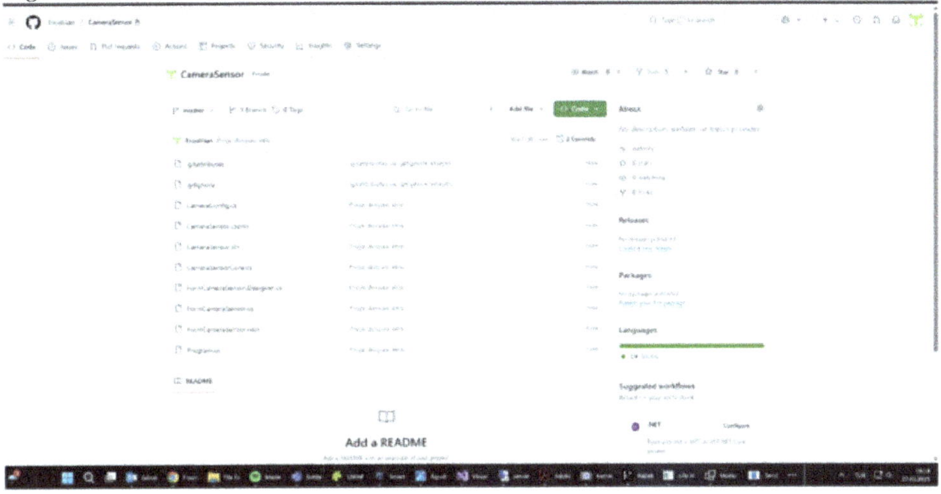

Figure 8 shows an overview of the Heat Sensor project's repository page on GitHub. This screen shows the interface where the project's source files are listed and version control is performed.

Figure 8. Heat sensor GitHub screenshot

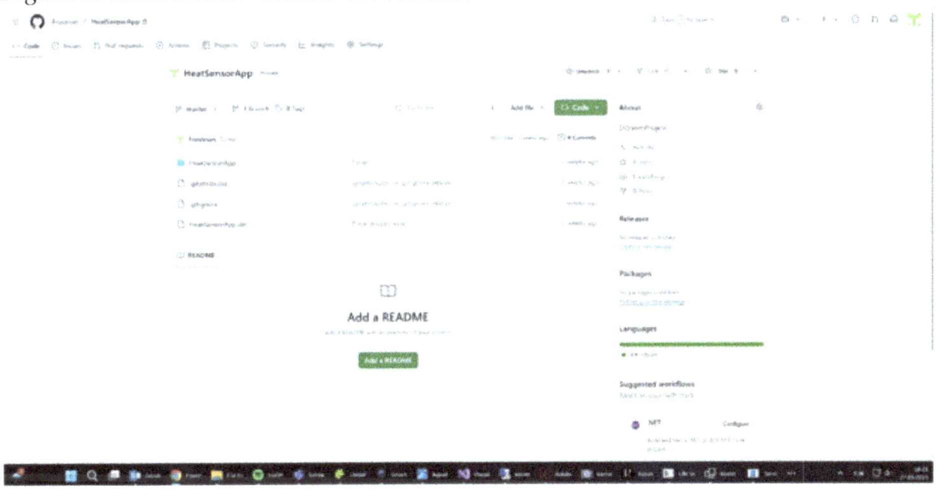

Figure 9 shows an overview of the GitHub repository page for the Motion Sensor project. This screen shows the platform interface where the project's source codes and related files are listed, and version control and collaboration processes are carried out. Figure 10 shows the GitHub repository of the Fingerprint sensor.

Figure 9. Motion sensor GitHub screenshot

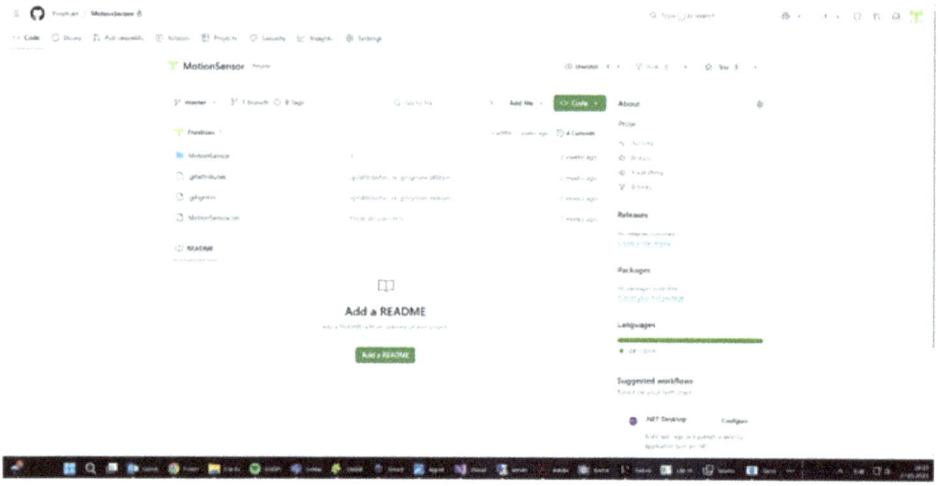

Figure 10. Fingerprint sensor GitHub screenshot

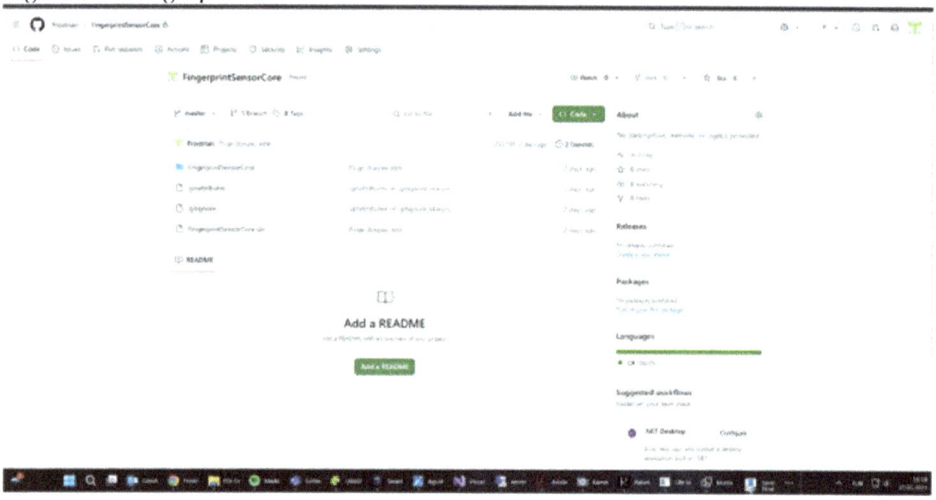

Comments on Application Environment and Software Components

Sentinel is an integrated system developed in C# language; device monitoring, behavior profiling and IDS module. The main purpose is to analyze the network traffic of connected devices and detect abnormal data behavior in real time and generate warnings. The application includes simulators of IoT devices. Each module provides data generation for a different sensor type via the MQTT protocol.

The modularity of the code structures supports the extensibility and reusability of the developed system.

Figure 11 shows a screenshot of various sensor simulators used in the developed IoT system being run and tested simultaneously. This structure allows the use of multiple simulators and offers significant advantages in the integration and performance tests of the system.

Figure 11. Test screenshot with different simulators

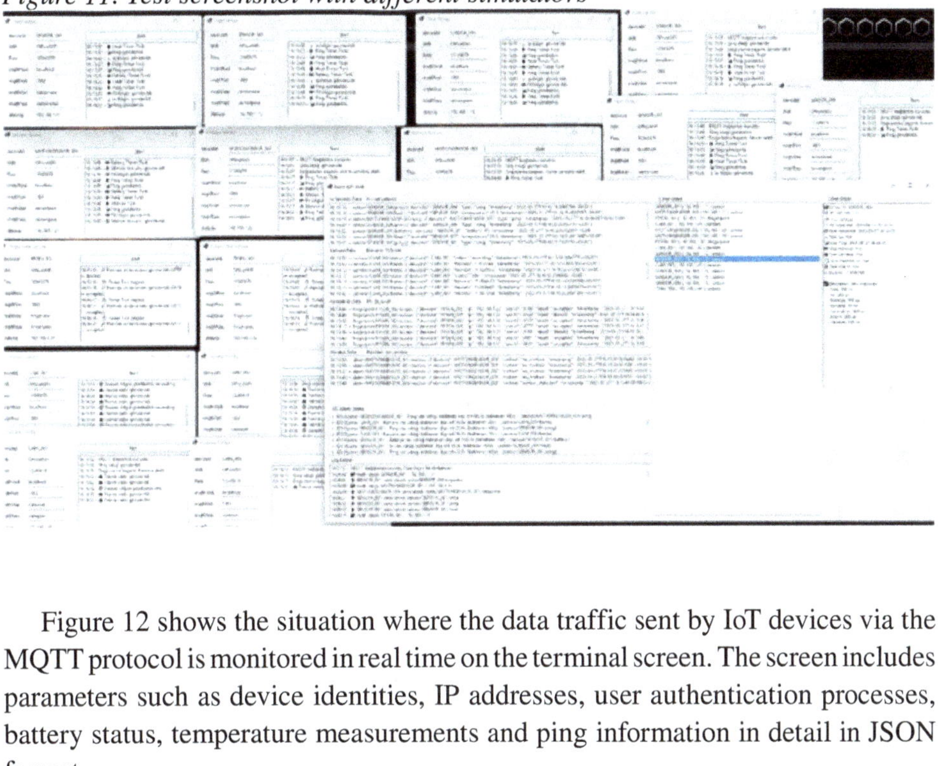

Figure 12 shows the situation where the data traffic sent by IoT devices via the MQTT protocol is monitored in real time on the terminal screen. The screen includes parameters such as device identities, IP addresses, user authentication processes, battery status, temperature measurements and ping information in detail in JSON format.

Figure 12. MQTT traffic screenshot

Evaluation on Findings Based on IDS Logs

Intrusion detection for different types of cyber attacks was performed in a simulated environment and the responses given by the system module were analyzed through screen outputs and log records.

Flood Attack Logs and Detection Mechanism

For flood attacks, sequential data transmission was provided to the selected devices (e.g. SENSOR_001) at short intervals. It was seen that the attack was initiated with the system's "All topics are being listened to" warning and active devices were monitored. When a large number of MQTT messages were detected, especially in a short time interval, the IDS module defined this as "abnormal traffic" and entered the process of blacklisting the device. This shows that the system is configured to be sensitive to traffic frequency.

Figure 13 shows the user interface of the attack simulator. By selecting different attack types on this screen, an attack simulation can be started with a certain number of packets, and the simulator's listening and response processes can be monitored in real time. The attack simulator is used as an effective tool for cybersecurity tests and IDS performance evaluations.

Figure 13. Attack simulator screenshot

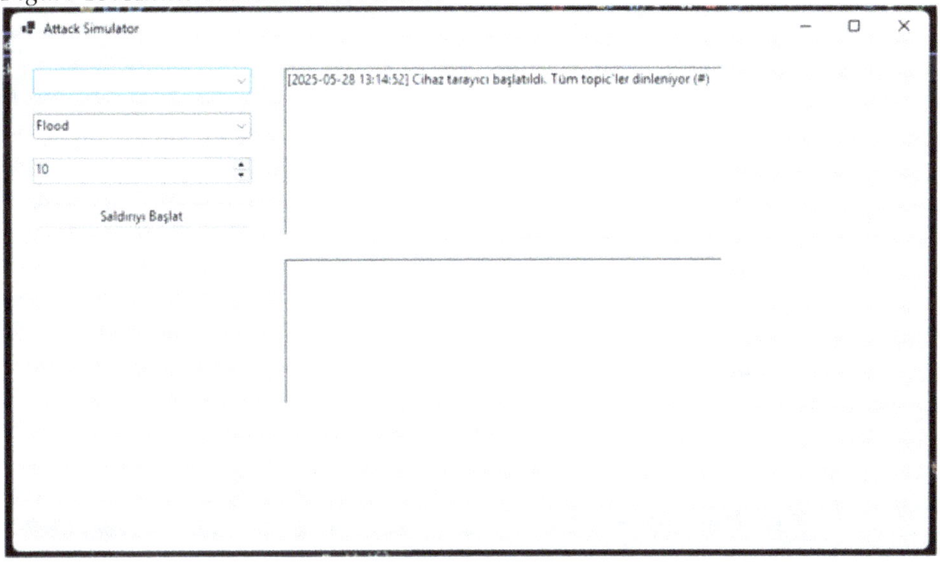

Figure 14 shows the Attack Simulator program's user interface and a screenshot at runtime. A specific type of attack simulation was started for the selected sensor on this screen, and the simulator's sending of the relevant attack packets and receiving feedback from the system were monitored in real time. The real-time feedback provided by the simulator plays an important role in evaluating the performance and durability of attack detection systems.

Figure 14. Attack simulator usage screen screenshot

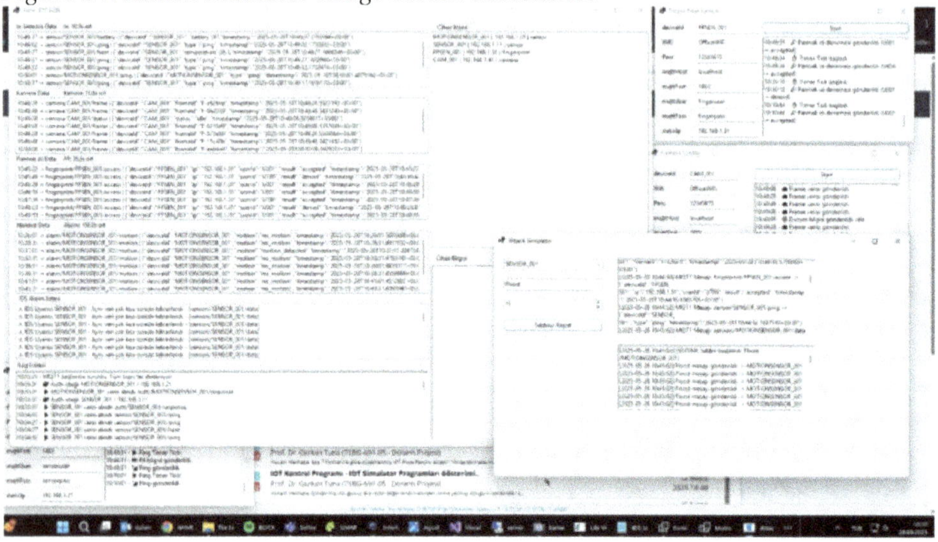

Figure 15 shows the homepage of the Attack Simulator project on GitHub. This screen presents the platform interface where the project's source files and management tools are listed, and version control and collaboration processes are managed.

Figure 15. Attack simulator GitHub screenshot

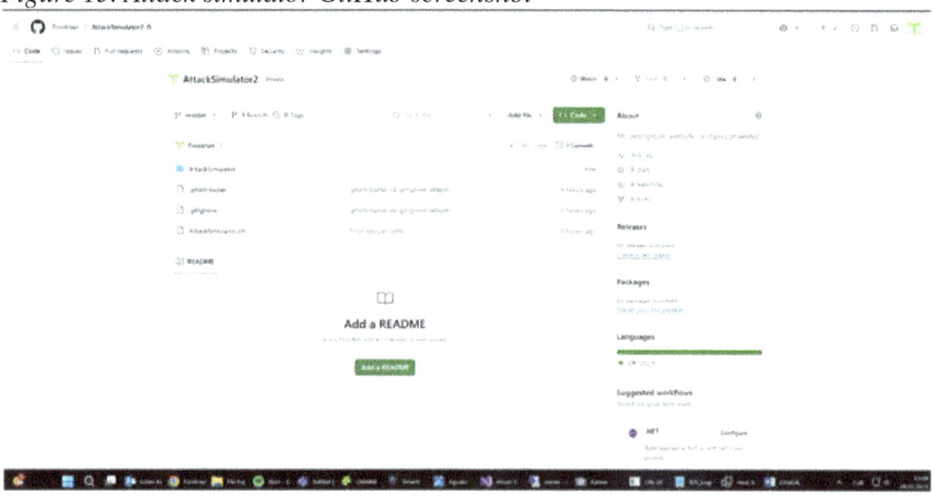

Tampering events occurred in the data coming from the fingerprint sensor. Figure 16 shows the sensor device ID, detected anomaly reasons, and event timestamps in detail in JSON format. These log records are critical to assessing the system's ability to detect attacks and manipulations.

Figure 16. Fingerprint data tempering log screenshot

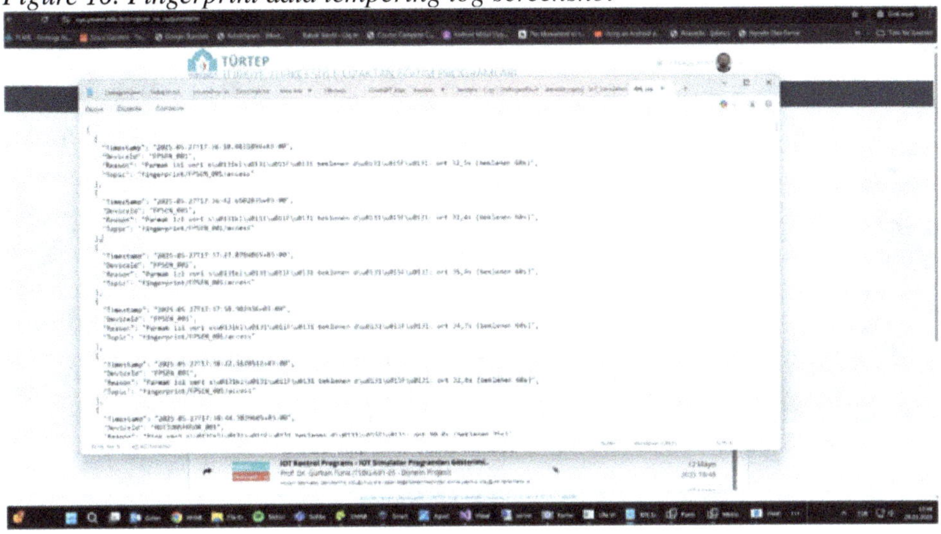

Figure 17 shows log records related to spoofing control performed on data coming from devices in the IoT system are shown. On the screen, device identities, IP addresses, log levels and detected spoofing events are detailed in JSON format. These logs were used to evaluate the effectiveness of the system's authentication mechanism and its resistance to security threats.

Figure 17. Spoof control log screenshot

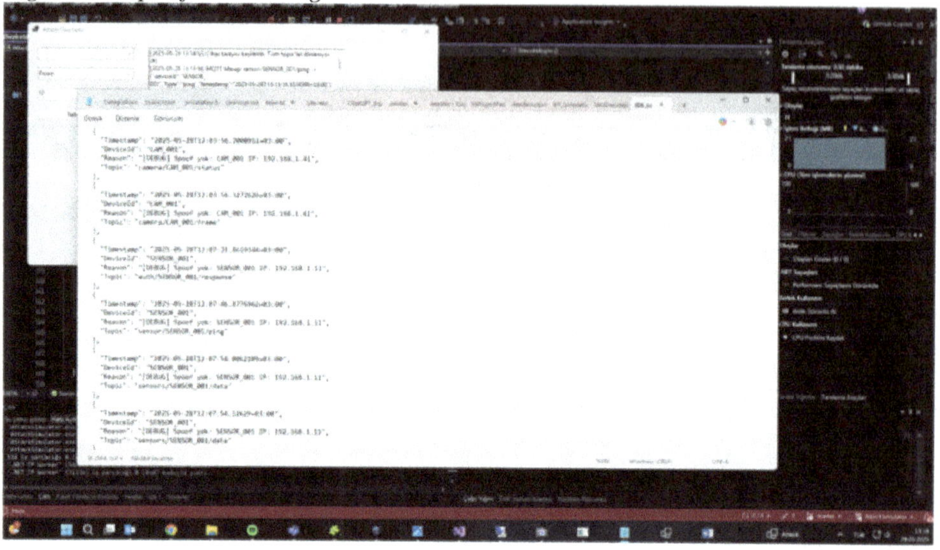

Figure 18 shows the blacklist management interface of the system developed against flood attacks and screenshots of the records created at the time of the attack. The image shows the process of the system detecting and blacklisting the attacker IP addresses and the related alarm and log records in detail. These screens provide important data in analyzing the effectiveness of the attack detection and prevention mechanism.

Figure 18. Flood-black list screenshots

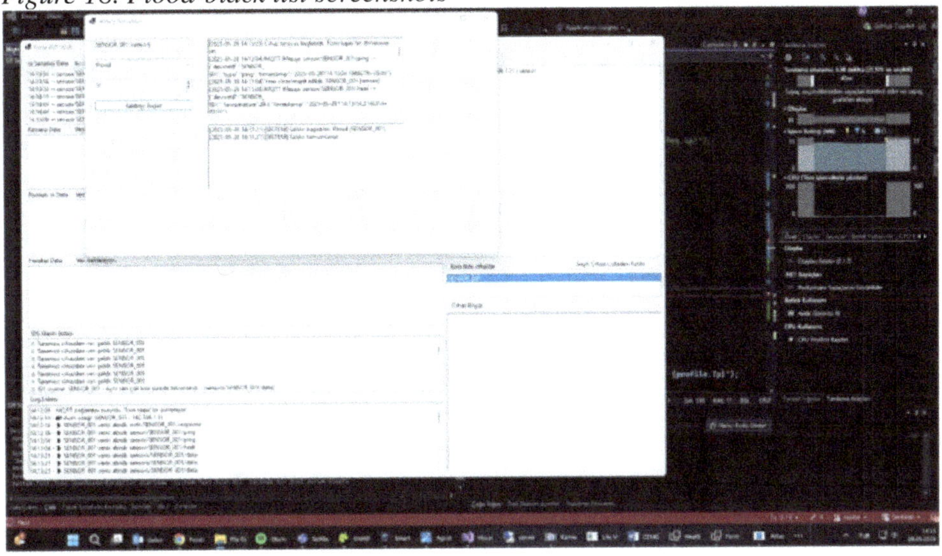

Data Tampering Attack and IDS Response

In the data integrity interference scenario, the attacker sent measurement values from the device named SENSOR_001 that the system did not expect. In the log records, it was seen that values such as "-491\u00BBC" at the end of the data outputs were marked as "outside the expected range" by the system and the IDS successfully detected these deviations. In the attack simulation screen, there were explanations such as "Data Tampering attack was initiated" and "Data other than the expected value was transmitted", which proves that the IDS provides a control mechanism based on basic value range checking.

Spoofing Attack and Undetected Collisions

In this type of attack, data from different IP addresses with the same device ID was analyzed by the IDS. In the images, it was seen that data was sent from different addresses other than the registered IP address of the device. However, in some scenarios, this situation was not clearly marked as an "identity conflict" by the IDS. The logs only stated "data came from the device". This situation shows that the IDS system is deficient in behavior-based or IP matching analysis.

Detection of Unidentified Devices

Data transmissions from devices that had not been previously identified to the Sentinel system (e.g. CAM_001 or SENSOR_999) were logged by the system as "unidentified devices". Such data is directly indicated in the IDS logs with different coloring or a "new device detected" message; thus, it is understood that the system uses a verification mechanism based on the device database.

IDS Success Rate According to Detected Attack Types

In this section, the performance of the IDS component according to attack types is analyzed.

Table 1. IDS success rate according to detected attack types

Attack Type	Detection Success	Explanation
Flood	High	Anomaly detection against traffic density is successful, the device is blacklisted.
Data Tampering	Medium-High	Measurements that deviate from the expected value are effectively detected.
Spoofing (ID Copy)	Low-Medium	Device ID conflicts cannot be detected in some cases.
Unidentified Device Access	High	According to the device database, unidentified devices are successfully blocked.

According to the data given in Table 1, the system is more effective in frequency-based and threshold value control-based attack types; however, it remains weak in behavioral or fake identity attacks.

The Role of Timestamp and Frequency Monitoring in Anomaly Detection

One of the most powerful aspects of the IDS system is its ability to control the frequency of MQTT messages via their timestamps. In particular, the frequency of messages transmitted in seconds was determined in the data, and this was compared with a threshold value and marked as anomalies.

In addition, the log generation method of the system was examined by evaluating the measurement data according to both content and transmission time. This shows that the IDS system has not only content-based but also temporal-based analysis capabilities.

Figure 19 shows a screenshot of a simulation of a data tempering attack on an IoT system. The data coming from the relevant sensors were detected by the system and processed into the alarm list. This screen provides basic data for analyzing the system's data integrity protection and anomaly detection capabilities.

Figure 19. Data tempering attack screenshot

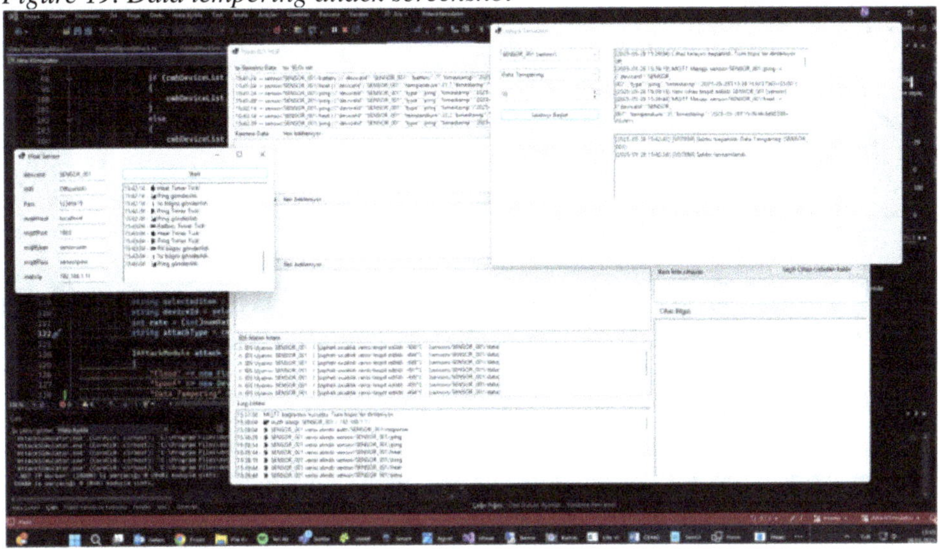

Figure 20 shows that there are log records regarding data tampering events detected in the data coming from IoT sensors. The records shown in JSON format include details such as device identities, timestamps when tampering was detected, and erroneous data values. These logs were used to evaluate the effectiveness of the system's data security mechanisms and its response to abnormal data entries.

Figure 20. Data tempering log screenshot

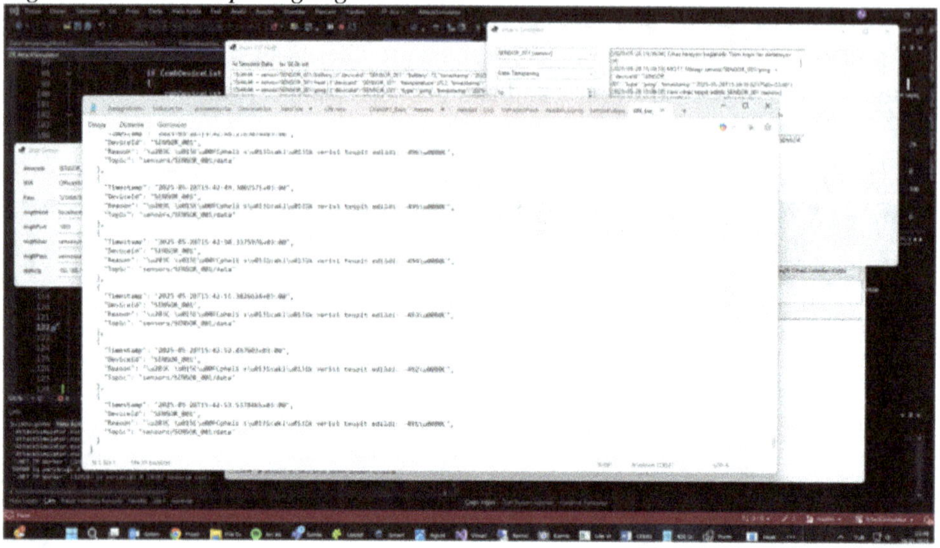

Figure 21 shows a screenshot of the ability of the IDS system to detect unidentified devices and report these situations. The screen shows that unidentified devices are detected among the devices connected to the system and an alarm is generated by the system. In addition, real-time data streams received from sensors and related log records are presented in detail.

Figure 21. IDS-unidentified device control screenshot

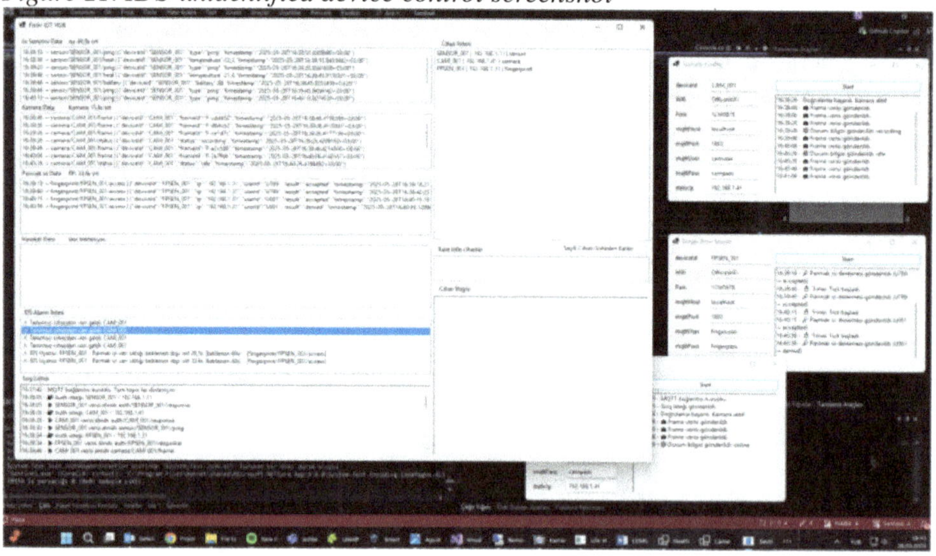

Figure 22 shows the log records created after the detection of devices that are not defined in the IDS system. The logs contain device identities, the cause of the detected event, and related data topic information. These records document the system's ability to track and report unknown devices on the network and its sensitivity to security breaches.

Figure 22. Unidentified device log screenshot

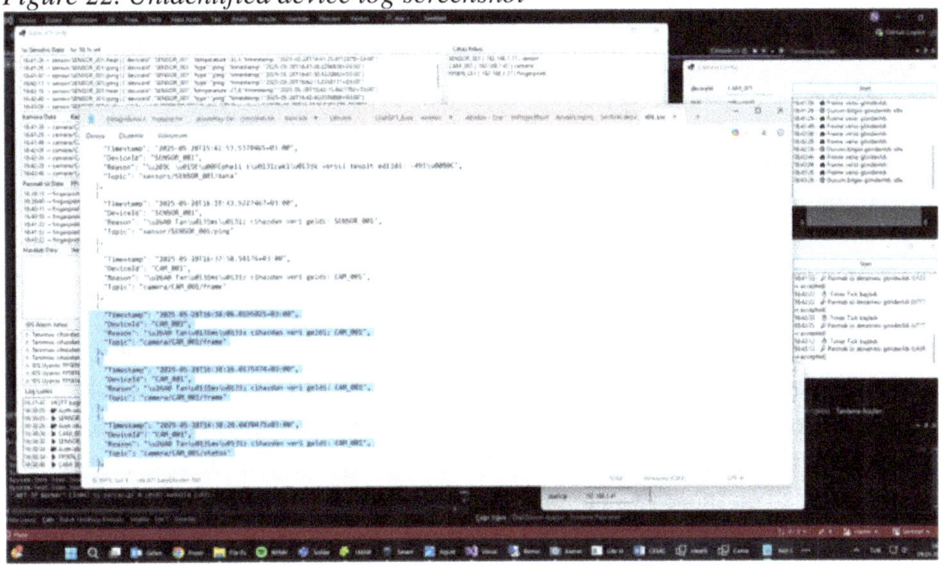

Discussion and Recommendations

It was observed that the IDS module could detect anomalies with high precision based on time-stamped frequency analysis. In flood attack scenarios, a large number of message transmissions in short time intervals were marked as “traffic density” by the system. The device was blacklisted and dropped into the log system. On the other hand, it was been observed that the system remains weak in more sophisticated spoofing attacks, and data coming from different IP addresses with the same device ID were processed only as data transmission and were not classified as anomalies. It was understood that IP mismatches could not be sufficiently analyzed by the system in terms of behavior. This can be solved by behavioral profiling or machine learning supported modules.

The IDS module was shown to be successful against data tempering attacks to a certain extent. The logs showed that the system could detect measurements outside the expected value range as anomalies and issue warnings based on “value deviations". However, these warnings were based on absolute threshold values, and the lack of contextual analysis was occasionally caused false positives.

One of the strengths of the system was its response to unidentified devices. It was seen that the system instantly identified transmissions from devices that were not registered in the database, generated IDS alerts, blocked or quarantined these devices. This feature demonstrated the effectiveness of the system's device-based verification layers.

The findings of the study can be summarized with the following key conclusions:

- Flood attacks can be detected with high accuracy thanks to time-based traffic control.
- Data tampering attacks can be detected using fixed threshold checks; however, false positive results may be produced due to context-independent evaluations.
- Against spoofing attacks, the IDS module remained weak because there is not enough matching between the IP address and the device ID.
- Unidentified devices are successfully blocked by the system's database-based authentication mechanism.
- Timestamp and frequency monitoring constitute the basic building blocks of IDS performance, and the anomaly detection process is successfully carried out based on these two parameters.

The following features are recommended to integrate into the application.

- Considering the availability of existing threat detection mechanisms, new behavioral profiling algorithms must be integrated into the IDS module.
- Instead of fixed threshold values, dynamic threshold determination mechanisms that can perform contextual analysis must be developed.
- The false positive rate must be reduced by implementing a scoring system that scores anomalies against data integrity attacks.
- The system must be able to track multiple IP connections more effectively by matching device ID with IP address.
- To further optimize the IDS response time, an event-driven process structure could be preferred.
- There is a need to add cameras and high-bandwidth devices into the network traffic analysis module.
- User traceability can be increased by creating time series-based visual analysis panels regarding device behavior.
- The simulation environment must be made compatible with different network topologies.

- To increase MQTT message security, TLS/SSL encryption options must be integrated into the system.
- Visual analysis of the sources of attacks can be provided by creating a real-time threat map on system logs.

CONCLUSION

In this study, the behaviors of IoT devices (heat sensors, motion sensors, fingerprint readers and IP cameras) against various types of cyber attacks were analyzed using a control and attack detection system called Sentinel developed in an MQTT-based IoT simulation environment. The software application was developed using C# and works integrated with Mosquitto MQTT Broker.

In the experimental test scenarios, it was understood that the proposed system was successful in terms of timestamping, frequency analysis and data integrity checks; however, it was limited in more advanced attack types such as identity cloning and behavior manipulation. The system is a potential solution in terms of real-time security monitoring in distributed networks with its capabilities such as online behavior profiling, device type-based analysis and visualization of MQTT traffic. In general, the results show attack detection capacity of the system, cyber security resistance of the devices and time-based behavioral analysis potential. In line with these results, it is recommended that machine learning-based behavioral analysis, anomaly scoring, and contextual alarm thresholds be integrated into the IDS system in future studies. Such approaches will contribute to the more accurate identification of fake identities, complex attack patterns, and abnormal behaviors.

ACKNOWLEDGMENT

Some parts of this chapter were produced from the first author's master's project. This research received no specific grant from any funding agency in the public, commercial, or not-for-profit sectors.

REFERENCES

Abdullah, A., Hamad, R., & Abdulrahman, M. (2019). CyberSecurity: A review of IoT security issues, challenges and techniques. *IEEE Information Security and Cyber Forensics*, *4*(1), 12–20.

Abomhara, M., & Køien, G. M. (2015). Cyber security and the internet of things: Vulnerabilities, threats, intruders and attacks. *Journal of Cyber Security and Mobility*, *4*(1), 65–88. DOI: 10.13052/jcsm2245-1439.414

Ahmad, W., Rasool, A., Javed, A. R., Baker, T., & Jalil, Z. (2021). Cyber security in IoT-based cloud computing: A comprehensive survey. *Electronics (Basel)*, *11*(1), 16. DOI: 10.3390/electronics11010016

Asaad, R. R., & Saeed, V. A. (2022). Cyber security: Threats, vulnerabilities, challenges and proposed solution. *Applied Computing Journal*, *4*(1), 1–8. DOI: 10.52098/acj.202260

Aslan, Ö., Aktuğ, S. S., Ozkan-Okay, M., Yilmaz, A. A., & Akin, E. (2023). A comprehensive review of cybersecurity vulnerabilities, threats, attacks, and solutions. *Electronics (Basel)*, *12*(6), 1333. DOI: 10.3390/electronics12061333

Avcı, İ. (2022). Akıllı evlerde IoT teknolojileri ve siber güvenlik. *Avrupa Bilim ve Teknoloji Dergisi*, *21*(4), 88–97. DOI: 10.31590/ejosat.1080228

Aydın, Ö., & Yükçü, S. (2020). Siber saldırı önlemede blokzinciri teknolojisinin fayda-maliyet analizi. *Manas Sosyal Araştırmalar Dergisi*, *9*(2), 145–162.

Butun, I., Österberg, P., & Song, H. (2019). Security of the Internet of Things: Vulnerabilities, attacks, and countermeasures. *IEEE Communications Surveys and Tutorials*, *22*(1), 616–644. DOI: 10.1109/COMST.2019.2953364

Conti, M., Dehghantanha, A., Franke, K., & Watson, S. (2018). Internet of Things security and forensics: Challenges and opportunities. *Future Generation Computer Systems*, *78*, 544–546. DOI: 10.1016/j.future.2017.07.060

Fidaner, Y. S., & Savaşçıhabeş, A. (2024). Examining IoT Applications in the Perspective of Security Technologies in 5G Networks. *Orclever Proceedings of Research and Development, 12*(1), 99–110. DOI: 10.56038/oprd.v5i1.599

Ghelani, D., Hua, T. K., & Koduru, S. K. R. (2022). Cyber Security Threats, Vulnerabilities, and Security Solutions Models in Banking. *Authorea Preprints*. DOI: 10.22541/au.166385206.63311335/v1

Gönen, S., Yılmaz, E. N., & Şanoğlu, S. (2021). Dijital dönüşümün vebası: Siber suçlar ve güvenlik. *Akademik Derleme, 6*(1), 112–129.

Islam, M. R., & Aktheruzzaman, K. M. (2020). An analysis of cybersecurity attacks against internet of things and security solutions. *Journal of Computer and Communications, 8*(4), 19–28. DOI: 10.4236/jcc.2020.84002

Kaşif, A. (2020). Design of Privacy and Security Centric Home Automation System [Master's thesis]. Bursa Uludağ University, Türkiye.

Kimani, K., Oduol, V., & Langat, K. (2019). Cyber security challenges for IoT-based smart grid networks. *International Journal of Critical Infrastructure Protection, 25*, 36–49. DOI: 10.1016/j.ijcip.2019.01.001

Liu, Y., Xiao, Y., Li, S., Liang, W., & Chen, C. P. (2019). Cyber security and privacy issues in smart grids. *IEEE Communications Surveys and Tutorials, 14*(4), 981–997. DOI: 10.1109/SURV.2011.122111.00145

Lone, A. N., Mustajab, S., & Alam, M. (2023). A comprehensive study on cybersecurity challenges and opportunities in the IoT world. *Security and Privacy, 6*(2), e318. DOI: 10.1002/spy2.318

Naik, S., & Maral, V. (2017). Cyber security—IoT. In *2nd IEEE International Conference on Communication and Electronics Systems* (pp. 159–164). IEEE.

Neshenko, N., Bou-Harb, E., Crichigno, J., Kaddoum, G., & Ghani, N. (2019). Demystifying IoT security: A survey on IoT vulnerabilities and attacks. *IEEE Communications Surveys and Tutorials, 21*(3), 2702–2733. DOI: 10.1109/COMST.2019.2910750

Parashar, A. (2023). Cybersecurity threats in the Internet of Things (IoT). *IOSR Journal of Computer Engineering, 25*(5), 1–8.

Pehlivanoğlu, M. K., Kuyucu, A., & Kaya, R. (2023). IoT veri kümelerinde makine öğrenmesi tekniklerine dayalı saldırı tespiti. *Avrupa Bilim ve Teknoloji Dergisi, 25*(110), 312–325.

Raimundo, R. J., & Rosário, A. T. (2022). Cybersecurity in the internet of things in industrial management. *Applied Sciences (Basel, Switzerland), 12*(3), 1598. DOI: 10.3390/app12031598

Razaque, A., Amsaad, F., Jaro Khan, M., Hariri, S., Chen, S., Siting, C., & Ji, X. (2019). Survey: Cybersecurity vulnerabilities, attacks and solutions in the medical domain. *IEEE Access : Practical Innovations, Open Solutions, 7*, 168774–168797. DOI: 10.1109/ACCESS.2019.2950849

Sakal, M. (2022). Endüstri 4.0 ve Güvenlik. In *Endüstri 4.0 ve İşletne Yönetimi* (Eds.Mehmet Marangoz & Hatice Hicret Özkoç, Beta Yayınevi. Saylam, R. (2020). Applicability of the Internet of Things (IoT) in military environment and a proposed acceptance model [Master's thesis]. Atatürk University, Türkiye.

Sevinç, M., & Avcı, İ. (2024). IoT cihazlarında insan hatasından kaynaklanan güvenlik açıklarının analizi. *Mühendislik Bilimleri ve Tasarım Dergisi, 12*(1), 57–68.

Siwakoti, Y. R., Bhurtel, M., & Rawat, D. B. (2023). Advances in IoT security: Vulnerabilities, attacks, and countermeasures. *IEEE Internet of Things Journal, 10*(4), 3456–3473. DOI: 10.1109/JIOT.2023.3252594

Tawalbeh, L., Muheidat, F., Tawalbeh, M., & Quwaider, M. (2020). IoT privacy and security: Challenges and solutions. *Applied Sciences (Basel, Switzerland), 10*(12), 4102. DOI: 10.3390/app10124102

Tok, M. S. (2019). Nesnelerin internetinde botnetler: Mirai zararlı yazılımı üzerine bir çalışma. *[Unpubllished master's thesis]. Gazi University, Ankara, Türkiye.*

Tulgar, M., Zaim, A. H., & Aydın, M. A. (2022). Ulusal bilgi ve iletişim güvenliği rehberi: IoT güvenliği için bir uygulama örneği. İstanbul Ticaret Üniversitesi. *Fen Bilimleri Dergisi, 21*(2), 99–115.

Türker, G. F., & Tanyeri, K. (2024). Blokzincir teknolojisi ile IoT sistemlerinin veri güvenliğinin sağlanması. Gazi Üniversitesi Fen Bilimleri Dergisi C Serisi, 8(1), 45-59.

Whitmore, A., Agarwal, A., & Da Xu, L. (2015). The Internet of Things—A survey of topics and trends. *Information Systems Frontiers, 17*(2), 261–274. DOI: 10.1007/s10796-014-9489-2

Yalçın, N., & Çakır, S. (2024). IoT ağlarında saldırı tespiti için federe öğrenme. *Bilgisayar Bilimleri ve Mühendisliği Konuları, 10*(2), 50–67.

Yıldız, B. (2022). *Privacy Preserving Data Analysis For Information Systems* [Unpublished doctoral dissertation]. Dokuz Eylül University. İzmir, Türkiye.

Yılmaz, E. N., Gönen, S., Şanoğlu, S., Karacayilmaz, G., & Özbi̇ri̇nci̇, Ö. (2021). Endüstri 4.0'ın gelişim sürecinde unutulan bileşen: Siber güvenlik. *DÜBİTED*, *4*(2), 87–103. DOI: 10.29130/dubited.905340

Yin, X. C., Liu, Z. G., Nkenyereye, L., & Ndibanje, B. (2019). Toward an applied cyber security solution in IoT-based smart grids. *Sensors (Basel)*, *19*(22), 4952. DOI: 10.3390/s19224952 PMID: 31739400

Yurttaş, H. G., & Güzel, A. (2023). Kablosuz ağa bağlı tıbbi cihazlarda siber güvenlik açıkları ve çözüm önerileri. *Uluslararası Sağlık Yönetimi ve Stratejileri Araştırma Dergisi*, *5*(1), 45–60.

Zarpelão, B. B., Miani, R. S., Kawakani, C. T., & de Alvarenga, S. C. (2017). A survey of intrusion detection in Internet of Things. *Journal of Network and Computer Applications*, *84*, 25–37. DOI: 10.1016/j.jnca.2017.02.009

Zeybek, M., & Yılmaz, E. N. (2019). Nesnelerin İnterneti: Risk temelli yaklaşım. *Denetişim*, *3*(2), 12–29.

ADDITIONAL READING

Awad, A. I., & Abawajy, J. H. (2022). *Security and privacy in the Internet of things: architectures, techniques, and applications. IEEE Press.* John Wiley & Sons, Inc., DOI: 10.1002/9781119607755

Dhondge, K. (2021). *Lifecycle IoT security for engineers.* Artech House.

Liyanage, M., Braeken, A., Kumar, P., & Ylianttila, M. (2020). *IoT security: advances in authentication.* John Wiley & Sons, Inc. DOI: 10.1002/9781119527978

Tripathy, B. K., & Anuradha, J. (2018). *Internet of things (IoT): technologies, applications, challenges and solutions.* Taylor & Francis, CRC Press.

KEY TERMS AND DEFINITIONS

Internet of Things: Internet of things describes devices with sensors, processing ability, software and other technologies that connect and exchange data with other devices and systems over the Internet or other communication networks.

JavaScript Object Notation (JSON): JSON is an open standard file format and data interchange format that uses human-readable text to store and transmit data objects consisting of name–value pairs and arrays. It is a commonly used data format with diverse uses in electronic data interchange, including that of web applications with servers.

MQTT: MQTT is a lightweight, publish-subscribe, machine-to-machine network protocol for message queue/message queuing service. It is designed for connections with remote locations that have devices with resource constraints or limited network bandwidth, such as in the Internet of Things.

Sensor: A sensor is a device that detects and responds to some type of input from the physical environment.

Compilation of References

Abdulhameed, A. A., Ssemakula, J. B., & Abbas, O. O., AbdulAzeez, A. A., & Ahmed, A. A. (2025, January). Design and implementation of a voiceprint recognition system for hybrid security and authentication in a university organizational infrastructure. *International Journal of Engineering Technology Research & Management, 9*(1), 140–149.

Abdullah, A., Hamad, R., & Abdulrahman, M. (2019). CyberSecurity: A review of IoT security issues, challenges and techniques. *IEEE Information Security and Cyber Forensics, 4*(1), 12–20.

Abdul, Z. K., & Al-Talabani, A. K. (2022, November). Mel Frequency Cepstral Coefficient and its applications: A review. *IEEE Access : Practical Innovations, Open Solutions, 10*, 122136–122158. DOI: 10.1109/ACCESS.2022.3223444

Abomhara, M., & Køien, G. M. (2015). Cyber security and the internet of things: Vulnerabilities, threats, intruders and attacks. *Journal of Cyber Security and Mobility, 4*(1), 65–88. DOI: 10.13052/jcsm2245-1439.414

Accurso, A. W. (2023, October). Digital voiceprinting is not ready for court. *Criminal Legal news, 6*(10), 47. https://www.criminallegalnews.org/news/2023/oct/1/digital-voiceprinting-not-ready-court

Acien, A., Morales, A., Monaco, J. V., Vera-Rodriguez, R., & Fierrez, J. (2021). TypeNet: Deep learning keystroke biometrics. *IEEE Transactions on Biometrics, Behavior, and Identity Science, 4*(1), 57–70. DOI: 10.1109/TBIOM.2021.3112540

ACLU Illinois. (2025, May 15). *Biometric Information Privacy Act (BIPA)*. https://www.aclu-il.org/sites/default/files/field_documents/protect_bipa_5.8.24.pdf

Adjerid, I., & Kelley, K. (2018). The role of biometric authentication in reducing identity fraud. *Journal of Cybersecurity, 4*(1), 1–10. DOI: 10.1093/cybsec/tyy002

Agrawal, V., Hazratifard, M., Elmiligi, H., & Gebali, F. (2023). Electrocardiogram (ECG)-based user authentication using deep learning algorithms. *Diagnostics (Basel)*, *13*(3), 439. Advance online publication. DOI: 10.3390/diagnostics13030439 PMID: 36766544

Ahmad, W., Rasool, A., Javed, A. R., Baker, T., & Jalil, Z. (2021). Cyber security in IoT-based cloud computing: A comprehensive survey. *Electronics (Basel)*, *11*(1), 16. DOI: 10.3390/electronics11010016

Ahmed, M., & Salem, A.-B. (2021). INTELLIGENT SYSTEM FOR HUMAN AUTHENTICATION USING FUSION OF DORSAL HAND, PALM AND FINGER VEINS. *Int. J. Intell. Comput. Inf. Sci.*, *21*(2), 140–154. DOI: 10.21608/ijicis.2021.73726.1087

Akbari, H., Yuan, L., Qian, R., Chuang, W. H., Chang, S. F., Cui, Y., & Gong, B. (2021). Vatt: Transformers for multimodal self-supervised learning from raw video, audio and text. *Advances in Neural Information Processing Systems*, *34*, 24206–24221.

Alay, N., & Al-Baity, H. H. (2020). Deep learning approach for multimodal biometric recognition system based on fusion of iris, face, and finger vein traits. *Sensors (Basel)*, *20*(19), 5523. DOI: 10.3390/s20195523 PMID: 32992524

Ali, M. L., Monaco, J. V., Tappert, C. C., & Qiu, M. (2017). Keystroke biometric systems for user authentication. *Journal of Signal Processing Systems for Signal, Image, and Video Technology*, *86*(2-3), 175–190. DOI: 10.1007/s11265-016-1114-9

Ali, M., Shaikh, N. A., & Mahoto, N. A. (2023). Deep learning-based presentation attack detection for face recognition systems: A comprehensive survey. *IEEE Access : Practical Innovations, Open Solutions*, *11*, 42357–42385. DOI: 10.1109/ACCESS.2023.3271428

Alimovski, E. (2019). Data augmentation with GAN for robust face recognition system based on deep learning [Unpublished master's thesis]. İstanbul Sabahattin Zaim University.

Aliyu, I., Bomoi, M. A., & Maishanu, M. (2022). A comparative study of eigenface and Fisherface algorithms based on opencv and sci-kit libraries implementations. *International Journal of Information Engineering and Electronic Business*, *14*(3), 38–44. DOI: 10.5815/ijieeb.2022.03.04

Alqahtany, S. S., & Syed, T. A. (2024). ForensicTransMonitor: A Comprehensive Blockchain Approach to Reinvent Digital Forensics and Evidence Management. *Information (Basel)*, *15*(2), 109. DOI: 10.3390/info15020109

Alsaadi, H. A., & Tubaishat, A. (2015). Internet of Things: Features, challenges, and vulnerabilities. *International Journal of Advanced Computer Science and Applications*, *6*(11), 128–133.

Altyar, S. S., Hussein, S. S., & Mohammed, M. J. (2022, March). Human recognition by utilizing voice recognition and visual recognition. *International Journal of Nonlinear Analysis and Applications*, *13*(1), 343–351. DOI: 10.22075/ijnaa.2022.5501

Alver, C. (2007). Voice biometrics in financial services. *The Journal of Financial Services Technology*, *1*(1), 75–81.

Amin, S., & Di Iorio, C. (2025). A Review of Several Keystroke Dynamics Methods. *arXiv preprint arXiv:2502.16177.*

Ammour, B., Boubchir, L., Bouden, T., & Ramdani, M. (2020). Face–Iris multimodal biometric identification system. *Electronics (Basel)*, *9*(1), 85. DOI: 10.3390/electronics9010085

Amos, B., Ludwiczuk, B., & Satyanarayanan, M. (2016). OpenFace: A general purpose face recognition library with mobile applications. CMU Technical Report.

Anderson, R., & Murdoch, S. J. (2024). Security engineering for biometric systems in the age of AI: Challenges and opportunities. *IEEE Security and Privacy*, *22*(1), 45–58. DOI: 10.1109/MSEC.2024.3356789

Anwar, S., & Barnes, N. (2019). Real image denoising with feature attention. In *Proceedings of the IEEE/CVF International Conference on Computer Vision* (pp. 3155–3164). DOI: 10.1109/ICCV.2019.00325

Aponte-Novoa, F. A., Orozco, A. L. S., Villanueva-Polanco, R., & Wightman, P. (2021). The 51% Attack on Blockchains: A Mining Behavior Study. *IEEE Access : Practical Innovations, Open Solutions*, *9*, 140549–140564. DOI: 10.1109/ACCESS.2021.3119291

Arora, G., Singh, A., Nigam, A., Pandey, H. M., & Tiwari, K. (2022). FKPIndexNet: An efficient learning framework for finger-knuckle-print database indexing to boost identification. *Knowledge-Based Systems*, *239*(108028), 108028. DOI: 10.1016/j.knosys.2021.108028

Asaad, R. R., & Saeed, V. A. (2022). Cyber security: Threats, vulnerabilities, challenges and proposed solution. *Applied Computing Journal*, *4*(1), 1–8. DOI: 10.52098/acj.202260

Asadi Shirzi, M., & Kermani, M. R. (2024). Real-Time Point Recognition for Seedlings Using Kernel Density Estimators and Pyramid Histogram of Oriented Gradients. *Actuators*, *13*(3), 81–81. DOI: 10.3390/act13030081

Aslan, Ö., Aktuğ, S. S., Ozkan-Okay, M., Yilmaz, A. A., & Akin, E. (2023). A comprehensive review of cybersecurity vulnerabilities, threats, attacks, and solutions. *Electronics (Basel)*, *12*(6), 1333. DOI: 10.3390/electronics12061333

Avcı, İ. (2022). Akıllı evlerde IoT teknolojileri ve siber güvenlik. *Avrupa Bilim ve Teknoloji Dergisi*, *21*(4), 88–97. DOI: 10.31590/ejosat.1080228

Aw, C., & O'Leary, C. (2025, July 28). *Cambodia moves to enact comprehensive data privacy law*. Hogan Lovells. https://www.hoganlovells.com/en/publications/cambodia-moves-to-enact-comprehensive-data-privacy-law

Aydın, Ö., & Yükçü, S. (2020). Siber saldırı önlemede blokzinciri teknolojisinin fayda-maliyet analizi. *Manas Sosyal Araştırmalar Dergisi*, *9*(2), 145–162.

Bäckström, T., Räsänen, O., Zewoudie, A., Pérez Zarazaga, P., Koivusalo, L., Das, S., Gómez Mellado, E., Mansali, N. B., Ramos, D., & Vali, M. H. (2022). *Introduction to speech processing* (2nd ed.). Aalto University., https://speechprocessingbook.aalto.fi

Badrinath, G. S., Nigam, A., & Gupta, P. (2011). An efficient finger-knuckle-print based recognition system fusing SIFT and SURF matching scores. In Vol. 7043, pp. 374–387). Lecture Notes in Computer Science. Springer., DOI: 10.1007/978-3-642-25243-3_30

Baig, A. F., & Eskeland, S. (2021). Security, Privacy, and Usability in Continuous Authentication: A Survey. *Sensors (Basel)*, *21*(17), 5967. DOI: 10.3390/s21175967 PMID: 34502865

Baig, Z. A., Szewczyk, P., Valli, C., Rabadia, P., Hannay, P., Chernyshev, M., Johnstone, M., Kerai, P., Ibrahim, A., Sansurooah, K., Syed, N., & Peacock, M. (2017). Future challenges for smart cities: Cyber-security and digital forensics. *Digital Investigation*, *22*, 3–13. DOI: 10.1016/j.diin.2017.06.015

Baines, K. N. (2011). *Dorsal knuckle crease pattern: Classification and inter-observer consistency* (Master's thesis, University of Dundee). University of Dundee Research Portal. https://discovery.dundee.ac.uk/files/90057557/Nicoll_Baines_msc_2011.pdf

Baker, C. P., Sundberg, J., Purdy, S. C., & Rakena, T. O. (2024, July). Female adolescent singing voice characteristics: An exploratory study using LTAS and inverse filtering. *Logopedics, Phoniatrics, Vocology*, *49*(2), 83–95. DOI: 10.1080/14015439.2022.2140455 PMID: 36322641

Baltruaitis, T., Ahuja, C., & Morency, L. P. (2019). Multimodal machine learning: A survey and taxonomy. *IEEE Transactions on Pattern Analysis and Machine Intelligence*, *41*(2), 423–443. DOI: 10.1109/TPAMI.2018.2798607 PMID: 29994351

Bansal, A., & Jain, A. K. (2019). Deep learning for biometric recognition: Trends and challenges. *IEEE Transactions on Pattern Analysis and Machine Intelligence*, *43*(10), 3461–3481.

Bansal, P., & Ouda, A. (2024). Continuous Authentication in the Digital Age: An Analysis of Reinforcement Learning and Behavioral Biometrics. *Computers*, *13*(4), 103. DOI: 10.3390/computers13040103

Baptista, D. (2025, July 14). *Will Mexico's new biometric ID card harm digital privacy?* Context News. https://www.context.news/digital-rights/will-mexicos-new-biometric-id-card-harm-digital-privacy

Basheer Aljbour, K. S., Aleshoush, M., El-Manaseer, S., Aloun, D. M., Alkayid, J. H. H., & Namous, A. A. A. Y. A. (2025). In A. Hannoon and A. Mahmood (Eds.), *Intelligence-Driven Circular Economy: Regeneration Towards Sustainability and Social Responsibility, Volume 2*. From Studies in Computational Intelligence 1174 (pp. 343-353), Springer. DOI: 10.1007/978-3-031-74220-0_26

Basu, N., Bali, A. S., Weber, R., Rosa-Aguilar, C., Edmond, G., Martire, K. A., & Morrison, G. S. (2022, October). Speaker identification in courtroom contexts – Part I: Individual listeners compared to forensic voice comparison based on automatic-speaker-recognition technology. *Forensic Science International*, *341*, 111499. DOI: 10.1016/j.forsciint.2022.111499 PMID: 36283276

Behind the PDPL: Navigating Argentina's Personal Data Protection Law. (2024, February 2). Pandectes. https://pandectes.io/blog/behind-the-pdpl-navigating-argentinas-personal-data-protection-law

Benziane, S. H., & Benyettou, A. (2017). Dorsal hand vein identification based on binary particle swarm optimization. *Journal of Information Processing Systems*, *13*(2), 268–283. DOI: 10.3745/JIPS.03.0065

Benziane, S., & Benyettou, A. (2011). An introduction to biometrics. *International Journal of Computer Science and Information Security*, *9*(4), 40–47.

Benziane, S., & Benyettou, A. (2013). Biometric technology based on hand vein. *Oriental Journal of Computer Science and Technology*, *6*(4), 401–412.

Best-Rowden, L., Han, H., Otto, C., Klare, B. F., & Jain, A. K. (2014). Unconstrained face recognition: Identifying a person of interest from a media collection. *IEEE Transactions on Information Forensics and Security*, *9*(12), 2144–2157. DOI: 10.1109/TIFS.2014.2359577

Betts, S. (2021, January 30). Rockland man charged with making false marine distress call. *Portland Press Herald*. https://www.pressherald.com/2021/01/29/rockland-man-charged-with-making-false-marine-distress-call

Bhagat, P. K., Singh, A., & Choudhary, P. (2023). Blockchain-based audit trail system for biometric authentication: Design and implementation. *Journal of King Saud University. Computer and Information Sciences*, *35*(8), 101642. DOI: 10.1016/j.jksuci.2023.101642

Bharadwaj, S., Kirchbacher, M., Ras, Z. W., & Harrison, M. (2014). *Biometric quality: A review of fingerprint, iris, and face. EURASIP Journal on Image and Video Processing, 2014*(34). DOI: 10.1186/1687-5281-2014-34

Bhardwaj, A., & Goundar, S. (2020). Keyloggers: Silent cybersecurity weapons. *Network Security*, *2020*(2), 14–19. DOI: 10.1016/S1353-4858(20)30021-0

Bhargava, B., & Yu, L. (2017). Biometric authentication in cloud computing: Challenges and solutions. *Computer*, *50*(6), 60–69.

Bhavya, D. N., & Chethan, H. K. (2020, December). A System to Detect the Faces in Unconstrained Environment. In *International Conference on Innovations in Bio-Inspired Computing and Applications* (pp. 196-205). Cham: Springer International Publishing.

Biel, L., Pettersson, O., Philipson, L., & Wide, P. (2001). ECG analysis: A new approach in human identification. *IEEE Transactions on Instrumentation and Measurement*, *50*(3), 808–812. DOI: 10.1109/19.930458

Biometric Information Privacy Act (BIPA), 740 ILCS 14/1 (2008).

Biswas, P. S., & Mishra, S. D. (2024). Novel approach for personal identification using dorsal knuckle crease patterns: A pilot study. *Mapana Journal of Sciences*, *23*(4), 83–97. DOI: 10.12723/mjs.71.5

Biswas, S., & Mishra, A. (2024). Emerging biometric modalities for disaster victim identification. *Journal of Forensic Sciences*, *69*(2), 311–320.

Bitter, A. (2023, January 6). Whole Foods must pay $300,000 over accusations that a system meant to track productivity captured warehouse workers' voices without their consent. *Business Insider.* https://www.businessinsider.com/whole-foods-pay-300k-recording-warehouse-employees-voices-2023-1

Black, S.. (2009). Forensic analysis of the hand in child abuse investigations. *Journal of Forensic Sciences, 54*(1), 20–25. PMID: 19018931

Bodapati, J. D., Veeranjaneyulu, N., & Shareef, S. N. (2024). Deep learning based facial recognition in forensic investigation: State-of-the-art review. *Forensic Science International Digital Investigation, 48*, 301615. DOI: 10.1016/j.fsidi.2024.301615

Bodepudi, A., Reddy, M., Gutlapalli, S. S., & Mandapuram, M. (2019). Voice recognition systems in the cloud networks: Has it reached its full potential? *Asian Journal of Applied Science and Engineering, 8*(1), 51–60. DOI: 10.18034/ajase.v8i1.12

Borak, M. (2025, July 18). *Mexico makes biometric identifier mandatory for all citizens.* Biometric Update. https://www.biometricupdate.com/202507/mexico-makes-biometric-identifier-mandatory-for-all-citizens

Bowyer, K. W., Hollingsworth, K., & Flynn, P. J. (2008). Image understanding for iris biometrics: A survey. *Computer Vision and Image Understanding, 110*(2), 281–307. DOI: 10.1016/j.cviu.2007.08.005

Boyko, N., Basystiuk, O., & Shakhovska, N. (2018). Performance evaluation and comparison of software for face recognition based on dlib and opencv library. 2018 IEEE Second International Conference. DOI: 10.1109/DSMP.2018.8478556

Bozik, T. (2019). Liveness detection in face recognition systems [Unpublished master's thesis]. Yıldız Technical University.

Brown, C. S. (2015). Investigating and Prosecuting Cyber Crime: Forensic Dependencies and Barriers to Justice. *International Journal of Cyber Criminology*, •••, 55–119.

Budnik, R. A., & Evpak, E. G. (2024). Human voice: Legal protection challenges. *Legal Issues in the Digital Age, 5*(4), 28–45. DOI: 10.17323/2713-2749.2024.4.28.45

Buolamwini, J., & Gebru, T. (2018). Gender shades: Intersectional accuracy disparities in commercial gender classification. *Proceedings of the Conference on Fairness, Accountability and Transparency*, 77–91.

Buresh, D. L. (2023, March/April). BIPA: The Illinois' Biometric Information Privacy Act what every Illinois attorney should know. *DuPage County Bar Association Brief, 35*(5), 6-13. https://www.dcba.org/mpage/v35-Donald-L-Buresh

Burt, C. (2024, November 8). *Biometric privacy law in Texas close enough to BIPA to protect Match.* Biometric Update. https://www.biometricupdate.com/202411/biometric-privacy-law-in-texas-close-enough-to-bipa-to-protect-match

Butler, S. (2025, January 15). *Digital vs. analog audio: The key differences and do they matter?* Yahoo Tech. https://tech.yahoo.com/audio/articles/digital-vs-analog-audio-key-150015487.html

Butun, I., Österberg, P., & Song, H. (2019). Security of the Internet of Things: Vulnerabilities, attacks, and countermeasures. *IEEE Communications Surveys and Tutorials, 22*(1), 616–644. DOI: 10.1109/COMST.2019.2953364

Cambodia: Data protection laws of the world. (2025, January 20). DLA Piper. https://www.dlapiperdataprotection.com/?c=KH

Cappelli, R., Ferrara, M., & Franco, A. (2021). Fingerprint image reconstruction from standard templates. *IEEE Transactions on Pattern Analysis and Machine Intelligence, 43*(7), 2324–2337. DOI: 10.1109/TPAMI.2020.2974446 PMID: 17627039

Carrell, S. (2009, May 7). Scottish paedophile ring found guilty of more than 50 charges. *The Guardian.* https://www.theguardian.com/uk/2009/may/07/scottish-paedophile-ring-guilty-child-abuse

Cavoukian, A. (2013). *Privacy by design: The 7 foundational principles.* Information and Privacy Commissioner of Ontario.

Cebe, M., Erdin, E., Akkaya, K., Aksu, H., & Uluagac, S. (2018). Block4Forensic: An Integrated Lightweight Blockchain Framework for Forensics Applications of Connected Vehicles. *IEEE Communications Magazine, 56*(10), 50–57. DOI: 10.1109/MCOM.2018.1800137

Chattopadhyay, S., & Sukul, B. (2012). Identification from dorsal finger pattern: A new approach. *Medicine, Science, and the Law, 52*(1), 17–21. DOI: 10.1258/msl.2011.010145 PMID: 22399024

Cheng, K. H. M., Xu, C., & Zhao, G. (2023). Advancing 3D finger knuckle recognition via deep feature learning. *arXiv.* https://arxiv.org/abs/2301.02934

Chen, J., Miao, W., Wu, W., Yang, L., & Yuan, H. (2024). Secure Relative Detection in (Forensic) Database with Homomorphic Encryption. *Lecture Notes in Computer Science, 14955*, 410–422. DOI: 10.1007/978-981-97-5131-0_35

Chen, L., Wang, X., & Zhang, D. (2023). Privacy-preserving biometric authentication using homomorphic encryption: A practical implementation. *IEEE Transactions on Information Forensics and Security, 18*, 3456–3470. DOI: 10.1109/TIFS.2023.3287654

Choras, R. S. (2020). Multimodal Biometrics for Person. *Security and Privacy From a Legal, Ethical, and Technical Perspective*, 177.

Chouchane, O., Brossier, B., Gamboa, J. E. G., Lardy, T., Tak, H., Ermis, O., Kamble, M. R., Patino, J., Evans, N., Önen, M., & Todisco, M. (2021, August). Privacy-preserving voice anti-spoofing using secure multi-party computation. *Interspeech, 2021*, 856–860. Advance online publication. DOI: 10.21437/Interspeech.2021-983

Claypoole, T. F. (2021, September 28). *Voice analysis complicates personal privacy.* Womble Bond Dickinson. https://www.womblebonddickinson.com/us/insights/blogs/voice-analysis-complicates-personal-privacy

Colorado General Assembly. (2024). *HB24-1130 Privacy of biometric identifiers & data.* https://leg.colorado.gov/bills/hb24-1130

Comparing privacy laws: GDPR vs. Personal Data Protection Act. (2023). OneTrust DataGuidance. https://www.dataguidance.com/sites/default/files/gdpr_v._personal_data_protection_act_argentina.pdf

Conklin, A., Dietrich, G., & Walz, D. (2004). Password-based authentication: A system perspective. *37th Annual Hawaii International Conference on System Sciences (pp.* 10-*pp). IEEE.*

Conti, M., Dehghantanha, A., Franke, K., & Watson, S. (2018). Internet of Things security and forensics: Challenges and opportunities. *Future Generation Computer Systems*, *78*, 544–546. DOI: 10.1016/j.future.2017.07.060

Dakshinamurthy, S. (2018). *The classification of knuckle creases and their prevalence within a population: Inter-observer and intra-observer validation* (Unpublished master's thesis). School of Science and Engineering, University of Dundee.

Dan, L., Hackman, L., Rinaldi, V., & Trucco, E. (2023). The persistence of knuckle creases during finger flexion for the identification of perpetrators from digital images of their hands. *Forensic Science International*, *348*, 111603. DOI: 10.1016/j.forsciint.2023.111603 PMID: 36842847

Daugman, J. (2004). How iris recognition works. *IEEE Transactions on Circuits and Systems for Video Technology*, *14*(1), 21–30. DOI: 10.1109/TCSVT.2003.818350

Deng, J., Guo, J., Xue, N., & Zafeiriou, S. (2019). Arcface: Additive angular margin loss for deep face recognition. In *Proceedings of the IEEE/CVF conference on computer vision and pattern recognition* (pp. 4690-4699). DOI: 10.1109/CVPR.2019.00482

Deng, Y., & Zhong, Y. (2015). Keystroke dynamics user authentication using advanced machine learning methods. *Recent Advances in User Authentication Using Keystroke Dynamics Biometrics, GCSR*, *2*, 23–40.

Deskus, C., & Fattal, J. R. (2019, October 17). *A Fourth Amendment framework for voiceprint database searches.* Just Security. https://www.justsecurity.org/66571/a-fourth-amendment-framework-for-voiceprint-database-searches

Dholi, P. R., & Chaudhari, K. P. (2012). Typing pattern recognition using keystroke dynamics. *International Conference on Advances in Information Technology and Mobile Communication (pp.* 275–280*)*. Springer.

Diviya, P., Logapriya, K., Febiyana, G. N., Sivashankari, M., & Kumar, R. D. (2015). Identification of suspects using finger knuckle patterns in biometric fusions. *International Journal of Research in Computer Science*, *2*(2), 5–9. https://researchscript.in/wp-content/uploads/2015/03/IJRCS020202.pdf

Does the Fifth Amendment protect biometrics? (2022, March 2). Purdue Global Law School. https://www.purduegloballawschool.edu/blog/constitutional-law/fifth-amendment-biometrics

Doran, B. M., Fordyce, T. S., Valdetero, J. M., & Pestine, Z. (2024, August 14). *BIPA update: Illinois limits liability and clarifies electronic consent for biometric data collection.* Greenberg Traurig. https://www.gtlaw.com/en/insights/2024/8/bipa-update-illinois-limits-liability-and-clarifies-electronic-consent-for-biometric-data-collection

Durina, M. (2021, December). 90 years later: Revisiting the Lindbergh kidnapping case. *Journal of the American Society of Questioned Document Examiners*, *24*(2), 3–14. DOI: 10.69525/jasqde.276

Dykstra, J., & Riehl, D. (2012). Forensic Collection of Electronic Evidence from Infrastructure-as-a-Service Cloud Computing. *Richmond Journal of Law and Technology*, 1-27.

Dykstra, J., & Sherman, T. (2012). Acquiring forensic evidence from infrastructure-as-a-service cloud computing: Exploring and evaluating tools, trust, and techniques. *Digital Investigation*, *9*, S90–S98. DOI: 10.1016/j.diin.2012.05.001

Egypt: Data protection laws of the world. (2024, January 19). DLA Piper. https://www.dlapiperdataprotection.com/index.html?t=law&c=EG

El-Abed, M., Hemdan, E. E.-D., & El-Sayed, A. (2024). Multi-modal biometric recognition system using deep learning fusion techniques. *Neural Computing & Applications*, *36*(4), 2145–2163. DOI: 10.1007/s00521-023-09156-x

Elgohary, H. M., Darwish, S. M., & Elkaffas, S. M. (2022). Improving Uncertainty in Chain of Custody for Image Forensics Investigation Applications. *IEEE Access : Practical Innovations, Open Solutions, 10*, 14669–14679. DOI: 10.1109/ACCESS.2022.3147809

Erdogmus, N., & Marcel, S. (2014). Spoofing face recognition with 3D masks. *IEEE Transactions on Information Forensics and Security, 9*(7), 1084–1097. DOI: 10.1109/TIFS.2014.2322255

Espejel, E., & Garcia Gómez, F. Mosquera, A., Oacheco Mendoza, J. J., Luna, W., Arellano, C. A., & Torres Montalvo, R. (2025, March 7). *Mexico enacts new data protection regime.* White & Case. https://www.whitecase.com/insight-alert/mexico-enacts-new-data-protection-regime

European Union. (2016). General Data Protection Regulation (GDPR). Regulation (EU) 2016/679.

Evershed, N., & Taylor, J. (2023, March 16). AI can fool voice recognition used to verify identity by Centrelink and Australian tax office. *The Guardian.* https://www.theguardian.com/technology/2023/mar/16/voice-system-used-to-verify-identity-by-centrelink-can-be-fooled-by-ai

Ezazul Islam, Md., Rafiqul Islam, Md., Chetty, M., & Lim, S., & Mehmood Chadhar. (2023). User authentication and access control to blockchain-based forensic log data. *EURASIP Journal on Information Security*20231DOI: 10.1186/s13635-023-00142-3

Fábián, I., & Gulyás, G. G. (2021). A comparative study on the Privacy Risk of face recognition libraries. *Acta Cybernetica (Szeged), 25*(2), 233–255. DOI: 10.14232/actacyb.289662

Fast Fourier Transformation FFT: Basics. (n.d.). NTI Audio. https://www.nti-audio.com/en/support/know-how/fast-fourier-transform-fft

Fayaz, F. A., Mohi-Ud-Din, S., Batool, I., Kaur, S., & Rashid, M. (2019). *novel face recognition based examine Authentication system using Python D-lib.* IEEE Conference on Image Information Processing.

Federal Bureau of Investigation. (2013, June 3). *Speaker recognition.* Biometric Center of Excellence. https://ucr.fbi.gov/fingerprints_biometrics/biometric-center-of-excellence/files/speaker-recognition.pdf

Federal Trade Commission. (2013, January 17). 16 CFR Part 312, Children's Online Privacy Protection Rule ("COPPA"). *Federal Register, 78*(12), 3972–4014. https://www.ftc.gov/system/files/2012-31341.pdf

Fernandes, R., Colaco, R. M., Shetty, S., & H., R. M. (2020). A New Era of Digital Forensics in the form of Cloud Forensics: A Review. *Inventive Research in Computing Applications*, 422-427.

Fidaner, Y. S., & Savaşçıhabeş, A. (2024). Examining IoT Applications in the Perspective of Security Technologies in 5G Networks. *Orclever Proceedings of Research and Development*, *12*(1), 99–110. DOI: 10.56038/oprd.v5i1.599

Fierrez, J., Morales, A., Vera-Rodriguez, R., & Camacho, D. (2021). Multiple classifiers in biometrics. Part 1: Fundamentals and review. *Information Fusion*, *75*, 136–150. DOI: 10.1016/j.inffus.2021.04.008

FindBiometrics. (2019). Major biometric data breach hits Suprema's BioStar 2 platform. Retrieved from https://findbiometrics.com

Frye Standard. (2022, December). Cornell School of Law Legal Information Institute. https://www.law.cornell.edu/wex/frye_standard

Fudge, T. P. (n.d.). *The International Phonetic Alphabet.* Fudgelink: Voice lessons. https://fudgelink.com/IPA

Galbally, J., Cappelli, R., Lumini, A., Gonzalez-de-Rivera, G., Maltoni, D., Fierrez, J., Ortega-Garcia, J., & Maio, D. (2010). An evaluation of direct attacks using fake fingers generated from ISO templates. *Pattern Recognition Letters*, *31*(8), 725–732. DOI: 10.1016/j.patrec.2009.09.032

Galbally, J., Marcel, S., & Fierrez, J. (2023). Image quality assessment for fake biometric detection: Application to iris, fingerprint, and face recognition. *IEEE Transactions on Image Processing*, *32*, 2164–2178. DOI: 10.1109/TIP.2023.3265421 PMID: 26270913

Gao, C., Yang, Z., Teoh, A. B. J., & Zhu, M. (2024). Beyond first-order: A multi-scale approach to finger knuckle print biometrics. *arXiv preprint arXiv:2406.19672*. https://doi.org//arXiv.2406.19672DOI: 10.48550

Garg, M., Arora, A. S., & Gupta, S. (2020). A novel feature biometric fusion approach for iris, speech and signature. Inform. W Technol. *Mater.*, *20*(2), 63–71.

Geitgey, A. (2019). Face recognition documentation.

Genelza, G. G. (2024, October 29). A systematic literature review on AI voice cloning generator: A game-changer or a threat? *Journal of Emerging Technologies*, *4*(2). Advance online publication. DOI: 10.57040/ag587791

Gennaro, M. (2024, February 28). Privacy class action over unauthorized voiceprint collection to proceed against Meta. *Courthouse News Service.* https://www.courthousenews.com/privacy-class-action-over-unauthorized-voiceprint-collection-to-proceed-against-meta

Ghelani, D., Hua, T. K., & Koduru, S. K. R. (2022). Cyber Security Threats, Vulnerabilities, and Security Solutions Models in Banking. *Authorea Preprints.* DOI: 10.22541/au.166385206.63311335/v1

Gomez-Barrero, M., Rathgeb, C., Galbally, J., Busch, C., & Fierrez, J. (2022). Protected facial biometric templates based on local ranking. *IEEE Transactions on Information Forensics and Security, 17*, 1082–1095. DOI: 10.1109/TIFS.2022.3157171

Gönen, S., Yılmaz, E. N., & Şanoğlu, S. (2021). Dijital dönüşümün vebası: Siber suçlar ve güvenlik. *Akademik Derleme, 6*(1), 112–129.

Goyal, N. K. (2025). Weaponizing Misinformation and Cyberattacks on Supply Chains. In Bukhari, S., & Zafar, I. (Eds.), *Mass Media and Impact of Fake News on Supply Chains* (pp. 177–210). IGI Global Scientific Publishing., DOI: 10.4018/979-8-3373-2175-2.ch007

Goyal, N. K. (2025e). Security and Privacy in IoT, Fog, and Blockchain Networks. In Sathio, A., Rind, M., Awan, S., & Junejo, A. (Eds.), *Energy-Efficient Deep Learning Approaches in IoT, Fog, and Green Blockchain Revolution* (pp. 371–398). IGI Global Scientific Publishing., DOI: 10.4018/979-8-3373-0300-0.ch015

Goyal, N. K., Dandotiya, M., Kumari, M., Kaloria, S., & Anushya, A. (2025d). AI-Powered Haptic Devices for Patient Rehabilitation. In Mamodiya, U., Kalam, A., Bossoufi, B., Awad, M., & Somwanshi, D. (Eds.), *Integrating AI With Haptic Systems for Smarter Healthcare Solutions* (pp. 143–170). IGI Global Scientific Publishing., DOI: 10.4018/979-8-3373-2307-7.ch007

Goyal, N. K., Dandotiya, M., Kumari, M., Sheikh, M. F., & Sharma, S. (2025b). Emerging Job Roles in the Tourism Sector Leveraging Artificial Intelligence. In Silic, M., Luthia, M., Ahmad, F., & Chakraborty, P. (Eds.), *Empowering Sustainable Performance and Competitive Advantage in Tourism* (pp. 89–126). IGI Global Scientific Publishing., DOI: 10.4018/979-8-3373-4546-8.ch004

Goyal, N. K., Dandotiya, M., Qurashi, J. A., Kumari, M., & Sharma, S. (2025c). AI-Driven Predictive Analytics for Disease Management. In Ouaissa, M., Ouaissa, M., Imad, M., Qurashi, J., & Farooq, M. (Eds.), *Utilizing AI of Medical Things for Healthcare Security and Sustainability* (pp. 231–276). IGI Global Scientific Publishing., DOI: 10.4018/979-8-3373-0690-2.ch008

Goyal, N. K., Kumari, M., Kushwah, V., Dandotiya, M., & Vijay, A. (2025a). Evaluating blockchain security: Trade-offs in consensus mechanisms. *IET Conference Proceedings*, 2024(38), 80–88. DOI: 10.1049/icp.2025.0775

Grubb, B. (2014). Australian Taxation Office uses 'voiceprint' technology to speed up calls. *The Sydney Morning Herald*. https://www.smh.com.au/technology/australian-taxation-office-uses-voiceprint-technology-to-speed-up-calls-20140909-10edli.html

Gu, H., Yi, J., Wang, C., Ren, Y., Tao, J., Yan, X., Chen, Y., & Zhang, X. (2024, October 28). Utilizing speaker profiles for impersonation audio detection. [Conference proceedings]. *32nd ACM International Conference on Multimedia*. pp. 1961-1970. DOI: 10.1145/3664647.3681602

Gully, A., Harrison, P., Hughes, V., Rhodes, R., & Wormald, J. (2022, February 14). How voice analysis can help solve crimes. *Frontiers for Young Minds*, *10*, 702664. DOI: 10.3389/frym.2022.702664

Gunasekaran, K., Raja, J., & Pitchai, R. (2019). Deep multimodal biometric recognition using contourlet derivative weighted rank fusion with human face, fingerprint and iris images. *Automatika (Zagreb)*, *60*(3), 253–265. DOI: 10.1080/00051144.2019.1565681

Gündüz, G., & Cedimoğlu, İ. H. (2019). *Derin öğrenme algoritmalarını kullanarak görüntüden cinsiyet tahmini*. Sakarya University Journal of Computer and Information Sciences, 2(1).

Gu, X., Angelov, P. P., Zhang, C., & Atkinson, P. M. (2018). A Massively Parallel Deep Rule-Based Ensemble Classifier for Remote Sensing Scenes. *IEEE Geoscience and Remote Sensing Letters*, *15*(3), 345–349. DOI: 10.1109/LGRS.2017.2787421

Guy-Powell v Applebee's Restaurants LLC et al., case number 2022-CH-08365. Class Action. https://www.classaction.org/media/guy-powell-et-al-v-applebees-restaurants-llc-et-al.pdf

Hadid, A., Evans, N., Marcel, S., & Fierrez, J. (2021). Biometrics systems under spoofing attack: An evaluation methodology and lessons learned. *IEEE Signal Processing Magazine*, *38*(4), 20–30. DOI: 10.1109/MSP.2021.3077008

Haghighat, M., Abdel-Mottaleb, M., & Alhalabi, W. (2016). Discriminant correlation analysis: Real-time feature level fusion for multimodal biometric recognition. *IEEE Transactions on Information Forensics and Security*, *11*(9), 1984–1996. DOI: 10.1109/TIFS.2016.2569061

Hally, L. (2021, September 21). *Voiceprint hack*. Cyber Security: Luke Hally. https://www.lukehally.au/data-security-privacy/voiceprint-hack

Hameed, R. T., & Mohamad, O. A. (2023). Federated Learning in IoT: A survey on distributed decision making. *BJIoT, 2023*, 1–7. DOI: 10.58496/BJIoT/2023/001

Hammad, M., Zhang, S., & Wang, K. (2019). A novel two-dimensional ECG feature extraction and classification algorithm based on convolution neural network for human authentication. *Future Generation Computer Systems, 101*, 180–196. DOI: 10.1016/j.future.2019.06.008

Hansen, J. H. L., Sreenivasa Rao, K. S., D'Urso, P., & Kim, W. (2005). *Cepstral.* Science Direct Topics. https://www.sciencedirect.com/topics/computer-science/cepstral

Haque, M. A., Khan, N. Z., & Khatoon, G. (2015). Authentication through keystrokes: What you type and how you type. *2015 IEEE International Conference on Research in Computational Intelligence and Communication Networks (pp.* 257–261*)*. IEEE. DOI: 10.1109/ICRCICN.2015.7434246

Harland-Logan, S. (2023, November 14). *Exonerations: Guy Paul Morin.* Innocence Canada. https://www.innocencecanada.com/the-latest/exoneration/guy-paul-morin

Hasan, R. T. H., & Sallow, A. B. (2021). Face detection and recognition using OpenCV. *Journal of Soft Computing and Data Mining*, *2*(2), 86-97.

Hautamäki, R. G., Kanervisto, A., Hautamäki, V., & Kinnunen, T. (2018, June 26-29). *Perceptual evaluation of the effectiveness of voice disguise by age modification.* Speaker Odyssey 2018: The Speaker and Language Recognition, Les Sables d'Olonne, France (pp. 320-326). https://www.isca-archive.org/odyssey_2018/hautamaki18_odyssey.pdf

He, K., Zhang, X., Ren, S., & Sun, J. (2016). Deep residual learning for image recognition. In *Proceedings of the IEEE conference on computer vision and pattern recognition* (pp. 770-778).

How can speech recognition be used in law enforcement? (2023, May 23). Amberscript. https://www.amberscript.com/en/blog/how-can-speech-recognition-be-used-in-law-enforcement

Huang, J., & Ma, Y. (2018). Bat algorithm based on an integration strategy and Gaussian distribution. [Affiliation removed per APA rules.]. *Applied Soft Computing, 63*, 141–154. DOI: 10.1016/j.asoc.2017.11.040

Hussian, A., Murshed, F., Alandoli, M. N., & Aljafari, G. (2025). A Hybrid Deep Learning Approach for Secure Biometric Authentication Using Fingerprint Data. *Computers*, *14*(5), 178. DOI: 10.3390/computers14050178

İlkbahar, F., Ünal, S., & Karakaya, AT (2021). Akıllı Ev Sistemleri Üzerine Bir Model Önerisi. *AJIT-e*, *12*(45), 90–105.

Illinois General Assembly. (2008). *Civil Liabilities (740 ILCS 14/) Biometric Information Privacy Act.* https://www.ilga.gov/Legislation/ILCS/Articles?ActID=3004&ChapterID=57

Ingale, M., Cordeiro, R., Thentu, S., Park, Y., & Karimian, N. (2020). ECG biometric authentication: A comparative analysis. *IEEE Access : Practical Innovations, Open Solutions*, *8*, 117853–117866. DOI: 10.1109/ACCESS.2020.3004464

Introna, L., & Wood, D. (2004). Picturing algorithmic surveillance: The politics of facial recognition systems. *Surveillance & Society*, *2*(2/3), 177–198.

Islam, M. R., & Aktheruzzaman, K. M. (2020). An analysis of cybersecurity attacks against internet of things and security solutions. *Journal of Computer and Communications*, *8*(4), 19–28. DOI: 10.4236/jcc.2020.84002

Jain, A. K., Nandakumar, K., & Ross, A. (2016). 50 years of biometric research: Accomplishments, challenges, and opportunities. *Pattern Recognition Letters*, *79*, 80–105. DOI: 10.1016/j.patrec.2015.12.013

Jain, A. K., Ross, A., & Pankanti, S. (2006). Biometrics: A tool for information security. *IEEE Transactions on Information Forensics and Security*, *1*(2), 125–143. DOI: 10.1109/TIFS.2006.873653

Jain, A. K., Ross, A., & Prabhakar, S. (2004). An introduction to biometric recognition. *IEEE Transactions on Circuits and Systems for Video Technology*, *14*(1), 4–20. DOI: 10.1109/TCSVT.2003.818349

Jander, N., Hutter, N., Mueller, T., Immerz, A., Stritt, F., Traser, L., Spahn, C., & Richter, B. (2025, June 27). Protective factors for vocal health in teachers: The role of singing, voice training, and self-efficacy. *International Journal of Environmental Research and Public Health*, *22*(7), 1018. Advance online publication. DOI: 10.3390/ijerph22071018 PMID: 40724085

Jansen, F., Sánchez-Monedero, J., & Dencik, L. (2021, July). Biometric identity systems in law enforcement and the politics of (voice) recognition: The case of SiiP. *Big Data & Society*, *8*(2), 20539517211063604. Advance online publication. DOI: 10.1177/20539517211063604

Ja, R., Yang, Y., & Grundy, J. (2021). A systematic review of scheduling approaches on multi-tenancy cloud platforms. *Information and Software Technology*, *132*, 106478. DOI: 10.1016/j.infsof.2020.106478

Jasserand, C. (2024). Biometric Data, Within and Beyond Data Protection. In The Boundaries of Data (pp. 295–310). Amsterdam University Press. https://ssrn.com/abstract=4483962

Javed, A. R., Ahmed, W., Alazab, M., Jalil, Z., Kifayat, K., & Gadekallu, T. R. (2022). A comprehensive survey on computer forensics: State-of-the-Art, tools, techniques, challenges, and future directions. *IEEE Access : Practical Innovations, Open Solutions*, *10*(1), 11065–11089. DOI: 10.1109/ACCESS.2022.3142508

Jin, Z., Lim, M. H., & Teoh, A. B. J. (2023). Blockchain-based biometric authentication system with privacy preservation. *Journal of Network and Computer Applications*, *204*, 103415. DOI: 10.1016/j.jnca.2022.103415

Joseph, J. (2017). Technoprison: Technology and prisons. In Moriarty, L. J. (Ed.), *Criminal Justice Technology in the 21st Century* (3rd ed., pp. 172–205). Charles C. Thomas.

Jung, Y., & Virgil, E. D. (2024). Analysis of Legislative Framework Governing Biometric Data. *Procedia Computer Science*, *241*, 48–55. DOI: 10.1016/j.procs.2024.08.009

Kabakus, A. T. (2019). *An experimental performance comparison of widely used face Detection tools*. ADCA J. DOI: 10.14201/ADCAIJ201983512

Kabir, W., Ahmad, M. O., & Swamy, M. N. S. (2019). A multi-biometric system based on feature and score level fusions. *IEEE Access : Practical Innovations, Open Solutions*, *7*, 59437–59450. DOI: 10.1109/ACCESS.2019.2914992

Kadane, J. B. (2020). Knuckles and fingerprints: A comparison and case study. *Journal of Forensic Identification*, *70*(3), 265–272. https://dr.lib.iastate.edu/bitstreams/5c66d011-8161-4b8b-bc68-b9b27cc4bb55/download

Kadri, F., Meraoumia, A., Bendjenna, H., & Chitroub, S. (2016, March). *Palmprint & iris for a multibiometric authentication scheme using Log-Gabor filter response. In 2016 international conference on information technology for organizations development (IT4OD)*. IEEE.

Kaplan, E., & Rahman, M. (2024). OpenCV ile Yüz Tespiti. *IJANSER*, *8*(2), 530–534.

Kaşif, A. (2020). Design of Privacy and Security Centric Home Automation System [Master's thesis]. Bursa Uludağ University, Türkiye.

Kaur, H., & Khanna, P. (2024). Digital forensics in biometric systems: Challenges and research opportunities. *Forensic Science International Digital Investigation*, *49*, 301723. DOI: 10.1016/j.fsidi.2024.301723

Khan, M., Sen, A., Singh, O. P., Chaudhary, J. K., Soni, M., & Anbazhagan, K. (2024). Deep learning for biometric attack detection and recognition. In K. Santosh et al. (Eds.), Recent trends in image processing and pattern recognition (RTIP2R 2023) (Vol. 2026, pp. [insert page range]). Springer. DOI: 10.1007/978-3-031-53082-1_29

Khan, F. A., & Khan, M. K. (2025). Generative AI and deepfake detection in biometric systems. *Cognitive Computation, 17*(1), 112. DOI: 10.1007/s12559-025-10469-3

Khare, P., & Srivastava, S. (2023, June). Enhancing security with voice: A comprehensive review of AI-based biometric authentication systems. *International Journal of Research and Analytical Reviews, 10*(2), 398–403. https://ijrar.org/papers/IJRAR23B4732.pdf

Kimani, K., Oduol, V., & Langat, K. (2019). Cyber security challenges for IoT-based smart grid networks. *International Journal of Critical Infrastructure Protection, 25*, 36–49. DOI: 10.1016/j.ijcip.2019.01.001

Kimery, A. (2025, May 23). *Move in House to block state AI laws draws bipartisan fire; Senate support questionable.* Biometric Update. https://www.biometricupdate.com/202505/move-in-house-to-block-state-ai-laws-draws-bipartisan-fire-senate-support-questionable

Kim, J., Kim, H., & Huh, J.-H. (2022). A study on IoT-based biometric authentication system security threats and countermeasures. *Applied Sciences (Basel, Switzerland), 12*(10), 4956. DOI: 10.3390/app12104956

Kimura, S., Schaumann, B. A., Plato, C. C., & Kitagawa, T. (1990). Embryological development and prevalence of digital flexion creases. *The Anatomical Record, 226*(2), 249–257. DOI: 10.1002/ar.1092260214 PMID: 2301741

Kindt, E. J. (2013). *Privacy and data protection issues of biometric applications: A comparative legal analysis.* Springer. DOI: 10.1007/978-94-007-7522-0

Klasén, L., Fock, N., & Forchheimer, R. (2024). The invisible evidence: Digital forensics as key to solving crimes in the digital age. *Forensic Science International, 362*, 112133–112133. DOI: 10.1016/j.forsciint.2024.112133 PMID: 39089208

Korshunov, P., & Marcel, S. (2023). Vulnerability assessment and detection of morphing attacks in face recognition systems. *IEEE Transactions on Biometrics, Behavior, and Identity Science, 5*(2), 234–248. DOI: 10.1109/TBIOM.2023.3245678

Kose, N., & Dugelay, J.-L. (2013). On the vulnerability of face recognition systems to spoofing mask attacks. In *Proceedings of the IEEE International Conference on Acoustics, Speech and Signal Processing (ICASSP)* (pp. 2357–2361). IEEE. DOI: 10.1109/ICASSP.2013.6638076

Kramer, R. S. S., Jones, A. L., & Gous, G. (2021). Individual differences in face and voice matching abilities: The relationship between accuracy and consistency. *Applied Cognitive Psychology*, *35*(1), 192–202. DOI: 10.1002/acp.3754

Krishna Prakasha, K., & Sumalatha, U. (2025, January). Privacy-preserving techniques in biometric systems: Approaches and challenges. *IEEE Access : Practical Innovations, Open Solutions*, *13*, 32584–32616. DOI: 10.1109/ACCESS.2025.3541649

Krizhevsky, A., Sutskever, I., & Hinton, G. E. (2012). ImageNet classification with deep convolutional neural networks. *Advances in Neural Information Processing Systems*, *25*, 1097–1105.

Kulkarni, S. S., & Rout, R. D. (2012). Secure biometrics: Finger knuckle print. *International Journal of Advanced Research in Computer and Communication Engineering*, *1*(10), 852–854. https://www.ijarcce.com/upload/december/21-Secure%20Biometrics.pdf

Kumar Singh, K., & Barde, S. (2023). A feasible adaptive fuzzy genetic technique for face, fingerprint, and palmprint based multimodal biometrics systems. *Journal of Current Science and Technology*, *14*(1). Advance online publication. DOI: 10.59796/jcst.V14N1.2024.1

Kumar, A. (2014). Importance of being unique from finger dorsal patterns: Exploring minor finger knuckle patterns in verifying human identities. *IEEE Transactions on Information Forensics and Security*, *9*(8), 1288–1298. DOI: 10.1109/TIFS.2014.2328869

Kumar, A., & Ravikanth, C. (2009). Personal authentication using finger knuckle surface. *IEEE Transactions on Information Forensics and Security*, *4*(1), 98–110. DOI: 10.1109/TIFS.2008.2011089

Kumar, A., & Zhou, Y. (2009). Human identification using knuckle codes. In *Proceedings of the IEEE Biometrics Symposium* (pp. 1–6). IEEE. DOI: 10.1109/BTAS.2009.5339021

Kumar, V. M., Phanideep, D. M., Purushotham, C., Ramesh, R., & Mohan, J. (2025). User authentication from ECG signals using 1D convolutional neural network. *Procedia Computer Science*, *258*, 1262–1273. DOI: 10.1016/j.procs.2025.04.360

Kushwaha, S. S., Joshi, S., Singh, D., Kaur, M., & Lee, H.-N. (2022). Systematic Review of Security Vulnerabilities in Ethereum Blockchain Smart Contract. *IEEE Access : Practical Innovations, Open Solutions*, *10*, 6605–6621. DOI: 10.1109/ACCESS.2021.3140091

Kutlugün, M. A. (2023). *REAL-TIME FACE RECOGNITION USING EXTENDED CONVOLUTIONAL NEURAL NETWORKS [Unpublished doctoral thesis]*. İstanbul Sabahattin University.

Kwon, O., Bae, S., & Shin, B. (2020). Understanding the adoption intention of AI through the ethics lens. In T. X. Bui (Ed.), Proceedings of the 53rd Annual Hawaii International Conference on System Sciences, HICSS 2020 (pp. 4972-4981). (Proceedings of the Annual Hawaii International Conference on System Sciences; Vol. 2020-January). IEEE Computer Society. DOI: 10.24251/HICSS.2020.611

Kyi, A. S. S., & Lin, K. Z. (2019, November). Detecting voice features for criminal case. *Proceedings of the 2019 International Conference on Advanced Information Technologies (ICAIT)*, Yangon, Myanmar, 212-216. DOI: 10.1109/AITC.2019.8921212

Lee, C. K., Chang, C. C., Johar, A., & Wong, P. F. (2013). Fingerprint changes and verification failure among patients with hand dermatitis. *JAMA Dermatology*, *149*(3), 294–299. DOI: 10.1001/jamadermatol.2013.1425 PMID: 23682364

Li, J., Lei, H., Alavi, A. H., & Wan, G.-G. (2019). Elephant herding optimization: Variants, hybrids, and applications. [Note: standardized to journal publication; confirm if a different version is required.]. *Mathematics*, *7*(12), 1–30. DOI: 10.3390/math7121140

Li, M., Lal, C., Conti, M., & Hu, D. (2021). LEChain: A blockchain-based lawful evidence management scheme for digital forensics. *Future Generation Computer Systems*, *115*, 406–420. DOI: 10.1016/j.future.2020.09.038

Lin, J., Yu, W., Zhang, N., Yang, X., & Zhao, W. (2017). A survey on Internet of Things: Architecture, enabling technologies, security and privacy, and applications. *IEEE Internet of Things Journal*, *4*(5), 1125–1142. DOI: 10.1109/JIOT.2017.2683200

Lin, S., Li, C.-T., & Kot, A. C. (2020). Multi-domain adversarial feature generalization for person re-identification. *IEEE Transactions on Image Processing*, *30*, 1596–1607. DOI: 10.1109/TIP.2020.3046864 PMID: 33382653

Li, S., & Jain, A. K. (Eds.). (2015). *Encyclopedia of Biometrics* (2nd ed.). Springer. DOI: 10.1007/978-1-4899-7488-4

Liu, Y., Stehouwer, J., Jourabloo, A., & Liu, X. (2021). Deep tree learning for zero-shot face anti-spoofing. *IEEE/CVF Conference on Computer Vision and Pattern Recognition*, 4680-4689. DOI: 10.1109/CVPR46437.2021.00465

Liu, Y., Xiao, Y., Li, S., Liang, W., & Chen, C. P. (2019). Cyber security and privacy issues in smart grids. *IEEE Communications Surveys and Tutorials*, *14*(4), 981–997. DOI: 10.1109/SURV.2011.122111.00145

Li, X., Zheng, W. S., Wang, X., Xiang, T., & Gong, S. (2015). Multi-scale learning for low-resolution person re-identification. In *Proceedings of the IEEE international conference on computer vision* (pp. 3765-3773).

Lone, A. N., Mustajab, S., & Alam, M. (2023). A comprehensive study on cybersecurity challenges and opportunities in the IoT world. *Security and Privacy*, *6*(2), e318. DOI: 10.1002/spy2.318

Lugovaya, T. S. (2005). *Biometric human identification based on electrocardiogram* (Master's thesis). Electrotechnical University "LETI," Saint Petersburg, Russian Federation.

Lungova, V., & Thibeault, S. L. (2020, October). Mechanisms of larynx and vocal fold development and pathogenesis. *Cellular and Molecular Life Sciences : CMLS*, *77*(19), 3781–3795. DOI: 10.1007/s00018-020-03506-x PMID: 32253462

Luo, C., Song, C., & Zhang, Z. (2020). Generalizing person re-identification by camera-aware invariance learning and cross-domain mixup. In *Proceedings of the European Conference on Computer Vision* (pp. 224–241). Springer. DOI: 10.1007/978-3-030-58555-6_14

Lyon, D. (2007). *Surveillance studies: An overview*. Polity Press.

Makowski, D., Pham, T., Lau, Z. J., Brammer, J. C., Lespinasse, F., Pham, H., Schölzel, C., & Chen, S.-A. (2021). NeuroKit2: A Python toolbox for neurophysiological signal processing. *Behavior Research Methods*, *53*(4), 1689–1696. DOI: 10.3758/s13428-020-01516-y PMID: 33528817

Malik, G. (2024). Biometric Authentication — Risks and advancements in biometric security systems. *Journal of Computer Science and Technology Studies*, *6*(3), 159–180. DOI: 10.32996/jcsts.2024.6.3.14

Manisha, & Kumar, N. (2020). Cancelable biometrics: A comprehensive survey. *Artificial Intelligence Review, 53*(5), 3403–3446. https://doi.org/DOI: 10.1007/s10462-019-09753-8

Manral, B., Somani, G., & Conti, M. (2019). A Systematic Survey on Cloud Forensics Challenges, Solutions, and Future Directions. *ACM Computing Surveys*, •••, 1–38.

Martinez, A. (2025, June 27). Colorado's biometric privacy law takes effect July 1: Are you ready? *Forbes*. https://www.forbes.com/sites/alonzomartinez/2025/06/27/colorados-biometric-privacy-law-takes-effect-july-1-are-you-ready

Massachusetts Appellate Courts. (2024). *Commonwealth v Thanh Du SJC-13557*. https://www.ma-appellatecourts.org/docket/SJC-13557

Mayhew, S. (2021, October 5). *One in 10 New Zealanders have a voiceprint.* Biometric Update. https://www.biometricupdate.com/201210/one-in-10-new-zealanders-have-a-voiceprint

Mbimbi, B., Murray, D., & Wilson, M. (2024). IoT Forensics-Based on the Integration of a Permissioned Blockchain Network. *Blockchains, 2*(4), 482–506. DOI: 10.3390/blockchains2040021

McConvey, J. R. (2025, February 5). *Amazon can't block BIPA voice biometrics case even though federal case failed.* Biometric Update. https://www.biometricupdate.com/202502/amazon-cant-block-bipa-voice-biometrics-case-even-though-federal-case-failed

Meden, B., Rot, P., Terhörst, P., Damer, N., Kuijper, A., Scheirer, W. J., Ross, A., Peer, P., & Struc, V. (2021). Privacy–Enhancing Face Biometrics: A Comprehensive Survey. *IEEE Transactions on Information Forensics and Security, 16*, 4147–4183. DOI: 10.1109/TIFS.2021.3096024

Meingast, M., Roosta, T., & Sastry, S. (2007). Security and privacy issues with health care information technology. *Proceedings of the 28th IEEE EMBS Annual International Conference*, 5453–5458.

Miller, D. R. (1973, March). Criminal law: Voiceprint evidence is being heard. *Florida Law Review, 25*(3), 10. https://scholarship.law.ufl.edu/flr/vol25/iss3/10

Minaee, S., Abdolrashidi, A., Su, H., Bennamoun, M., & Zhang, D. (2023). Biometrics recognition using deep learning: A survey. *Artificial Intelligence Review, 56*(8), 8647–8695. DOI: 10.1007/s10462-022-10237-x

Mirjalili, V., Raschka, S., & Ross, A. (2018, October). Gender privacy: An ensemble of semi adversarial networks for confounding arbitrary gender classifiers. 2018 IEEE 9th International Conference on Biometrics, Applications and Systems (BTAS). DOI: 10.48550/arXiv.1807.11936

Mirjalili, S., & Lewis, A. (2016). The whale optimization algorithm. *Advances in Engineering Software, 95*, 51–67. DOI: 10.1016/j.advengsoft.2016.01.008

Mirjalili, S., Mirjalili, S. M., & Lewis, A. (2014). Grey wolf optimizer. *Advances in Engineering Software, 69*, 46–61. DOI: 10.1016/j.advengsoft.2013.12.007

Monahan, T. (2011). Surveillance as governance: Social inequality and the pursuit of democratic surveillance. *Surveillance & Society, 9*(3), 1–17. DOI: 10.24908/ss.v9i1/2.4099

Monrose, F., Reiter, M. K., & Wetzel, S. (1999). Password hardening based on keystroke dynamics. *Proceedings of the 6th ACM Conference on Computer and Communications Security (pp.* 73–82*)*.

Moore, K. L., Dalley, A. F., & Agur, A. M. R. (2018). *Clinically oriented anatomy* (8th ed.). Lippincott Williams & Wilkins.

Moussa, A. F. (2021). Electronic evidence and its authenticity in forensic evidence. *Egyptian Journal of Forensic Sciences*, *11*(1), 20. Advance online publication. DOI: 10.1186/s41935-021-00234-6

Mtibaa, A., Petrovska-Delacrétaz, D., Boudy, J., & Ben Hamida, A. (2021). Privacy-preserving speaker verification system based on binary I-vectors. *IET Biometrics*, *10*(3), 233–245. DOI: 10.1049/bme2.12013

Munjal, K., & Bhatia, R. (2022). A systematic review of homomorphic encryption and its contributions in healthcare industry. *Complex & Intelligent Systems*, *9*(4), 3759–3786. Advance online publication. DOI: 10.1007/s40747-022-00756-z PMID: 35531323

Naidu, B. R., & Babu, M. S. P. (2018). Biometric authentication data with three traits using compression technique, HOG, GMM and fusion technique. *Data in Brief*, *18*, 1976–1986. DOI: 10.1016/j.dib.2018.03.115 PMID: 29900333

Naik, S., & Maral, V. (2017). Cyber security—IoT. In *2nd IEEE International Conference on Communication and Electronics Systems* (pp. 159–164). IEEE.

Nanni, L., & Lumini, A. (2009). A multi-matcher system based on knuckle-based features. *Neural Computing & Applications*, *18*(1), 87–91. DOI: 10.1007/s00521-007-0160-4

Nasser, T. (2024, February 23). *Are data protection regulations all the same?* Right-Hand Cybersecurity. https://right-hand.ai/blog/are-data-protection-regulations-all-the-same

Neshenko, N., Bou-Harb, E., Crichigno, J., Kaddoum, G., & Ghani, N. (2019). Demystifying IoT security: A survey on IoT vulnerabilities and attacks. *IEEE Communications Surveys and Tutorials*, *21*(3), 2702–2733. DOI: 10.1109/COMST.2019.2910750

Nguyen, D. C., Hosseinalipour, S., Love, D. J., Pathirana, P. N., & Brinton, C. G. (2022). Latency Optimization for Blockchain-Empowered Federated Learning in Multi-Server Edge Computing. *IEEE Journal on Selected Areas in Communications*, *40*(12), 3373–3390. DOI: 10.1109/JSAC.2022.3213344

Nguyen, D. T., Pham, T. D., Lee, Y. W., & Park, K. R. (2023). Deep learning-based enhanced presentation attack detection for iris recognition by combining features from local and global regions based on NIR camera sensor. *Sensors (Basel)*, *23*(5), 2726. DOI: 10.3390/s23052726 PMID: 36904928

NIST. (2022). Face Recognition Vendor Test (FRVT). National Institute of Standards and Technology. Retrieved from https://www.nist.gov

Office of the Victorian Information Commissioner. (n.d.). Biometrics and privacy: Issues and challenges. https://ovic.vic.gov.au/privacy/resources-for-organisations/biometrics-and-privacy-issues-and-challenges

Öğüt, G. E. (2019). Comparison of CPU and GPU based acceleration techniques in machine learning and deep learning [Unpublished master's thesis]. Muğla Sıtkı Koçman University.

Oloyede, M. O., Hancke, G. P., & Myburgh, H. C. (2024). A review on face liveness detection algorithms. *IEEE Access : Practical Innovations, Open Solutions*, *12*, 15234–15257. DOI: 10.1109/ACCESS.2024.3362891

Omar, I. A., Jayaraman, R., Debe, M. S., Salah, K., Yaqoob, I., & Omar, M. (2021). Automating Procurement Contracts in the Healthcare Supply Chain using Blockchain Smart Contracts. *IEEE Access : Practical Innovations, Open Solutions*, *9*(1), 1–1. DOI: 10.1109/ACCESS.2021.3062471

Omotosho, L., Ogundoyin, I., Adebayo, O., & Oyeniyi, J. (2021). An enhanced multimodal biometric system based on convolutional Neural Network. *Journal of Engineering Studies and Research*, *27*(2), 73–81. DOI: 10.29081/jesr.v27i2.276

Oregon Department of Justice. Consumer Protection. (n.d.). *Consumer privacy: Oregon Consumer Privacy Law.* https://www.doj.state.or.us/consumer-protection/id-theft-data-breaches/privacy

Oregon Department of Justice. Media. (2024, July 2). *Oregonians gain significant new consumer privacy protections.* https://www.doj.state.or.us/media-home/news-media-releases/oregonians-gain-significant-new-consumer-privacy-protections

Osorio-Roig, D., Rathgeb, C., Drozdowski, P., & Busch, C. "Stable Hash Generation for Efficient Privacy-Preserving Face Identification", <i>IEEE Transactions on Biometrics, Behavior, and Identity Science</i>, vol. 4, no. 3, IEEE, pp. 333–348, 2022. DOI: 10.1109/TBIOM.2021.3100639

Özdemir, M. F. (2020). Comparison of face recognition algorithms in the embedded systems [Unpubllished master's thesis]. Afyon Kocatepe University.

Pan, S., Hoque, S., & Deravi, F. (2022). An Attention-Guided Framework for Explainable Biometric Presentation Attack Detection. *Sensors (Basel), 22*(9), 3365. DOI: 10.3390/s22093365 PMID: 35591055

Parashar, A. (2023). Cybersecurity threats in the Internet of Things (IoT). *IOSR Journal of Computer Engineering, 25*(5), 1–8.

Parmar, D. G. (2025, April 17). AI voice cloning: how a Bollywood veteran set a legal precedent. WIPO Magazine. https://www.wipo.int/web/wipo-magazine/articles/ai-voice-cloning-how-a-bollywood-veteran-set-a-legal-precedent-73631

Pasupuleti, M. K. (2025). *AI-Enabled Multimodal Biometrics: Advancing Security with Facial.* Voice, and Behavioral Integration., DOI: 10.62311/nesx/77579

Patel, V. M., Chellappa, R., Chandra, D., & Barbello, B. (2021). Continuous user authentication on mobile devices: Recent progress and remaining challenges. *IEEE Signal Processing Magazine, 38*(4), 49–61. DOI: 10.1109/MSP.2016.2555335

Patel, V. M., Gang, H., & Chellappa, R. (2015). Cancelable biometrics: A review. *IEEE Signal Processing Magazine, 32*(5), 54–65. DOI: 10.1109/MSP.2015.2434151

Patil, V. K., Birajdar, G. S., Chaudhari, S. M., & Gandhamal, A. N. (2023). Comparative approach for face Detection in Python, OpenCV and Hardware, *JETIR, 10*(7), 675-681.

Patro, K. K., Prakash, A. J., Rao, M. J., & Kumar, P. R. (2022). An efficient optimized feature selection with machine learning approach for ECG biometric recognition. *Journal of the Institution of Electronics and Telecommunication Engineers, 68*(4), 2743–2754. DOI: 10.1080/03772063.2020.1725663

Payne, A., Austin, T., & Clemons, A. M. (2024, June). Beyond the front yard: The dehumanizing message of accent-altering technology. *Applied Linguistics, 45*(3), 553–560. DOI: 10.1093/applin/amae002

Pehlivanoğlu, M. K., Kuyucu, A., & Kaya, R. (2023). IoT veri kümelerinde makine öğrenmesi tekniklerine dayalı saldırı tespiti. *Avrupa Bilim ve Teknoloji Dergisi, 25*(110), 312–325.

Penti, R. S., & Adams, A. A. (2025, July 14). *AI and tech under the One Big Beautiful Bill Act: Key restrictions, risks, and opportunities.* Ropes & Gray. https://www.ropesgray.com/en/insights/alerts/2025/07/ai-and-tech-under-the-one-big-beautiful-bill-act-key-restrictions-risks-and-opportunities

People v. King. (1968, October 10). Cal. App. 2d California Courts of Appeal Cases, *266.* https://law.justia.com/cases/california/court-of-appeal/2d/266/437.html

Pineau, E., Murray, M., & Chmielewski, D. (2025, July 31). Voice actors push back as AI threatens dubbing industry. *Reuters.* https://www.reuters.com/business/media-telecom/voice-actors-push-back-ai-threatens-dubbing-industry-2025-07-30

Platt, M., & McBurney, P. (2023). Sybil in the Haystack: A Comprehensive Review of Blockchain Consensus Mechanisms in Search of Strong Sybil Attack Resistance. *Algorithms, 16*(1), 34. DOI: 10.3390/a16010034

Polsinelli. (2024). *Global data protection law guide: USA – Illinois.* Multilaw. https://multilaw.com/Multilaw/Multilaw/Data_Protection_Laws_Guide/DataProtection_Guide_USA_Illinios.aspx

Prakash, A. J., Patro, K. K., Samantray, S., Pławiak, P., & Hammad, M. (2023). A deep learning technique for biometric authentication using ECG beat template matching. *Information (Basel), 14*(2), 65. Advance online publication. DOI: 10.3390/info14020065

Pravallika, P., & Satya Prasad, K. (2016, August). SVM classification for fake biometric detection using image quality assessment: Application to iris, face and palm print. In Proceedings of the 2016 International Conference on Inventive Computation Technologies (ICICT) (pp. –). IEEE. DOI: 10.1109/INVENTIVE.2016.7823189

Pugliese, A. (2020). Deepfake detection using convolutional neural networks. *IEEE Transactions on Neural Networks and Learning Systems, 31*(10), 3769–3780.

Pustovoitov, P. S., Seilova, N. A., & Gnatiuk, A. S. (2024, March 15). Analysis of voice impersonation fraud: Risks, cases and defense strategies. *International Journal of Information and Communication Technology, 5*(1), 122–134. DOI: 10.54309/IJICT.2024.17.1.010

Qasim, I. N. Q. (2023). *Using facial recognition software to analyze the emotions of individuals with special needs* [Unpublished master's thesis]. Selcuk University.

Qian, J., Du, H., Hou, J., Chen, L., Jung, T., & Li, X.-Y. (2018, November 4). Hidebehind: Enjoy voice input with voiceprint unclonability and anonymity. *SenSys '18: Proceedings of the 16th ACM Conference on Embedded Networked Sensor Systems* (pp. 82-94), Shenzhen, China. DOI: 10.1145/3274783.3274855

Quinn-Barabanov, J., Berman, E., & Reid, S. (2024, March 29). *BIPA 2.0? Washington's new privacy law creates private litigation and AG enforcement risk for businesses.* Steptoe. https://www.steptoe.com/en/news-publications/bipa-20-washingtons-new-privacy-law-creates-private-litigation-and-ag-enforcement-risk-for-businesses.html

Quinn, P., Ellyne, E., & Yao, C. (2024). Will the GDPR Restrain Health Data Access Bodies Under the European Health Data Space (EHDS)? *Computer Law & Security Review, 54*, 105993–105993. DOI: 10.1016/j.clsr.2024.105993

Quinton, C., & Lipton, B. (2022, September 1). The catalog of carceral surveillance: Voice recognition and surveillance. *Prison Legal News, 33*(9), 27–28.

Rabbi, B. M. S., Haque, E., Hui, H. S., & Taassum, N. (2024, October). *From impersonation to authentication: techniques for identifying deep fake voices.* [Thesis, Brac University]. http://hdl.handle.net/10361/25249

Raimundo, R. J., & Rosário, A. T. (2022). Cybersecurity in the internet of things in industrial management. *Applied Sciences (Basel, Switzerland), 12*(3), 1598. DOI: 10.3390/app12031598

Ramachandra, R., Raja, K., Venkatesh, S., & Busch, C. (2024). Presentation attack detection methods for face recognition systems: A comprehensive survey. *ACM Computing Surveys, 56*(4), 1–38. DOI: 10.1145/3630260

Rao, I. S., Kiah, M. L. M., Hameed, M. M., & Memon, Z. A.Iqra Sadia Rao. (2024). Scalability of blockchain: A comprehensive review and future research direction. *Cluster Computing, 27*(5), 5547–5570. Advance online publication. DOI: 10.1007/s10586-023-04257-7

Rao, M. R., & Yarlagadda, P. K. D. V. (2022). Ethical implications of facial recognition technology: A comprehensive review. *Technology in Society, 68*, 101885.

Ratha, N. K., Connell, J. H., & Bolle, R. M. (2021). Enhancing security and privacy in biometrics-based authentication systems. IBM Journal of Research and Development, 65(3), 8:1-8:15. DOI: 10.1147/JRD.2021.3067559

Ratha, N. K., Connell, J. H., & Bolle, R. M. (2001). An analysis of minutiae matching strength. *Proceedings of the International Conference on Audio- and Video-Based Biometric Person Authentication*, 223–228. DOI: 10.1007/3-540-45344-X_32

Rathgeb, C., & Uhl, A. (2011). A survey on biometric cryptosystems and cancelable biometrics. *EURASIP Journal on Multimedia and Information Security, 2011*(1), 3. Advance online publication. DOI: 10.1186/1687-417X-2011-3

Razaque, A., Amsaad, F., Jaro Khan, M., Hariri, S., Chen, S., Siting, C., & Ji, X. (2019). Survey: Cybersecurity vulnerabilities, attacks and solutions in the medical domain. *IEEE Access : Practical Innovations, Open Solutions, 7*, 168774–168797. DOI: 10.1109/ACCESS.2019.2950849

Regulation 2016/679 OJ L 119 of the European Parliament and of the Council of 27 April 2016 on the protection of natural persons with regard to the processing of personal data and on the free movement of such data, and repealing Directive 95/46/EC (General Data Protection Regulation). Official Journal of the European Union, 59, 1-88. https://eur-lex.europa.eu/legal-content/EN/TXT/PDF/?uri=CELEX:32016R0679

Reis, O., Eneh, N. E., Ehimuan, B., Anyanwu, A., Olorunsogo, T., & Abrahams, T. O. (2024). Privacy Law challenges in the Digital Age: A global review of legislation and enforcement. Int. j. *Appl. Res. Soc. Sci.*, *6*(1), 73–88. DOI: 10.51594/ijarss.v6i1.733

Reynolds, J. C., & Weber, J. W. (1979, October 1). The admissibility of spectrographic voice identification in the state courts. *The Journal of Criminal Law & Criminology*, *70*(3), 349–354. DOI: 10.2307/1142577

Rezgui, D., & Lachiri, Z. (2016). ECG biometric recognition using SVM-based approach. *IEEJ Transactions on Electrical and Electronic Engineering*, *11*(S1), S94–S100. DOI: 10.1002/tee.22241

Robinson, J. (2023, August). *Daubert Standard.* Cornell School of Law Legal Information Institute. https://www.law.cornell.edu/wex/Daubert_standard

Rodgers, W., Yeung, F., Odindo, C., & Degbey, W. Y. (2021). Artificial intelligence-driven music biometrics influencing customers' retail buying behavior. *Journal of Business Research*, *126*, 401–414. DOI: 10.1016/j.jbusres.2020.12.039

Ross, A., & Jain, A. K. (2003). Information fusion in biometrics. *Pattern Recognition Letters*, *24*(13), 2115–2125. DOI: 10.1016/S0167-8655(03)00079-5

Rubin, H. (2025, January). It's my data and I'll protect it if I want to: A proposal to standardize biometric data privacy laws between the United States and European Union. *Vanderbilt Journal of Transnational Law*, *58*(1), 261–304.

Ruiz Gara. (2018, July). *SOCOFing: Sokoto Coventry Fingerprint Dataset* [Data set]. Kaggle. https://www.kaggle.com/datasets/ruizgara/socofing2

Rukhiran, M., Wong-In, S., & Netinant, P. (2023). IoT-based biometric recognition systems in education for identity verification services: Quality assessment approach. *IEEE Access : Practical Innovations, Open Solutions*, *11*, 22767–22787. DOI: 10.1109/ACCESS.2023.3253024

Saadi, Z. M., Sadiq, A. T., Akif, O. Z., & Farhan, A. K. (2024). A survey: Security vulnerabilities and protective strategies for graphical passwords. *Electronics (Basel)*, *13*(15), 3042. DOI: 10.3390/electronics13153042

Sadikan, S. F. N., Ramli, A. A., & Fudzee, M. F. M. (2019). A survey paper on keystroke dynamics authentication for current applications. *AIP Conference Proceedings, 2173*(1), 020010. DOI: 10.1063/1.5133925

Safavipour, M. H., Doostari, M. A., & Sadjedi, H. (2023). Deep hybrid multimodal biometric recognition system based on features-level deep fusion of five biometric traits. *Computational Intelligence and Neuroscience, 2023*(1), 6443786. DOI: 10.1155/2023/6443786 PMID: 37469627

Saini, M., & Kumar Kapoor, A. (2016). Biometrics in Forensic Identification: Applications and Challenges. *Journal of Forensic Medicine, 1*(2). Advance online publication. DOI: 10.4172/2472-1026.1000108

Saini, P., & Khera, R. (2021). Aadhaar: Biometric surveillance or identity empowerment? *Economic and Political Weekly, 56*(10), 21–24.

Sakal, M. (2022). Endüstri 4.0 ve Güvenlik. In *Endüstri 4.0 ve İşletne Yönetimi* (Eds.Mehmet Marangoz & Hatice Hicret Özkoç, Beta Yayınevi. Saylam, R. (2020). Applicability of the Internet of Things (IoT) in military environment and a proposed acceptance model [Master's thesis]. Atatürk University, Türkiye.

Sambasivan, N.. (2021). Re-imagining algorithmic fairness in India and beyond. *Proceedings of the ACM Conference on Fairness, Accountability, and Transparency*, 1–18. DOI: 10.1145/3442188.3445896

Sangeeta, N., & Nam, S. Y. (2023). Blockchain and Interplanetary File System (IPFS)-Based Data Storage System for Vehicular Networks with Keyword Search Capability. *Electronics (Basel), 12*(7), 1545. DOI: 10.3390/electronics12071545

Saray Çetinkaya, T., & Sertbaş, A. (2022). Derin Öğrenme Algoritmalarının GPU ve CPU Donanım Mimarileri Üzerinde Uygulanması ve Performans Analizi: Deneysel Araştırma. *Avrupa Bilim Ve Teknoloji Dergisi*, (33), 10–19. DOI: 10.31590/ejosat.937936

Schenkel, T., Ringhage, O., & Branding, N. (2019). A comparative study of facial recognition techniques with focus on low computational power.

Schneier, B. (2015). *Data and Goliath: The hidden battles to collect your data and control your world.* W. W. Norton & Company.

Sedláček, M. (2018, November 27). *Using voice biometrics in the world of the GDPR*. Phonexia. https://www.phonexia.com/blog/using-voice-biometrics-world-gdpr

Sevinç, M., & Avcı, İ. (2024). IoT cihazlarında insan hatasından kaynaklanan güvenlik açıklarının analizi. *Mühendislik Bilimleri ve Tasarım Dergisi, 12*(1), 57–68.

Shadman, R., Wahab, A. A., Manno, M., Lukaszewski, M., Hou, D., & Hussain, F. (2025). Keystroke dynamics: Concepts, techniques, and applications. *ACM Computing Surveys*, *57*(11), 1–35. DOI: 10.1145/3733103

Sharma, D., & Selwal, A. (2023). A comprehensive review on GAN-based face forgery detection techniques. *Multimedia Tools and Applications*, *82*(18), 27605–27651. DOI: 10.1007/s11042-023-14625-9

Shinde, K., & Kayte, C. (2023). Multimodal deep learning based score level fusion using face and fingerprint. In *Advances in Intelligent Systems Research* (pp. 140–152). Atlantis Press International BV., DOI: 10.2991/978-94-6463-196-8_13

Shtepliuk, I., Montelius, K., Eriksson, J., & Puglisi, D. (2025). Adaptive Machine Learning for Electronic Nose-Based Forensic VOC Classification. *Advancement of Science*, *12*(36), 04657. DOI: 10.1002/advs.202504657 PMID: 40552384

Siddika, S. (2025, May 20). *Privacy-preserving biometrics: Enhancing security with Anonybit*. LinkedIn. https://www.linkedin.com/pulse/privacy-preserving-biometrics-enhancing-security-anonybit-siddika-6vw1c

Simmons, D. (2017, May 19). BBC fools HSBC voice recognition security system. *BBC*. https://www.bbc.com/news/technology-39965545

Simonyan, K., & Zisserman, A. (2014). Very deep convolutional networks for large-scale image recognition. *arXiv*. https://arxiv.org/abs/1409.1556

Singh, M., Singh, R., & Ross, A. (2022). A comprehensive overview of biometric fusion. *Information Fusion*, *82*, 103–121. DOI: 10.1016/j.inffus.2022.02.006

Sirivarshitha, A. K., Sravani, K., Priya, K. S., & Bhavani, V. (2023). An approach for face Detection and face recognition using OpenCV and face recognition libraries in Python. 2023 9th International Conference on Advanced Computing and Communication Systems (ICACCS), 1, 1274-1278. DOI: 10.1109/ICACCS57279.2023.10113066

Siwakoti, Y. R., Bhurtel, M., & Rawat, D. B. (2023). Advances in IoT security: Vulnerabilities, attacks, and countermeasures. *IEEE Internet of Things Journal*, *10*(4), 3456–3473. DOI: 10.1109/JIOT.2023.3252594

Smith, J. (2024, August 29). High court to decide if covert police recording violates wiretap law. *Commonwealth Beacon*. https://commonwealthbeacon.org/courts/high-court-to-decide-if-covert-police-recording-violates-wiretap-law

Snoek, C. G. M., Worring, M., & Smeulders, A. W. M. (2005). Early versus late fusion in semantic video analysis. In *Proceedings of the 13th Annual ACM International Conference on Multimedia* (pp. 399–402). DOI: 10.1145/1101149.1101236

Solove, D. J. (2006). A taxonomy of privacy. *University of Pennsylvania Law Review*, *154*(3), 477–560. DOI: 10.2307/40041279

Songara, D., Poonia, A., Gupta, G., & Sharma, A. (2022). A model for fingerprint liveness detection enabled by M-SSO heuristic algorithm using deep learning strategies. *Cybernetics and Systems*, *55*(8), 1793–1828. DOI: 10.1080/01969722.2022.2146848

Standring, S. (Ed.). (2016). *Gray's anatomy: The anatomical basis of clinical practice* (41st ed.). Elsevier Limited.

Stanley, J. (2014, October 16). *On the creation of giant voiceprint databases*. American Civil Liberties Union. https://www.aclu.org/news/privacy-technology/creation-giant-voiceprint-databases

Stanojević, M., Milenković, I., Starčević, D., & Stanojević, B. (2016, May). Optimization of thresholds in serial multimodal biometric systems. In *2016 6th International Conference on Computers Communications and Control (ICCCC)* (pp. 140-146). IEEE. DOI: 10.1109/ICCCC.2016.7496752

Stevens, C. A., Carey, J. C., Shah, M., & Bagley, G. P. (1988). Development of human palmar and digital flexion creases. *The Journal of Pediatrics*, *113*(1 Pt 1), 128–132. DOI: 10.1016/S0022-3476(88)80546-8 PMID: 3385521

Stylios, I., Skalkos, A., Kokolakis, S., & Karyda, M. (2022). BioPrivacy: A behavioral biometrics continuous authentication system based on keystroke dynamics and touch gestures. *Information and Computer Security*, *30*(5), 687–704. DOI: 10.1108/ICS-12-2021-0212

Sun, Y., Chen, Y., Wang, X., & Tang, X. (2014). Deep learning face representation by joint identification-verification. *Advances in Neural Information Processing Systems*, •••, 27.

Suvika, K. V., Sai, G. L., Jayachandran, J., & SA, K. A. (2024). Continuous user authentication through Keystroke Dynamics. *GIJET, 10.*

Tabassi, E., Wilson, C., & Watson, C. (2024). NIST Special Publication 800-76-3: Biometric specifications for personal identity verification. National Institute of Standards and Technology. DOI: 10.6028/NIST.SP.800-76-3

Tafiadis, D., Tatsis, G., Ziavra, N., & Toki, E. I. (2017). Voice data on female smokers: Coherence between the Voice Handicap Index and acoustic voice parameters. *AIMS Medical Science*, *4*(2), 151–163. DOI: 10.3934/medsci.2017.2.151

Taigman, Y., Yang, M., Ranzato, M. A., & Wolf, L. (2014). Deepface: Closing the gap to human-level performance in face verification. In *Proceedings of the IEEE conference on computer vision and pattern recognition* (pp. 1701-1708). DOI: 10.1109/CVPR.2014.220

Talreja, V., Valenti, M. C., & Nasrabadi, N. M. (2021). Deep hashing for secure multimodal biometrics. *IEEE Transactions on Information Forensics and Security, 16*, 1306–1321. DOI: 10.1109/TIFS.2020.3033189

Tawalbeh, L., Muheidat, F., Tawalbeh, M., & Quwaider, M. (2020). IoT privacy and security: Challenges and solutions. *Applied Sciences (Basel, Switzerland), 10*(12), 4102. DOI: 10.3390/app10124102

Tene, O., & Polonetsky, J. (2013). Big data for all: Privacy and user control in the age of analytics. *Northwestern Journal of Technology and Intellectual Property, 11*(5), 239–273.

Texas Attorney General. (n.d.). *Biometric Identifier Act: Texas Capture Or Use Of Biometric Identifier Act.* https://www.texasattorneygeneral.gov/consumer-protection/file-consumer-complaint/consumer-privacy-rights/biometric-identifier-act

Tok, M. S. (2019). Nesnelerin internetinde botnetler: Mirai zararlı yazılımı üzerine bir çalışma. *[Unpubllished master's thesis]. Gazi University, Ankara, Türkiye.*

Tolosana, R., Vera-Rodriguez, R., Fierrez, J., Morales, A., & Ortega-Garcia, J. (2022). Deepfakes and beyond: A survey of face manipulation and fake detection. *Information Fusion, 87*, 131–148. DOI: 10.1016/j.inffus.2022.07.015

Tombak, M. E. (2019). An application recognizing faces and applying Gaussian Blur in OpenCv and Python [Unpublished master's thesis]. Kırıkkale University.

Trabelsi, S., Samai, D., Meraoumia, A., & Bensid, K. (2020). Finger-knuckle-print recognition using deep convolutional neural network. In *Proceedings of the 1st International Conference on Communications, Control Systems and Signal Processing (CCSSP)* (pp. 1–6). IEEE. DOI: 10.1109/CCSSP49278.2020.9151531

Tremblay, P., Gagnon, L., Roy, J.-P., & Arseneault, A. (2023, November 9). Speech production in healthy older adults with or without amateur singing experience. *Journal of Speech, Language, and Hearing Research: JSLHR, 66*(11), 4332–4352. DOI: 10.1044/2023_JSLHR-23-00126 PMID: 37870784

Trivedi, C., Bhattacharya, P., Prasad, V. K., Patel, V., Singh, A., Tanwar, S., Sharma, R., Aluvala, S., Pau, G., & Sharma, G. (2024). Explainable AI for industry 5.0: Vision, architecture, and potential directions. *IEEE Open Journal of Industry Applications, 5*, 177–208. DOI: 10.1109/OJIA.2024.3399057

Tse, K. W., & Hung, K. (2020). User behavioral biometrics identification on mobile platform using multimodal fusion of keystroke and swipe dynamics and recurrent neural network. *2020 IEEE 10th Symposium on Computer Applications & Industrial Electronics (pp. 262–267). IEEE.*

Tulgar, M., Zaim, A. H., & Aydın, M. A. (2022). Ulusal bilgi ve iletişim güvenliği rehberi: IoT güvenliği için bir uygulama örneği. İstanbul Ticaret Üniversitesi. *Fen Bilimleri Dergisi*, *21*(2), 99–115.

Tuna, K. (2022). Non-Contact Fever Measurement With Deep Learning Based Face Detection System [Unpublished master's thesis]. Konya Technical University.

Türker, G. F., & Tanyeri, K. (2024). Blokzincir teknolojisi ile IoT sistemlerinin veri güvenliğinin sağlanması. Gazi Üniversitesi Fen Bilimleri Dergisi C Serisi, 8(1), 45-59.

Turner, J. (1973). Managing Digital Discovery in Criminal Cases. *The Journal of Criminal Law & Criminology*, •••, 237–312.

Uliyan, D. M., Sadeghi, S., & Jalab, H. A. (2020). Anti-spoofing method for fingerprint recognition using patch based deep learning machine. *Engineering Science and Technology, an International Journal*, *23*(2), 264–273. DOI: 10.1016/j.jestch.2019.06.005

Ullah, S., Ou, J., Xie, Y., Zhang, Y., & Huang, F. (2024). Compound facial expressions recognition approach using DCGAN and CNN. *Multimedia Tools and Applications*, *83*(38), 85703–85723. DOI: 10.1007/s11042-024-20138-6

Uludag, U., & Jain, A. K. (2024). Attacks on biometric systems: A case study in fingerprints. *IEEE Transactions on Information Forensics and Security*, *19*, 4512–4527. DOI: 10.1109/TIFS.2024.3389012

United States Attorney's Office. District of Massachusetts. (2025, August 12). *Thirteen individuals charged for operating transnational elder fraud scheme.* [Press release]. https://www.justice.gov/usao-ma/pr/thirteen-individuals-charged-operating-transnational-elder-fraud-scheme

Van der Walt, E., & Eloff, J. (2018). Using machine learning to detect fake identities: Bots vs. humans. *IEEE Access : Practical Innovations, Open Solutions*, *6*, 6540–6549. DOI: 10.1109/ACCESS.2018.2796018

van Drongelen, W. (2007). Continuous, discrete, and fast Fourier transform. In *Signal processing for Neuroscientists* (pp. 91–105). Elsevier., DOI: 10.1016/B978-012370867-0/50006-1

Vanderslice, R. (1969, February 1). *The "voiceprint" myth.* [Report]. University of Michigan Center for Research on Language and Language Behavior.

Vellani, N., & Common, D. (2025, March 20). Her grandson's voice said he was under arrest. This senior was almost scammed with suspected AI voice cloning. *Canadian Broadcasting Corporation.* https://www.cbc.ca/news/marketplace/marketplace-ai-voice-scam-1.7486437

Venkatesh, S., Ramachandra, R., Raja, K., & Busch, C. (2023). Face morphing attack detection using deep learning: A comprehensive review. *IEEE Access : Practical Innovations, Open Solutions, 11*, 89234–89258. DOI: 10.1109/ACCESS.2023.3305678

Verma, R., Dhanda, N., & Nagar, V. (2022). Application of Truffle Suite in a Blockchain Environment. *Proceedings of Third International Conference on Computing, Communications, and Cyber-Security,* 693–702. DOI: 10.1007/978-981-19-1142-2_54

Vidhyapriya, R., & Rose, L. (2019). Personal authentication mechanism based on finger knuckle print using Gabor filter and expectation-maximization algorithm. *Journal of Medical Systems, 43*(8), 232. DOI: 10.1007/s10916-019-1332-3 PMID: 31203468

Vinayagam, J., & Dilip, G. (2024). A two-step verification-based multimodal-biometric authentication system using KCP-DCNN and QR code generation. *Journal of Ambient Intelligence and Humanized Computing, 15*(12), 3973–3996. DOI: 10.1007/s12652-024-04872-1

Voiceprint concept supported by government-sponsored tests. (1971, June 1). Communications of the ACM, 14(6), 434–435. https://doi.org/ DOI: 10.1145/362604.1275495

Vyas, R., Williams, B. M., Rahmani, H., Boswell-Challand, R., Jiang, Z., Angelov, P., & Black, S. (2022). Ensemble-based bounding box regression for enhanced knuckle localization. *Sensors (Basel), 22*(4), 1569. DOI: 10.3390/s22041569 PMID: 35214467

Wagemann, J., Clements, O., & Figuera, R. M. (2018). Geospatial web services pave new ways for server-based on-demand access and processing of Big Earth Data. *International Journal of Digital Earth, 11*(1), 7–25. DOI: 10.1080/17538947.2017.1351583

Wang, Q., Zhang, P., Xiong, H., & Zhao, J. (2021). Face.evolve: A high-performance face recognition library.arXiv preprint. https://arxiv.org/abs/2107.08621

Wang, G., Shanker, S., Nag, A., Lian, Y., & John, D. (2024). ECG biometric authentication using self-supervised learning for IoT edge sensors. *IEEE Journal of Biomedical and Health Informatics, 28*(11), 6606–6618. DOI: 10.1109/JBHI.2024.3455803 PMID: 39250357

Wang, H., Wang, Y., Zhou, Z., Ji, X., Gong, D., Zhou, J., & Liu, W. (2018). Cosface: Large margin cosine loss for deep face recognition. In *Proceedings of the IEEE conference on computer vision and pattern recognition* (pp. 5265-5274). DOI: 10.1109/CVPR.2018.00552

Wang, Y., Cusick, M., Laila, M., Puech, K., Ji, Z., Hu, X., Wilson, M., Spitzer-Williams, N., Wheeler, M., & Ibrahim, Y. (2024, May 21). FairLENS: Assessing fairness in law enforcement speech recognition. ArXiv. https://arxiv.org/abs/2405.13166 DOI: 10.1109/BigData62323.2024.10825119

Washington State Legislature. (2017). *Title 19, Chapter 19.375: Biometric identifiers.* https://app.leg.wa.gov/RCW/default.aspx?cite=19.375

Westin, A. F. (1967). *Privacy and Freedom.* Atheneum.

What is forensic voice comparison? (n.d.). In Forensic voice comparison: The essential guide. Phonexia. https://www.phonexia.com/knowledge-base/forensic-voice-comparison-essential-guide

What is voice biometrics? (n.d.). Forensic voice comparison: The essential guide. Phonexia. https://www.phonexia.com/knowledge-base/voice-biometrics-essential-guide

Whitmore, A., Agarwal, A., & Da Xu, L. (2015). The Internet of Things—A survey of topics and trends. *Information Systems Frontiers*, *17*(2), 261–274. DOI: 10.1007/s10796-014-9489-2

Wilder, M. A., Scott, C., Pablos, N. P. G., Varady, R., & McEvoy, J. (2010). Adapting Across Boundaries: Climate Change, Social Learning, and Resilience in the U.S.–Mexico Border Region. *Annals of the Association of American Geographers*, *100*(4), 917–928. DOI: 10.1080/00045608.2010.500235

Woodard, D. L., & Flynn, P. J. (2005). Finger surface as a biometric identifier. *Computer Vision and Image Understanding*, *100*(3), 357–384. DOI: 10.1016/j.cviu.2005.06.003

Xiong, Q., Zhang, X., Xu, X., & He, S. (2021). A Modified Chaotic Binary Particle Swarm Optimization Scheme and Its Application in Face-Iris Multimodal Biometric Identification. *Electronics (Basel)*, *10*(2), 217. DOI: 10.3390/electronics10020217

Yalçın, N., & Çakır, S. (2024). IoT ağlarında saldırı tespiti için federe öğrenme. *Bilgisayar Bilimleri ve Mühendisliği Konuları*, *10*(2), 50–67.

Yang, X. S., & Deb, S. (2009). Cuckoo search via Lévy flights. In *Proceedings of the World Congress on Nature & Biologically Inspired Computing (NaBIC)* (pp. 210–214). IEEE. DOI: 10.1109/NABIC.2009.5393690

Yıldız, B. (2022). *Privacy Preserving Data Analysis For Information Systems* [Unpublished doctoral dissertation]. Dokuz Eylül University. İzmir, Türkiye.

Yılmaz, E. N., Gönen, S., Şanoğlu, S., Karacayilmaz, G., & Özbi̇ri̇nci̇, Ö. (2021). Endüstri 4.0'ın gelişim sürecinde unutulan bileşen: Siber güvenlik. *DÜBİTED, 4*(2), 87–103. DOI: 10.29130/dubited.905340

Yin, X. C., Liu, Z. G., Nkenyereye, L., & Ndibanje, B. (2019). Toward an applied cyber security solution in IoT-based smart grids. *Sensors (Basel), 19*(22), 4952. DOI: 10.3390/s19224952 PMID: 31739400

Yi, X., Hong, L., Zhong, E., Liu, N. N., & Rajan, S. (2014). Beyond clicks: Dwell time for personalization. *8th ACM Conference on Recommender Systems (pp.* 113–120*).*

Yurttaş, H. G., & Güzel, A. (2023). Kablosuz ağa bağlı tıbbi cihazlarda siber güvenlik açıkları ve çözüm önerileri. *Uluslararası Sağlık Yönetimi ve Stratejileri Araştırma Dergisi, 5*(1), 45–60.

Zahoor, E. M. S. (2023). Biometric encryption: Integrating artificial intelligence for robust authentication. *Dandao Xuebao/Journal of Ballistics, 35*(3), 25-33.

Zakaria, S., Manaf, S. M. A., Amron, M. T., & Suffian, M. T. M. (2023). Has the world of finance changed? A review of the influence of artificial intelligence on financial management studies. *Information Management and Business Review, 15*(4), 420–432. DOI: 10.22610/imbr.v15i4(SI)I.3617

Zarpelão, B. B., Miani, R. S., Kawakani, C. T., & de Alvarenga, S. C. (2017). A survey of intrusion detection in Internet of Things. *Journal of Network and Computer Applications, 84*, 25–37. DOI: 10.1016/j.jnca.2017.02.009

Zeybek, M., & Yılmaz, E. N. (2019). Nesnelerin İnterneti: Risk temelli yaklaşım. *Denetişim, 3*(2), 12–29.

Zhang, L., Zhang, L., & Zhang, D. (2009, November). Finger-knuckle-print: a new biometric identifier. *In 2009 16th IEEE international conference on image processing (ICIP) (pp. 1981-1984). IEEE.* DOI: 10.1109/ICIP.2009.5413734

Zhang, D., Li, J., & Shan, Z. (2020). Implementation of Dlib deep learning face recognition technology. *IEEE Conf. on Intelligent Robots*. DOI: 10.1109/ICRIS52159.2020.00030

Zhang, L., Li, L., Yang, A., Shen, Y., & Yang, M. (2017). Towards contactless palmprint recognition: A novel device, a new benchmark, and a collaborative representation based identification approach. *Pattern Recognition*, *69*, 199–212. DOI: 10.1016/j.patcog.2017.04.016

Zhang, X., & Yue, W. T. (2020). Integration of On-Premises and Cloud-Based Software: The Product Bundling Perspective. *Journal of the Association for Information Systems*, *21*, 44. DOI: 10.17705/1jais.00645

Zhao, G., Wu, Z., Gao, Y., Niu, G., Wang, Z. L., & Zhang, B. (2020). Multi-layer extreme learning machine-based keystroke dynamics identification for intelligent keyboard. *IEEE Sensors Journal*, *21*(2), 2324–2333. DOI: 10.1109/JSEN.2020.3019777

Zhou, Z., Dong, C., & Kumar, A. (2025). Towards explainable and unprecedented accuracy in matching challenging finger crease patterns. In *Proceedings of the IEEE/CVF Conference on Computer Vision and Pattern Recognition (CVPR)* (pp. 6212–6221). IEEE. DOI: 10.1109/CVPR52734.2025.00582

About the Contributors

Sarah Benziane is a distinguished researcher and university lecturer in computer science and distributed systems, currently based at the Computer Science Department of the University of Science and Technology in Oran, Algeria. With a strong academic foundation, she holds an engineering degree in Parallel and Distributed Systems, a Magister in Electronics specializing in Vision and Pattern Recognition, and a Ph.D. in Artificial Intelligence and Pattern Recognition. Her research has produced a wealth of publications primarily focused on biometrics, hand vein recognition, and the application of convolutional neural networks (CNNs) in image processing. In addition to her research contributions, Sarah Benziane has supervised numerous undergraduate and master's projects on topics ranging from biometric systems and irrigation optimization to intelligent asset management. She has also been actively involved in organizing national and international conferences, such as doctoral workshops and conferences on embedded and distributed systems, and has served on scientific committees for conferences focused on artificial intelligence and the Internet of Things (IoT). Sarah Benziane is the author of several academic books, including one on reactive navigation for mobile robots and another on computer networks. Her recent work includes innovative research on AI applications for weather prediction and anti-theft systems for agriculture. She is currently involved in several national and international projects related to the application of AI in smart industry automation and renewable energy systems, as well as in fundamental research in computer science.

***

Claritta AlSaneh is a Telecommunications and Networks student at the University of Balamand, passionate about AI, electronics and embedded systems. My work spans telecom and networks, and AI-driven solutions for real world problem with a strong focus on making technology more innovative, efficient, and inclusive for everyone.

Saloni Bahl is an Assistant Professor of Law at Vivekananda Global University,Jaipur

Hakim Bendjenna is an Associate Professor in the Department of Computer Science at Tebessa university-Algeria. He received his Ph.D. in Computer Science from Mentouri university- Algeria and Toulouse university- France. His research interests lie in the general field of software engineering and decision support with special focuses on requirement engineering. Dr Hakim has taught primarily introductory programming courses and courses in software engineering for over ten years. Dr. Hakim also frequently serves as a program committee member for various international conferences and workshops

Prachi Sharma Biswas is a Ph.D. Scholar in Forensic Science at Shri Vaishnav Institute of Forensic Science, Shri Vaishnav Vidyapeeth Vishwavidyalaya, Indore, Madhya Pradesh, India. Her research interests lie in the intersection of forensic science and biometrics, with a focus on human identification, anatomical trait analysis, and the development of systematic classification approaches. She has published scholarly work in peer-reviewed journals and contributed to book chapters in her field. Her academic pursuits also extend to biometric recognition, image processing, and digital forensics, reflecting an interdisciplinary approach that combines traditional forensic practices with emerging technological methods. She is committed to promoting scientifically robust, ethically responsible, and practically applicable methods that enhance accuracy and credibility in forensic investigations and judicial decision-making..

Bayram Cadıl has an MSc in cyber security from Ahmet Yesevi University, Türkiye. His research interests include machine learning, image processing, and cyber security.

Salim Chitroub is a distinguished Full Professor in the Telecommunications Department at the Faculty of Engineering, University of Science and Technology Houari Boumediene (USTHB) in Algiers, Algeria. He began his academic journey at USTHB, earning a Magister (M.S.) degree in satellite image processing, followed by a Ph.D. specializing in signal processing and pattern recognition, with a particular focus on remote sensing image classification. Throughout his career, Professor Chitroub has demonstrated exceptional expertise in a broad spectrum of fields, including signal and image processing, biometrics, cybersecurity, artificial intelligence, and data science. His pioneering research has significantly advanced the development of sophisticated classification techniques, utilizing neural networks, data fusion, and independent component analysis (ICA), especially in applications related to remote sensing and multimodal biometric systems. Professor Chitroub is the author of numerous peer-reviewed publications and has supervised many master's and doctoral theses, nurturing the next generation of researchers and engineers. His work has had a notable impact both in academia and industry, with many of his doctoral students progressing to successful careers in research and engineering. Beyond his research and teaching responsibilities, Professor Chitroub is actively engaged in the scientific community. He serves as a reviewer for various international journals and conferences and holds editorial roles on several scientific committees. Recently, he was appointed Senior Guest Editor for a special issue on Artificial Intelligence and Machine Learning Algorithms for Cybersecurity, published in the journal Security and Privacy (Wiley, 2023). His current research interests include the integration of artificial intelligence with cybersecurity frameworks, particularly in resource-constrained environments, and the enhancement of biometric authentication systems. Additionally, Professor Chitroub is exploring the reliability of big data in AI applications, emphasizing the need for more efficient and less energy-intensive data utilization methods.

Swati Dubey Mishra, a Gold Medalist, holds a Ph.D. in Criminology and Forensic Science from Dr. Harisingh Gour University, Sagar (M.P.). She qualified the UGC-NET and was awarded the Junior Research Fellowship by the Directorate of Forensic Science, Ministry of Home Affairs, Govt. of India, later elevated to Senior Research Fellow. With nearly 16 years of academic, research, and administrative experience in Forensic Science, she currently serves as Associate Professor and Head, Shri Vaishnav Institute of Forensic Science, Shri Vaishnav Vidyapeeth Vishwavidyalaya (SVVV), Indore. She has guided four Ph.D. scholars (five ongoing) and contributed extensively through publications in reputed journals and conference proceedings. Actively engaged in academic exchange, she has participated in over 75 national and international conferences, including the 13th International Congress of the Indo-Pacific Association of Law Medicine and Science, Dhaka (2019). Awards & Recognitions to her credit include the Junior Scientist Award (2015), Shining Rotarian Award (2023), Atal Alankaran Award (2024), and Champion Change Award (2024). Dr. Mishra is a life member of multiple professional bodies such as ISCA, FSDS, IFS, and the Indian Society of Criminology. She serves on editorial and advisory boards of reputed forensic journals and is a reviewer for leading academic publications. Beyond academics, she is socially active as a Rotarian since 2022 and serving as Rotaract Chair (RY 2024–25) for District 3040 and editor of the Rotary Club of Indore Uptown's bimonthly bulletin Unmesh.

Tamara Phillips Fudge is a full-time professor in the graduate technology programs at Purdue University Global. She has won fellowships and awards for innovation and teaching, and has taught a wide variety of topics, including web development, human-computer interaction, systems analysis and design, and those featuring documentation, diagramming, problem-solving, and presentation. Her career started with music degrees from Indiana University and Florida State University. She sang opera, oratorio, and in recital; her compositions have been heard on Public Radio, at various universities, and a state choral convention. She taught vocal and choral music, piano, pedagogy, foreign language diction, opera production, music theory, composition, and related courses in the traditional college classroom for 20-odd years. For 7 years, she was a weekend correspondent for the Quad-City Times (Davenport, IA). Following a brief stint as an agent/registered representative selling life and health insurance and variable products, she returned to school with a keen interest in technology and has since distinguished herself in online teaching, coordination of large projects, and curriculum development.

Nikhil Kumar Goyal is currently Working as Assistant Professor at Poornima University Jaipur

Deepak Gupta is a distinguished academician with over 20 years of teaching experience and expertise in Computer Science and Engineering. He holds a Ph.D. in Computer Engineering and has an extensive research portfolio, including many international journal papers, multiple conference proceedings, and several books. His research interests span machine learning, artificial intelligence, and data mining. Currently, Dr. Gupta serves as an Associate Professor and Head of Department of CSE (Emerging Branches) at ITM Gwalior, MP, India. He has also contributed significantly to the field through patents and copyrights and actively engages as a reviewer, conference chair and editor.

Kavita Kanwar is currently Working as Assistant Professor at Poornima University Jaipur

S. Keerthi earned her Ph.D. in Information and Communication Engineering from Anna University, Chennai, in 2023. She completed her M.E. at PSRR College of Engineering, in 2015 and her B.E. at SBC College of Engineering in 2009. With over nine years of academic experience, Dr. Keerthi has been serving as an Assistant Professor in the Department of Computer Science and Engineering at Dayananda Sagar College of Engineering, Bangalore, since 2021. Her Research Interests include Machine Learning, Deep Learning, Computer Networks, and Cloud Computing.

Cemgil Kurt has an MSc in cyber security from Ahmet Yesevi University, Türkiye. His current research interests include software development, cyber security and ERP systems.

Claudia Matar is an assistant professor at the University of Balamand, Issam Fares Faculty of Technology. She received the Bachelor and the Master of Science in Computer Engineering from the University of Balamand in 2003 and 2005 respectively, and the PhD in Industrial Engineering/field of study: Engineering and Quality Management from the Polytechnic University of Bucharest - Romania in 2014. She has a teaching experience in higher education for more than 15 years in several universities in Lebanon. She has been elected President of the Scientific Committee at the Order of Engineers and Architects-North Lebanon for years 2023/2024 and 2024/2025. Her research focuses on: Technology, Technology in Education, Augmented Reality, Quality Management.

Abdallah Meraoumia received his PhD degree in Signal and Image Processing from the University of Science and Technology of Houari Boumedienne, USTHB, Algiers, Algeria, in 2014. Currently, he is an associate professor at the Department of Electrical Engineering in the University of Tebessa, Algeria. His research interests are related to signal and image processing, pattern recognition, information security and biometrics. He has published research papers at national and international journals, conference proceedings as well as chapters of books.

Srinivasan Nagaraj is currently working in Chitanya Bharathi Institute of Technology, Proddatur as Professor in Computer Science and Engineering Department. He completed his M.Tech(CSE) from Acharya Nagarjuna University and was awarded with Ph.D. from Annamalai University in 2023. He published more than 20 research papers in National and International journals, he published 5 patents to his credit and also participated in 15 National and International Conferences. He taught all Computer science and Machine Learning subjects for Undergraduate and Post graduate students. He has a stint of 24 years of experience in various reputed colleges and universities at various academic and administrative positions.

Raghu Nangunuri is an experienced academic and developer with a 20-year tenure as an Assistant Professor in the Department of Computer Science and Engineering at Kamala Institute of Technology & Science (KITS Singapur). He is currently pursuing a Ph.D. (part-time) from GLOCAL University, having completed his pre-Ph.D. requirements. He holds an M.Tech in Software Engineering from JNTU Hyderabad (2013) and a B.Tech in CSE from JNTU Hyderabad (2005). **Roles & Administrative Leadership** Beyond teaching, Raghu plays a multifaceted role in institutional administration. He serves as the Google Workspace Admin,

Admin for the institute's social media accounts, and the Single Point of Contact (SPOC) for several key initiatives, including the JNTUH Biometric Attendance System, EDUSKILLS, NPTEL-SWAYAM, Coursera, Infosys SpringBoard, and AWS Academy. He is also the AICTE IDEA LAB Co-Coordinator and a Nodal Officer for both AICTE ASLAP-PARAKH and the Malaviya Mission (NEP2020). **Technical Expertise & Development** Raghu possesses a strong technical skillset in web development and cloud computing. His proficiency spans programming languages such as C, C++, Java, and Python, alongside web technologies including HTML, CSS, JavaScript, PHP, Django, and React. He has developed numerous official web applications for his institution, including: **Student Attendance & Profile Management Systems:** Web portals for managing student attendance, sending automated notifications to parents, and maintaining student academic profiles. **Institutional Websites:** He designed, developed, and deployed the official websites for KITS Engineering College and Jayaprakash Narayan College of Engineering. **Specialized Portals:** He created systems for grievance redressal, online admissions, alumni relations, and faculty profiles. **Research Projects:** He developed a "Fog Computing" research project focused on secure document storage with time-restricted access. **Awards & Achievements** His contributions have been recognized with several awards, including the "Niyukthi Aacharya" from the Bharat Education Excellence Awards (2025) and the "Fastest Project Drona" award from HYSEA-INFOSYS. He is a four-time recipient of the "Active SPOC Award" from NPTEL-SWAYAM. In August 2022, he mentored a student team that won the Smart India Hackathon Final, securing a 1 Lakh cash prize. **Research & Publications** Raghu is an active researcher with three published patents, including a utility patent for optimizing resource allocation in cloud computing environments using predictive analytics (2025). He has co-authored three books, including titles on *Cryptography and Cyber Security* and *Artificial Intelligence and Machine Learning*. Additionally, he has published multiple papers in international journals focusing on cloud computing security and data mining efficiency. **Certifications & Skills** He holds numerous professional certifications in areas such as Cloud Computing, Artificial Intelligence, Agile Software Development, and Python from reputable bodies like NPTEL, Infosys Springboard, Coursera, and AWS.

Aditya Ojha holds an M.Sc. in Mathematics & Computing from Banaras Hindu University (BHU), Varanasi, and a B.Sc. degree. His academic research has focused on the practical applications of computational intelligence, particularly within the field of secure authentication. His postgraduate dissertation addressed the "Nature-Inspired Feature Fusion Approach for ECG-based Biometric Authentication," reflecting work in nature-inspired algorithms, biometric security systems, and mathematical modeling for signal processing.

Soraia Oueida was born in Brazil. She holds a Ph.D. degree in Computer and Industrial Engineering from the Polytechnic University of Bucharest and a Master's degree in Computer Engineering from the University of Balamand. Oueida worked in both industry and academia. She is now holding the role of Assistant Professor in the Department of Computer Engineering at the American University of the Middle East, Kuwait. She is an IEEE member and her research interests include but are not limited to AI and Machine Learning, Workflows, Petri Nets, Smart Cities, Blockchain, and Healthcare. Her first book was published in 2020 with Taylor & Francis entitled, Modeling a New Computer Framework for Managing Healthcare Organizations: Balancing and Optimizing Patient Satisfaction, Owner Satisfaction, and Medical Resources.

Ravikumar R. Natarajan is an experienced academician with over 12+ years of teaching and research expertise across institutions in India, the UAE, and the Maldives. He currently serves as an Assistant Professor at Marwadi University, Rajkot, specializing in Machine Learning, Recommender Systems, and AI-driven applications in healthcare and security. He holds a Ph.D. in Machine Learning and Recommender Systems from Amity University, Jaipur, a Master of Engineering in CSE from VMRF University, Salem, and a Bachelor of Engineering in CSE from Anna University. He also has a MiniMBA from the Swiss eLearning Institute. Dr. Ravikumar has authored more than 8+ multiple SCIE/Scopus-indexed journal articles, 25+ IEEE/Springer conference papers, and co-authored a book on Artificial Intelligence. He has also published 9 patents and contributed 10+ book chapters in emerging domains such as Federated Learning, Blockchain, and Digital Twins for IGI Global, Wiley and Springer publisher. He actively mentors students through initiatives like the Research Activities Club (RAC) and Student Research Internship Training (SRIT), promoting innovation and research culture.

Aarthi S is a dedicated Assistant Professor in Computer Science with over six years of teaching experience. She is currently pursuing her Ph.D. at Marwadi University, Gujarat, and holds an M.Phil. and MCA from Bharathiar University, Tamil Nadu. Throughout her academic career, Aarthi has specialized in programming languages such as C++, Java, and C, and contributed to research in recommender systems and AI-related data privacy. She has published journal articles, contributed to book chapters, and actively participated in technical workshops. Known for her student mentorship and engagement in academic events, Aarthi remains committed to fostering innovation and excellence in computer science education.

Abhilash Arun Sapre is an Assistant Professor of Law at Gujarat National Law University,Gandhinagar

Someshwar Siddi is an esteemed Associate Professor at St. Martins Engineering College, India. With over 20 years of teaching experience at engineering colleges affiliated with JNTUH, he has made a lasting impact on UG and PG students. Holding a Doctorate degree from Rayalaseema University, Kurnool, Dr. Siddi has published more than 20 research papers in international journals and contributed to Scopus Book Chapters and Conference Proceedings. As a prolific inventor, he has co-invented over 10 Indian patents. Dr. Siddi's academic excellence has been recognized with the prestigious Best Teacher Award in 2025. His research pursuits and academic contributions continue to inspire and educate students and professionals alike.

Gurkan Tuna is a Professor at the Department of Computer Technologies of Trakya University (2006–. . ..). He has authored several papers in international conference proceedings and refereed journals, and has been actively serving as a reviewer for international journals and conferences. His current research interests include cyber security and machine learning.

C. Umarani is an Associate Professor with over 22 years of experience in teaching, research, and academic leadership in Computer Science and Applications. Demonstrates deep expertise in Cybersecurity, Artificial Intelligence, Networking, Software Engineering, and IT Governance, supported by strong programming and analytical skills. A strategic thinker and problem-solver with a distinguished record of academic excellence, evidenced by 129+ research publications, 8 patents, and extensive contributions to national and international conferences. Known for impactful student mentorship, fostering innovation, and cultivating collaborative, learner-centric environments. Committed to advancing institutional growth through ethical research practices, interdisciplinary engagement, and continuous professional development.

Index

A

B

C

D

E

F

S

V

W

www.ingramcontent.com/pod-product-compliance
Lightning Source LLC
LaVergne TN
LVHW061525290926
849162LV00004B/664